Brain Plasticity and Learning

"Jennifer offers an informed challenge to those working in education to re-frame the professional language and knowledge base around teaching and learning. She offers detailed examples of practice and situates these within a stance which affirms the humanity and uniqueness of educational relationship and decision."

—Dr. Rachel Lofthouse, *professor of teacher education at Leeds Beckett University School of Education, United Kingdom*

"Traditional educational systems have often neglected preparing educators in the application of the affective and social neurosciences in the deepened understanding of how our brains and bodies are impacted by adversity and trauma. Addressing brain and nervous system development and integrating this research and science into the developing educational worlds of our children and youth creates hopefulness and possibility. Jennifer Hawkins has shared a comprehensive exploration of the critical importance and impact of how brain science and neuroplasticity can contribute to the growth and the resiliency of our world's children, youth and communities."

—Dr. Lori Desautels, Assistant Professor, *Butler University's College of Education, Indianapolis, USA*

"By exploring the disconnect between the fields of neuroscience and education as it is traditionally conceived, Hawkins makes a compelling case for change. Published in the wake of a global pandemic, when the younger generation sacrificed so much educational opportunity to protect the health of the older, we cannot ignore the huge potential that is brain plasticity. Hawkins explores what harnessing that might look like for the teacher. Her insights, and their biologically informed underpinning, must be read by anyone interested in the potential of education to transform lives."

—Mary Meredith *is Head of Inclusion at Lincolnshire County Council, U.K. - Education and skills, Employment, Diversity equality, Conferences and Training*

"This is a book that is a 'must read' for any educator who wants to ensure that their students receive a quality education. Having the awareness of brain plasticity is one of the golden keys to avoiding putting glass ceilings on our ability and potential as human beings. I believe that Jennifer's insights, which are drawn from research and outstanding practices, will be transformational for schools and colleges."

—Dr. Neil Hawkes, (DPhil Oxford). *Founder of Values-based Education (VbE) Website; www.valuesbasededucation.com*

"Jennifer Hawkins has produced another book that brings the science of the brain to the classroom in a way that could make a positive difference to the lives of children. The book urges new thinking in the way we approach teaching, questioning some of our traditional practices and their impact. It is fascinating, unsettling and uplifting."

—Mick Waters, *author with Tim Brighouse of 'About Our Schools' (2021) Camarthen: Independent Thinking Press - to be published in autumn 2021*

"Jennifer A. Hawkins new book, Brain Plasticity and Learning: Implications for Educational Practice, is an important exploration of neuroplasticity and its critical role in the learning process. Hawkins takes the reader through the fascinating history of neuroplasticity and explains the tenets of neuroplasticity in a very accessible manner. Hawkins leaves the reader inspired by the brain's plastic nature, its diversity, how it drives behaviour and its promise that if we can understand the brain's malleable nature, we can create treatments to address a number of conditions. A book well worth reading."

—Barbara Arrowsmith-Young, *author of* 'The Woman Who Changed Her Brain' *(2012) London: Vintage, Random house*

"One of the most referenced and researched books I have read on the subject of neuroscience and education, Hawkins brings to the fore all that can no longer be ignored. Comprehensive, compelling and a call to action for all those engaged in education policy and practice that neuroscience can no longer be kept out of the classroom. This is the instruction book for the overhaul that education is yearning for."

—Lisa Cherry, *Author of* "Conversations That Make A Difference for Children and Young People" *(2021) Routledge, Speaker and Trainer on Trauma, Recovery and Resilience. Currently researching 'belonging'. @_lisacherry | www.lisacherry.co.uk*

"The dominant approach to children's behaviour in school has focused for centuries on performance, what can be seen, largely ignoring potential, what is possible, the fact that change is always happening. The recent and current science that Hawkins explores underpins the shift which is underway to bring children's potential for change and growth into the light. To be able to stand their emerging practice on the evidence, teachers and school leaders need a guiding hand through the forest of neuroscience and they have it here."

—Dr Geoffrey James, (Ph.D.) *Solution Support trainer and practitioner, author of* "Transforming behaviour in the classroom – a solution focused guide for new teachers" *(2016)* Sage *and* "Solutions Focused Coaching Workbook for Educators" *(2019) Singular thesolutionsfocusedcoach.com*

Jennifer Anne Hawkins

Brain Plasticity and Learning

Implications for Educational Practice

palgrave
macmillan

Jennifer Anne Hawkins
Liverpool Merseyside, Cheshire, UK

ISBN 978-3-030-83529-3 ISBN 978-3-030-83530-9 (eBook)
https://doi.org/10.1007/978-3-030-83530-9

This Palgrave Macmillan imprint is published by the registered company Springer Nature Switzerland AG.
The registered company address is: Gewerbestrasse 11, 6330 Cham, Switzerland

To fellow researchers on this subject, my family and friends

Brain Plasticity and Learning: Foreword

Jennifer Hawkins proposes in her introduction that this book is going to be a journey through 'eclectic phenomenological research' relating to the plasticity of human cognition in the learning journey, and this is a good description of the contents of the text.

In this text, she considers the plasticity of human cognition in understanding and managing the self, and how such concepts might be used to improve and justify teaching practice for the teacher; an ambitious breadth of focus. She reflects upon the need to recognise the influence of culture in its broadest construction, not only in the sense of the need for sensitivity to the norms and values of different ethnicities, but in the sense of how societies define special educational needs and for what reasons.

A major theme permeating the text is children's huge cognitive flexibility, which bestows a great potential for learning, and consequently raises the equal requirement for teachers to remain flexible. The curriculum in turn needs to reflect the significant impact of the environment upon learners' chances of success, and particularly to ensure that potential is not curtailed in attempts to force 'square pegs' into 'round holes.' In this context, she raises important questions for very narrowly framed curriculums; for example England's National Curriculum.

Having created this panoramic perspective, she subsequently considers problems resulting from narrow conceptions of assessment. In particular, she focuses upon how these may too quickly label children as being in need of remedial measures, whilst they might be supported to achieve more successfully in a system that frames learning in a less rigid fashion.

Within this context, she explores a plastic brain in a rigid system and makes some useful observations of problems that may result.

On a similar theme, she visits the blurred lines between learning and indoctrination and makes some observations that are timely for England's education system. She highlights the manner in which it currently appears to be losing its way, encasing children in a culturally narrow pedagogy which constructs learning as the rote memorisation of fixed 'facts.'

Her final chapter moves to the role of economics as the underpinning ethos for this system; how people are constructed as profit-making units, and how curriculum is constructed from the basis of what children need to know for society to extract the maximum profit from their contribution to an overwhelming national, rather than increasingly global economy; a short sighted policy where information travels around the world via the tap of a screen.

This is a book that ranges widely and encourages the reader to construct both human beings and their societies as highly flexible entities. It left me contemplating how we might reimagine education to permit our plastic brains more space to imagine, invent and create, and to harness multi-cultural information to underpin our education processes.

I have considered this issue within my own writing and research and constantly raised the question why education in England in particular seems to take so little account of the way that human beings think, develop and learn, but starts from the idea of what the government would like them to be. Over twenty years ago, Singer (1999, p. 61) asked why, instead of trying to force human beings into systems that do not suit their biology or psychology, why we do not create 'policies ... grounded on the best available evidence of what human beings are like.' This question is just as salient today, and Hawkins takes the reader through a journey that explores this question in a wide ranging, eclectic narrative.

Leeds, West Yorkshire, UK　　　　Pam Jarvis

References

Singer, P. (1999). *A Darwinian Left: Politics, Evolution and Cooperation*. Weidenfeld and Nicolson.

Preface

In this book I look at world trends affecting education and discuss issues involved in teacher, parent and learner experiences connected to brain plasticity, which in one way or another affect us all through life. I am a teacher with a lot of curiosity, who likes to challenge 'obvious' assumptions and uncover what may lie beneath. If you read my first book (2017), you will know my journey is a continuous one as I research and learn. Ideas in this book will connect you to research in different social contexts, stories about discoveries in neuroplasticity and clinical psychology, stories about learners and teacher explanations of learning. I link neuroscience and psychological research to practitioners' narrative evidence and look for connections around how human beings learn in different situations and settings. I looked at possible ways to understand how brain plasticity relates to teaching and learning.

In the process of writing this book I have read about, met and had conversations with teachers, therapists, learners, parents, social experts, authors and psychologists. However, there are still many more 'experts' on learning out there who are professionally and unofficially recognised. It is a fascinating complicated world and education is a rich and complex field. I explore available educational advice to find out if such information can shed light on why and when some educational approaches work or fail and in what context. This phenomenological research produced a body of data that synthesises, elucidates and demonstrates the wisdom shown by all kinds of teachers, parents and learners every day. As a teacher, I hope such research may inform us about how to be more successful in our everyday practice.

Some of you suggested books, websites and webcams and shared ideas in your blogs, on Twitter and LinkedIn and by email (over 1250 papers, blogs and books). Thank you—your varied data and analyses are thought provoking and insightful. The text and backup references may be helpful for other psychological, philosophical or social educational researchers pursuing their own agenda. There are links to psychology, philosophy, technology, politics, economics and sociology. I believe your research is important for the future. I apologise to those I have inevitably missed and look forward to your constructive criticism. The 'online' information was useful, however sadly the internet links referenced will vary in longevity and are always open to author editing or removal.

I have not been able to reference directly in the text all the references I have read and considered, but nevertheless they have informed my work and are included as backup data in the reference lists for each chapter. I hope the links and books suggested for additional reading in these specific areas may be useful for others as they plan, deliver and evaluate their own and other people's learning in different ways. Where possible I substantiated your ideas discussed by asking for unpublished written data or referenced your books, papers and articles. Some of you gave your time in person to research collaboratively chatting by phone, video link and in conversations in schools and at events. I am particularly grateful to those who shared their personal stories and thoughts with supporting data contributions.

The book deploys the research data as appropriate to different chapters, sections and themes. Inevitably in dealing with such a complex subject this involves a great deal of overlapping of related themes. There are many possibilities and so inevitably I tended to choose those I found were at the time of current concern in educational and public discussion. I started off by making connections between educational and psychological evidence looking towards further developing an active 'feelings' learning theory and made links to neurocognitive science. When I ended I discovered I had found many starting points which I could not possibly follow by myself! I hope others will follow up and research these new pathways.

Liverpool Merseyside, Cheshire Jennifer Anne Hawkins

Acknowledgements

Thank you to Andy Williams and Helen Pitt for sharing their management issues, friendship, their staff and pupils with me at Lunt's Heath Primary School. As always extra special thanks to David Lobb for his unfailing support, interest and encouragement. I would also like to thank my daughters Claire Teague and Lucy Jones and their families for their love and support. More thanks to Yvonne Metcalf and Regina Tsaliovich and all the data contributors and folk who messaged me on EduTwitter and LinkedIn.

Special Thanks to the Following Special Data Contributors

Chapter 1: 'Journey of Peace' (2019) by Joseph Critchlow, aged 14—St. Vincent's School for Sensory Impairment, Liverpool, United Kingdom.
Chapter 2: 'Background Knowledge' (2020) by Dr Anna Maria Rostomyan, Corporate Communications Specialist, Yerevan State University.
Chapter 3: 'Too Old to Suffer' (2014) Chris K. Pearson, Video and Poem Transcript, a poem about emotions for his daughter.
Chapter 5: 'How it feels to be a new parent of an autistic child' (2020) by Kirsty Henderson on Twitter, 23 July 2020, 20 tweets, 5 min read.
Chapter 6: 'Happiness' (2019) A poem from 'My Mind's House' by Dr Christine Challen, 20 August 2019.

Chapter 7: 'My Education Journey' (2021) by Muhammad Shehu Shuaibu, 20 January 2021.

Chapter 8: 'Confidence: Anything you can do once, you can do again better. Learning to invest through failure' (2019) by Sifu John, Wing Chun martial arts master, March 2019.

Contents

CHAPTER 1

The Discovery and Implications of Neuroplasticity

1.1 Introduction

In my first book I researched some people's individual responses to learning and pointed out that their feelings and emotions made sense to them. I found these phenomena helped to explain their thinking and behaviour when linked to their history and circumstances. At first sight it is obvious that human thinking works in this way. However, when I looked at academic literature, teaching management and even practice on the ground, the habitual 'disconnect' in thinking was evident. We assume we know how 'we think' and the role of feelings and emotions in our thoughts, but although we are often 'driven' by them, we still tend to dismiss them. This subject has never been *sufficiently* acknowledged as an area of inquiry by academics or even teacher leaders, practitioners and learners themselves. It seems as though a great many of us have never fully valued our own humanity and diversity.

In this book my thinking and aims are as follows. If there is a 'disconnect' in human thinking, if we only acknowledge our feelings when it suits us to do so and if we know they are present in most of our thinking. It is time to ask ourselves—what is their role and can neuroscience help us to

Special thanks to Joseph Critchlow, for 'Journey of Peace' (2019) Prizewinning Essay, pupil aged 14—St. Vincent's School for Sensory Impairment, Liverpool, United Kingdom.

J. A. Hawkins, *Brain Plasticity and Learning*,
https://doi.org/10.1007/978-3-030-83530-9_1

understand these processes better? I invite you to join me in looking for ways to understand how brain plasticity, feelings and emotions influence human learning and to find out whether this kind of research can inform our teaching. My method of research is an eclectic phenomenological one, explained in my first book (Hawkins, 2017). Research into emotional learning processes is a relatively new field in psychology. No one person can cover every possible aspect of the subject. However, I am interested in making a start to find out if it is possible to develop the learning theory previously discussed across a wider macro-system (Bronfenbrenner, 1979).

We need to check ideas out, understand and manage ourselves better if we are to develop our own neuroplasticity and survive the physical, social and environmental challenges of our age. In doing this research I suggest, therefore, we need to take a respectful and rational approach to other people's opinions and frames of reference.

Some incidental questions might be:

- Is there any evidence to prove that feelings and emotions create logical connections in the brain, and are there feelings that are not emotional?
- How and why do emotions add so significantly to important personal learning experiences? For example, survival, motivation, confidence, achievement, pleasure, creation, practical gain, a sense of well-being, demotivation, fear, disempowerment, hopelessness and decline.
- How can different understandings generate new solutions to learning difficulties, a particular compromise or a fresh idea?

I have positioned most of this researching discussion in the field of education hoping to discover ways we may be able to improve and justify our future work as teachers. However, a full debate informed by many other disciplines by people from many other sociocultural settings than my own is essential to developing our knowledge on this subject. There will be many valuable ideas and approaches in very different cultural contexts out there of which I am unaware. Historically grounded critical research by people of other countries, social and cultural groups is urgently needed. We have much to learn from others about social settings, nations, ethnic groupings and civilisations where people represent different and similar points of view across a range of cultures. *We are all different, but we are all part of the human family.*

Clinical psychology and neuroscience research discovered brain plasticity processes some time ago. It is taking time for us to understand the implications. We are accumulating a new body of more detailed physical, scientific and medical neuroplasticity information and narrative experience. As a species we have always searched for meanings, but we now have more information about how we think than ever before. We can take a more enlightened view and learn to understand ourselves in more intelligent ways—particularly with regard to our capacity for active and rapid assessment, reassessment and prediction. Our brain plasticity has helped and hindered adjustment in learning as human knowledge has developed. For example, we found new solutions, but sometimes convinced ourselves mistaken ideas were correct.

At last we are in a position to survey and reconsider educational, medical and psychological research analyses. The operational diversity of brain functions through bodily communication systems has the potential to be much better understood. It is scientific to recognise that *feelings and emotional intelligence* skills interacting and informing our thinking not only are dynamic, intensely personal and linked to changeable situations but are an important biological necessity. It is time to start seriously challenging established assumptions, reassessing, analysing and reimagining to create new areas of knowledge.

1.2 A Short Summary of the Historic Background

During the past 100 years brain and body neuroplasticity has become a leading field of study in the world. Scientists have taken a long and surprisingly varied route to its discovery. They eventually proved that brain plasticity is an essential electrochemical life force affecting and enabling all aspects of all of our lives. However, there are still many aspects of how it actually works being researched all over the world for many purposes. The term 'neuroplasticity' is derived from 'neuron,' a nerve cell, and 'plasticity' meaning malleable, modifiable, changeable, adaptable, alterable, fluid, mouldable or impressionable. Although 'neuroplasticity' is definitely proved, scientists still have much to learn about it.

Our recent realisation of the existence of brain plasticity changes many of our previous assumptions about ourselves. The full implications still aren't fully understood and have yet to be researched and related to other disciplines. Neural plasticity is demonstrated in real time through neural imaging. Even an awareness of its very existence may give us choices and possibilities for future self and group development. It can offer us different perspectives on ourselves. I am looking, from a teacher's point of view, at neuroscience and clinical psychology discoveries about plasticity that may be important for teaching and learning.

This is an area of knowledge that has potential for providing new perspectives on research and development in many social academic and practitioner disciplines. Neuroplasticity might be researched across a range of subjects in education. It could even help explain and justify existing educational approaches, beliefs and practices and help develop new ones. It might help us to understand ourselves better! I am researching this from my own teacher perspective as I invite you to form your own opinions. I am starting off by exploring some background information. We know that intelligence systems throughout our brains and bodies have operated over millennia adapting to complex life environments and cultures.

In the 1970s Richard Klein, a paleoanthropologist at Stanford University, suggested a genetic mutation in the human brain 40,000 years ago caused the appearance of the 'modern' version (2002). Since then archaeologists excavating and studying skulls in Africa have started to dispute this theory. They found artefacts that show evidence of 'symbolic behaviour' in earlier and yet earlier ages. For example, pigments made from red ochre, perforated shell beads and ostrich shells engraved with geometric designs in South Africa dated to more than 70,000 years ago and even back as early as 164,000 years ago. More and more anthropologists agree that modern cognition was probably in place when Homo sapiens first appeared. See Article Endnote.[1]

Although it is hard to understand intelligence from skull remains, there are clues like evidence of early bone surgery, even healed trepanning operations (incisions made to remove and replace parts of the skull crown) in ancient Egypt. According to Matthew Cobb, two Greek anatomists, Herophilus and Erasistratus living in Alexandria around and after the time of Aristotle, were known to have dissected the brain and nervous system. Herophilus is credited with describing the cortex (large brain lobes) and the cerebellum, spinal cord, nervous system, motor and optic nerves (2020). I recommend Cobb's book *The Idea of the Brain: A History* for its scholarship and erudition. We will refer to it at various points in this book.

Cobb says in his opinion there are two main problems in understanding the brain. The first is that within our knowable universe to date it is the most complex object we have found. The second is that in spite of the massive 'tsunami' of data produced from research around the world, brain science is really still in its infancy. His book focuses upon what we have already thought about this amazing organ through history, what we presently know and some ideas about the future. He says that in spite of all our accumulated knowledge we have a crisis in our understanding of our own mental health, our levels of consciousness and how some aspects of the modern computer science of 'deep learning' actually works.

The early scientist Descartes (1596–1650) discovered that the heart pumps blood around the body. This discovery may have encouraged the idea that the body is a 'machine' with separate parts that are joined together. Surgeons continued to develop the idea of 'localisation' in the nineteenth century, because they found different damaged areas of the brain under surgery in particular patients were responsible for particular disabilities. A further assumption was made that these injuries were likely to be permanent and the body incapable of recovery. Santiago Ramon y Cajal (1852–1934), a Spanish pathologist and neuroanatomist, studied the central nervous system and was one of the first neuroscientists to make detailed drawings of the microscopic structure of the brain.

By the end of the nineteenth century it was known that the functioning adult human brain has around 100 billion neurons within the brain and nervous systems. Cobb tells us that the idea of explaining bodily organs as machine-like has been popular since the beginning of Western science. Nature and the universe were believed to be vast 'mechanisms' which obeyed the then known laws of physics. This idea led to a number of common assumptions about ourselves as creatures. We believed the brain and body operated through fixed mechanical processes. We thought we had a set amount of potential intelligence at birth and if body parts failed, they might not perform their appointed tasks again. We assumed we would not recover their use. We often learned instead to adapt behaviour to compensate for their loss (Cobb, 2020; Doidge, 2007).

We now know our negative assumptions about ourselves sometimes prevent adaptation and recovery. Our initial response to the brain is to make analogies to explain it. I did this myself when I said that "*our brains are biological machines or engines that constantly adjust to conditions and adapt to events in order to help us survive*" (Hawkins, 2017). This is inadequate, because the brain cannot be properly understood by likening it to a non-biological creation. It seems better to think of the brain as a complex, living and evolving natural organism, which responds to physical and mental influences both common to our species and personal.

I am a narrative researcher and a lay person in this field and have found the stories compiled by Dr Norman Doidge, a psychiatrist and psychoanalyst in Canada and the United States helpful. He has recorded narrative information about the discovery of 'brain' plasticity from before the 1960s up to recent times. His first book is called *The Brain that Changes Itself* (2007), and his second book *The Brain's Way of Healing* (2015a) continues his explanations in story form. As he writes he gives an analysis of his own conclusions. Although not his direct focus, Doidge's narrative research on the discovery of neuroplasticity also clearly demonstrates the importance of feelings and emotion for how we learn.

Doidge travelled the world recording key events in the discovery of neuroplasticity talking to neuroscientists, psychologists, doctors and therapists. They gave him their technical conclusions together with their opinions and insights and those of their patients. Doidge records their often incidental discoveries and how they researched, tested and evidenced them in laboratories and clinics. These stories resonate with me as a teacher. I begin to understand the ways human physical biology affects us all. I can make connections between my own and other teachers' experiences and the learning of the people described. He tells us the brain's amazing ability to recover was first demonstrated in the work of Russian neuropsychologist Alexander Romanovich Luria (1902–1977).

Luria's severe injury reports gave analyses of the functioning of various brains and their evidenced ability to cope with a variety of specific disabilities. His job was to treat soldiers injured in the Second World War. He researched diverse neuropsychological conditions by assessing and documenting his patients' stories as they recovered from a variety of brain-connected injuries. Luria's books *Higher Cortical Functions in Man* (1962) and *The Working Brain* (1973) are still used as reference works. Incidentally, before the war Luria researched into linguistics looking at the psycho-semantics or attribution of human meanings to words. He was a friend of the educational theorist Vygotsky during the 1920s and 1930s.

At first neuroplasticity was denied and resisted by traditional scientists because of their fixed traditional assumptions. Edward Taub's (1931…) work was famously discredited because of public concern about experimenting on live animals. Fellow scientists eventually accepted his findings when he was reinstated and his work acknowledged. Along with the work of other such scientists, Taub's work and that of Michael Merzenich (1942…) using monkeys has led to a much better understanding of how to treat stroke, brain damage, paralysis and cerebral palsy in humans. This was the pathway taken by neuroscientists, which led to the discovery and demonstration of explicit ways brain plasticity is vital in regeneration and recovery of parts and functions of the human body (Cobb, 2020; Doidge, 2007). See Video Endnote.[2]

In North America the Behaviourist school discovered they could 'teach' a habitual learned reaction by stimulating response behaviours in animals and humans (Skinner et al., 1957). They sometimes discovered that it was relatively easy to produce predictable and fairly consistent reactions in human groups, particularly in social situations. Many of these could be learned and strengthened by a repetition process they called 'conditioning.' However, this only happened in particularly contrived and *conducive* artificial situations. At the time, scientists generally discounted feelings, complex motivations, desires and emotions and many considered these to be an 'inferior' kind of subjective experience.

The discovery of neuroplasticity was made not only through its mapping of live electric connection through imaging, but also through clinical medical therapy. Psychologists, therapists and doctors found out about the brain by working with live patients. Out in the field teachers have always known that conditioning is useful in teaching and many applied it appropriately and with kindness. However, behaviourists persisted in ignoring the right to informed choice, free will and the brain's own individuality, motivation and adaptability. They were not curious about the human mind in the ethical sense or interested in their 'subject's' opinions (Doidge, 2007).

Behaviourists found that the traditional scientific method of repetition under controlled conditions lends itself to successful manipulation of human behaviour. They saw that behaviour was often a response to a habitual stimulus, later on discovering behaviour is affected by the production of 'reward' chemicals such as endorphins and dopamine. The internal responses, subject's opinions, thinking and the human consequences were not investigated or recorded. In spite of evidence all around them to the contrary and perhaps because they had no traditional scientific means to record them, they seem to have decided these could not be 'scientifically' assessed.

Behaviourists tended not to acknowledge that different forms of coercion and/or conditioning played a part in their experiments, for example, social conformity, reward and punishment, fear of failure, perceptions of real or imaginary threat, ridicule and so on. The beneficiaries of their work have generally continued not to seriously consider or to deliberately ignore their subjects' feelings and emotions or the ethical consequences of their experiments on populations. For example, some of the socially damaging effects of political advertising and social media companies today.

The behaviourists' assumptions diverted attention away from some potentially fascinating fields of dynamic context-based psychological and physiological research for which they had laid the basis. Their discoveries about human learning requiring repetition and conditioning could have been less exploitative. They could have informed and validated interventions that aid learning, while maintaining and respecting human values. Their 'objective standpoint' and rationale was eventually superseded by the development of qualitative and mixed-method research, but this has yet to have its full effect on society. Such research and analyses would benefit from a context agreed system of informed dialectic inquiry—that is by both researchers and 'subjects.'

Taub is a behavioural psychologist who has developed his work in a more therapeutic direction. For example, a counter-intuitive treatment was devised later developed by Taub called 'constraint-induced movement therapy' helping to rehabilitate people who have developed a common condition called 'learned non-use' resulting from neurological injuries due to a stroke. *Our understanding is now profoundly changed.* Neuroscience research is telling us more about how our bodies and brains are capable of partial or even complete recovery of functionality even when we don't expect it. Our bodies have their own unique and similar organic systems for regeneration and for developing ability and intelligence throughout life.

It turns out our human beliefs, motivations, feelings and emotions are integral and crucial to this process. Research into neuroplastic learning by neuroscientists using imaging techniques are telling us about human abilities and deficits to explore further in relation to teaching. Random trials across different populations by collecting standardised statistics are of limited use, but the considerable benefits of researching with individuals in-depth are evident in Doidge's books. Meanwhile the ethics of experimentation upon animals rightly remains a debate of public concern.

1.3 Firing and Wiring with Neuroplasticity Points List at the End

Donald Hebb (1949) is credited with being the first to describe the ability of neurons to connect by 'firing and wiring.' Hebb's theory was that the changes in neuronal structures are caused by responding to experience, an idea apparently proposed by Sigmund Freud 60 years before. The detailed process of electrical connection between neurons was first demonstrated, observed and recorded under laboratory conditions by Norwegian physiologist Terje Lomo in 1973. Lomo and neuroscientist Timothy Bliss discovered that [when] synaptic cells were given an extra high-frequency electrical stimulus, they developed a long-term enhancement response. The cells when tested later responded again and again even when stimulated at a lower level. See Article Endnote.[3]

This was called 'long-lasting potentiation,' Timothy Bliss collaborated with Lomo and the two published a report in 1973. It has now become known as long-term potentiation or LTP and is perhaps the basis of human memory. It was eventually proved that plasticity was indisputable in childhood, but traditional scientists were still reluctant to accept that it is continuous throughout life (Cobb, 2020; Greenfield, 2014, Doidge 2007). This makes me wonder (perhaps naively) if we may suppose and prove that situations creating effects of strong sensory stimulus such as being in novel and exciting situations can create a stronger electrical 'buzz' that aids memory, for example, school trips, the arts and outdoor pursuits.

Specific conditions for development are acknowledged to be necessary for brain and body growth at specific stages. For example, exposure to light is necessary at a particular stage for the development of the physical 'apparatus' for sight. In the 1960s David Hubel and Torsten Wiesel examined the visual cortex in kittens and discovered that deprivation of light in one eye during early development caused blindness for life in that eye. This is interesting from a learning and teaching point of view. It proves that deprivation causes loss and it also proves there are indeed some physical windows of opportunity in developmental growth (Vygotsky, 1896–1934; Freud, 1856–1939).

Since we now know about plasticity, it may also be possible to stimulate those physical mental growth opportunities at later stages. Doidge mentions that goslings have an instinct to follow a moving creature or object in the absence of a mother goose for a short period straight after hatching. The process is referred to as 'imprinting' by Austrian zoologist Konrad Lorenz (1903–1989). *The implication for learning development is that mental and physical connections are impossible to separate in practice and so both should be taken into account when planning interventions.* We should, perhaps, use more practical 'holistic' approaches in teaching and more physiotherapeutic approaches in medicine.

Important scientists in the history of neuroscience focused on their own different projects, discovering different aspects of plasticity. Michael

Merzenich (1942–) is a key neuroscientist who performed many complicated surgical experiments upon monkeys' brains micro-mapping with microelectrodes. Although perhaps unacceptably invasive today, this proved more precise and more informative in speed and accuracy than even some current methods of brain scanning. For example, he had to make 500 separate insertions to map a monkey's hand to neuronal maps in the brain. This gives us an indication of the complexity of our 'body-brains.'

Merzenich discovered with colleagues much more than previously known about the complicated connections made by neuronal systems in the body. In one groundbreaking experiment he proved the process of adult brain plasticity by observing its effects in real time under laboratory conditions. He carried out a piece of research in which he sewed a monkey's two fingers together observing the neural connections in the brain controlling their separate movement. After several months the two neural nets for each finger became one. They adapted back into two in a similar way when the fingers were separated again. This demonstrates the action and facility of neural plasticity in the brain's degenerative and regenerative processes at any stage of life (Doidge, 2007, 2015a).

Doidge explains that the command and control centre for the human body is the central nervous system—the brain and spinal cord. This is the essential two-way communication highway to and from the peripheral nervous system in muscles and glands. There are about 80–100 billion neurons, which may receive and react to electrical signals. *These electrical signals can travel at between 2 and 200 miles per hour through* axons *to excite or inhibit different areas of the nervous system.* See Video Endnotes.[4,5]

Neurons can retain constant functions, but are also capable of changing their formations of connection (neural nets). Each neuron has a number of dendrites sustaining and connecting it to other neurons by means of axons, varying in length from microscopic to very long 'wires'—extending and carrying messages all over the body. Synapses are the microscopic spaces between dendrites and axons. Micro-transmitters are the chemical messengers, which cross synapses when stimulated and messaged through electrical stimuli. We are complicated animals indeed!

Merzenich finally realised the extraordinarily versatile nature of plasticity when he observed how cut off nerves could regenerate and 'grow back.' Regeneration was known about, but had never been observed so clearly in action and with so much complex definition. In a series of experiments Merzenich proved that nerves could be re-designated by the surgeon, but were then reassigned by the brain itself in order to reconnect and be directed by different brain regions. The brain was proved to be able to self-normalise its structure *in alternative ways*, reconfiguring connections and restoring itself to functionality.

Merzenich proved the body was not hard wired and body parts could no longer be thought of as separate static pieces of equipment because *the brain itself is able to reconstitute, redeploy and reuse them under the 'right' conditions.* These were breakthrough groundbreaking events and even some of Merzenich's fellow researchers questioned the findings at first. The experiments were eventually fully proved and accepted as irrefutable. The process is illustrated by the iconic phrase, *"neurons that fire together wire together"*—created by Carla Shatz.

Neuroscientists have now observed plastic processes as they happen in the living brain operating in real time in a whole variety of research situations. They have devised several ways to observe electrochemical neuronal activity using various neuro-scientific observation techniques. See Brain Scanning Methods Endnote List.[6] All of these technical methods of observation show that the human brain has the ability to create and eradicate, alter and develop neural maps or nets with complex electrochemical connections, that is, *'plasticity.'* Neural images clearly show the physical adaptability of the brain as it responds to situations it encounters. The central implication for social as well as medical researchers is the realisation that plastic regeneration and deterioration is an ongoing human biological process.

I find this interesting because my father lost feeling in one leg following an operation on his spine after the Second World War. He told me he could feel his sciatic nerve growing back and that he had started to feel his toes again about 20 years later. It seems this was perhaps due to brain reconfiguration and reconnection as well the nerve sensation itself 'growing' or reconnecting in the leg. These are difficult *feeling sensation* processes to imagine. They can only truly be evidenced by patient description in response to surgical intervention, medication and therapy, recovery outcomes and behaviour as well as neural imaging. The fact that he was motivated to go for long walks every day and stayed active into old age was probably a recovery factor.

Questions then arise for all of us as to how we can use this personal information. The evidence suggests the natural process of plasticity is much more effectively activated if the patient's mind is empowered and encouraged to collaborate in the process. This involves awareness, empathy and imagination on the part of the therapist (or surgeon) in the development of interventions, skills and knowledge. For example, a woman who played her violin as the surgeon removed her brain tumour so that her playing ability might be less likely to be affected. See Video Endnote.[7]

Modern science is a relatively young discipline, but its recently acquired 'traditional' ideas and attitudes still permeate Western lay cultures and encourage us to continue to misunderstand our own and other people's potential. We still assume that human brains are fixed and do not allow sufficiently for physical difference, regeneration, degeneration and self-recovery. For example, our expectation that medicines will necessarily be a solution to disability when this is not necessarily true. This causes us to neglect the importance of the need to research and develop alternative medicines and ancient organic folk remedies, physical therapies, operational strategies and social nurturing in an open-minded manner.

As well as specific medical treatments human recovery from illness and accidents depends to some extent on luck as well as individual biology, type and severity of the injury. However, the physical environments, social contexts, attitudes and beliefs of those involved—*significantly not only those of the patient but also of families and clinicians*—are more important than we thought. The ability of the brain to direct the body to make its own compensations and recover itself (under conducive and encouraging conditions) has not always been appreciated in modern traditionally based Western medicine—feeling, well-being and emotionality were not generally considered important in traditional medical practice.

Important neuroplastic reference points for research into learning and teaching:

In his scientific research with monkeys Merzenich discovered and proved:

- Neural maps change their borders, become greater and less detailed, move around the brain and can even disappear.
- Neurons tend to connect to one another when activated at the same moment—*"neurons that fire together wire together"*—a phrase created by Carla Shatz.
- Brain maps tend to organise themselves in groups that relate to common sequences of actions that frequently happen.
- Those neural maps that do not fire at the same time tend to be further away from each other—*"neurons that fire apart wire apart."*

- The brain responds plastically and adapts when a person is motivated to learn.
- When we start to learn a physical skill, we use a whole range of superfluous bodily movements, which gradually reduce with practice of the skill as the neural maps are embedded. We stop using irrelevant muscles and fine tune our bodies to the particular task more efficiently using fewer neural maps.
- In a neural map dealing with the sense of touch each neuron relates to a particular area of skin on the body. As the sense of touch becomes more careful and precise neurons relate to smaller areas of skin used and the neural map becomes more discriminatory and able to fine tune.
- The speed at which we think is itself plastic and variable.
- When we have learned to do a new learning task, the processing speed between those neurons connecting increases as we become more and more proficient.
- As we repeat a learning task the signals in neural maps tend to become stronger and clearer until they are established, but even so 'if you don't use it you will eventually lose it' as it withers, fades or becomes dormant.
- Lasting changes in brain plasticity only occur if a person is motivated to focus and pay close attention.
- Learning separate and different tasks *simultaneously* (as in multitasking) tends to be counterproductive for deep and long-term learning.

1.4 Neurodiversity Includes Neurodivergence and 'Disability'

Traditional cognitive scientists tried to learn about the brain by 'comparing,' in their terms 'normally' or 'typically' functioning people to those they considered to be experiencing a 'mental health difficulty' or 'physical disability.' They tended not to investigate particularly healthy or particularly intelligent individuals or those with extraordinary abilities and lives. This has meant that within the field of disability medicine, breakthroughs and progress have often been achieved by insights and persistence in the face of the dis-encouragement and ignorance of closed-minded 'experts.' For example, it is entirely possible for individuals to demonstrate exceptional human abilities in some areas in combination with a wide variety and degree of disability in others such as the famous English theoretical physicist and cosmologist Stephen Hawking (1942–2018). In the long, painful and fascinating process of the discovery of neuroplasticity, difference, human diversity, resilience and determination have been of paramount importance.

Neuroplasticity appears to consist of multiple natural biological, electrochemical and 'mini-engineering' processes we still do not fully comprehend. However, human perceptions, free will and agency are fundamental to its operation. Merzenich theorised that people with learning difficulties, psychological problems, disabilities, stroke or brain damage might be helped and he eventually went on to prove this through his work with patients (2013). Knowledge about brain plasticity is accumulating in fits and starts through the painstaking detailed work of neuro-therapists. Such research and discovery is often unrecognised initially but eventually informs the knowledge base. It is often only possible to gain an overview in retrospect. There is groundbreaking work going on as visually impaired student Joseph Critchlow proved when he won an award for his philosophical essay writing. See Video Endnote[8] and Special Data Contribution.[9]

Teachers' and scientists' own disability stories sometimes motivate original research and provide lessons in neuroplasticity. Their work with patients and students to develop new physiotherapy procedures has important implications for our understanding about ways to leverage solutions in all areas of the social and behavioural sciences. For example, Canadian teacher Barbara Arrowsmith-Young (1951–), who had multiple learning difficulties at birth and learned how to help herself. She has a Master's degree and is the founder of a school in Toronto which teaches children with learning difficulties. Along with colleagues she has developed the Arrowsmith Cognitive Exercises Program. See Website Article.[10]

The motivations for her lifelong work came originally from her own journey of discovery in overcoming her own severe learning disabilities (Arrowsmith-Young, 2012). She was born with severe multiple learning disabilities from birth. She was physically uncoordinated, read and wrote everything backwards, had trouble processing concepts in language and continuously got lost. According to Doidge (2007), she has an exceptional auditory and visual memory and 'remarkably developed' frontal lobes enabling her to process information when she has absorbed it. Her brain was 'asymmetrical' and her abilities coexisted with 'areas of retardation.' After many struggles with a difficult compensatory education, which taught her to hide or mask her problems, Arrowsmith-Young worked out how to pass exams.

She read academic papers 20 times, learning the complete contents using her remarkable memory through repetition and rote. However, this was still not a solution and she was still depressed and frustrated by her difficulties. Looking for answers as an adult, she found useful clues to her condition from Luria's writing on the recovery of damaged brains. She was particularly impressed with the case of a wounded soldier called Zazetsky. Luria (1902–1977) told how a bullet lodged in Zazetsky's brain affected the juncture between three major perceptual areas. These were the temporal lobe (sound and language), the parietal lobe (spatial relationships and the senses) and the occipital lobe (visual images).

She realised she had similar experiences and problems in her life to those Zazetsky talked about. Suddenly she knew she was not the only person to experience these specific problems. Still looking for a remedy at 28, she came across Mark Rosenzweig's (1922–2009) research on neuroplastic growth in rats. He had discovered in post-mortems that rats living in stimulating environments had more 'developed' brains. He proved that brain-stimulating activity can produce physical change—increasing brain capacity (1987, 1962). Arrowsmith-Young decided to work on stimulating her own brain. She worked on her spatial recognition, sound and language and visual imagery recognition centres. She devised flash cards to read a variety of clock faces which she practised repeatedly to recognise, gradually increasing difficulty.

Her identification of her specific problems and her realisation about neuroplasticity had helped her create a way to rebuild and 'kick start' her own neural connectivity. The intervention she devised and developed turned out to have *remedied abilities in other of her 'problem' perceptual areas*, which had caused her to live much of her childhood *'in a fog.'* She continues to tailor and develop programmes for students with specific learning difficulties to this day. In spite of some opposition from traditional experts, they continue to prove their worth with students. See Chap. 3. The Plastic Brain and Its Educational Development.

South African John Pepper provided Doidge with an important long-term case study that illustrates similar findings about neuroplasticity in a different case of disability. Pepper was diagnosed with Parkinson's disease in his 30s—a virtual death sentence at the time. He found out that meaningful walking by taking enjoyable nature walks every day kept his symptoms at bay. This occupied and challenged his body *and mind* better than repetitive routine exercises in the gym. Although it was not a cure, this special type of activity was useful to him. Presumably it stimulated his brain circulation, body and central nervous system helping him fight the effects of his condition.

Doidge references Pepper's story in both his books and describes his achievements and difficulties in great detail. Pepper wrote his own book of his experiences and was in his 90s at the time of Doidge's second book (2015). He visited Pepper several times, looked at his medical records and talked to his medical consultant, who explained Pepper had at no time been cured of his illness. His symptoms always returned if, for any reason, he could not take his special walks every day. Pepper was accused of being a fraud by two doctors in the Parkinson's organisation to which he belonged. These doctors were unable to accept he could counteract the symptoms of this 'normally' degenerative disease.

Although Pepper did not recommend it, the doctors believed other patients would give up their medication to follow him, leading to their inevitable degeneration. They refused to take an action research approach to this new information, which had not been included in their training.

Pepper's walks stimulated his nervous system with feeling sensations of mind and body out of doors on the sea shore. He studied the birds and penguins on his walks and scrambled over rocks. His neural connectivity was stimulated by physical sensations, exercise and mental feelings in combination, facilitating the maintenance of good health. The intervention worked for him in a personal way far more effectively than medication. This daily work helped him to maintain active neural pathways and remain healthy for as long as he carried it out.

Vilayanur Ramachandran (1951–), a leading Asian Indian American clinical neurologist, says he finds out more about how normal brains function by helping patients who have unusual problems. He finds these cases of more value than 'normal' brains because they reveal differences and capabilities that challenge conventional assumptions. They have informed him about neurological 'oddness' and so helped him understand more about 'ordinary' brain function as well as specific disability (2003, 2002). His groundbreaking research involves visual perception, phantom limbs, synaesthesia, autism, body integrity, 'identity disorder' and mirror therapy. See Video Endnote.[11]

Ramachandran in a series of groundbreaking experiments collaborated with accident victims with a variety of medical conditions causing chronic pain. Amputees who experience phantom limb pain and 'sensory ghost' experiences as a kind of biological 'virtual reality' show that feelings and perceptions are an integral component in motivating the brain to 'rewire' itself. Abandoning the traditional medical model of chemical pain relief, Ramachandran used mirrors to help patients' brains revisualise their lost limbs. They were then able to 'learn' afresh to reset their neural maps to accept the loss. This stopped the pain. He discovered that supporting them to adjust dormant bodily 'memories' of pain helped the brain to 'reboot' itself. See Video Endnotes.[12,13]

Michael Moskowitz, psychotherapist and neuroscience researcher was informed by his personal experience of chronic pain after a traumatic skiing accident. In studying the effects of neuronal activity on pain, his own experiences helped him to find ways to help his patients weaken pain by focusing on the mind. Neuroscientists' increasing awareness of new interventions involving brain plasticity is making them appreciate the value of engaging the co-operation and collaboration of the 'subject' in their research. Unique case narratives constantly contribute information to the body of knowledge about plasticity. The study of the brain through imaging is not sufficient in itself. When investigating some problems, qualitative feedback data and scientific testing are also informative. The two research approaches are not mutually exclusive and may complement each other. See Article Endnote.[14]

As we have seen above, Doidge related personal stories of disabled people and the experts who found ways to help them. As a psychiatrist and psychotherapist, his journey informs his and our understanding about the possibilities for neuroplastic healing. There is a great deal of straightforward logical effort, feeling and insight involved in these stories, both on the part of the medical therapist and the patient as the work is done through trial and error. He shows how empowering the patient enables therapies to be developed for patients previously thought to be beyond recovery (2007, 2015a). Patients were encouraged to do various repeat exercises, creating and reviving neural pathways by practice and their interactions with the world. They benefited from following, initiating and persevering in healing behaviours. Research approaches such as clinical analysis, traditional medical observation, chemical medication, surgical and therapeutic intervention, mixed-method data analysis and patient narrative feedback were all found to be useful.

The power of the mind to heal is evident as proved by the now commonly accepted placebo effect used in randomised trials. New techniques can help such conditions like Parkinson's disease, learning difficulties, chronic pain and paralysis. Practical therapies can be tailored and adjusted to support a specific individual's needs. Feelings and emotional systems of bodily interaction are involved in triggering these neuroplastic recovery processes. These interventions produced positive results as feelings of limitation, hopelessness and dependency were overcome. In his latest video, he concludes that neuroplasticity is the brain's power to form and organise synaptic connections in response to learning, environmental experiences, disease and injury. See Video Endnote.[15]

Janet Zadina (2015) explains some of the difficulties in the emerging role of neuroscience in education reform. Bridging the gap between education and psychology has been difficult. Neuroscientists have felt they were better qualified to speak to teachers about neuroscience than educationalists. It is over 25 years since brain-based learning was initiated by teachers to make inferences from findings in neuroscience in relation to classroom practice. Some psychologists have felt this was a 'bridge too far' because the practitioners were lacking in scientific understanding and referred to the making of 'untenable leaps'. This has caused a separation in practice. It is still difficult for teachers to get training and support to action research and apply this knowledge. Zadina points out that we need to recognise and fund educational neuroscience research in order to improve training, research and practice so that information can *flow both ways.*

Currently collaborative co-operation between psychologists and teachers is urgently needed for the benefit of unsupported special needs pupils and also those in mainstream who have learning and well-being issues. Taking a fixed inflexible approach, reducing all research to norms between individuals and across populations has limitations in usefulness and appears to call into question the idea, entertained by some professionals that randomised tests are the only truly valid method of social research. Of course this very much depends on the research question, which if based on assumption may not reveal new information that isn't pre-envisaged by the researcher. Ignoring variations, variables and exceptions may be particularly wrong in social sciences practice when developing new ways forward—*more discussion and collaboration is needed.*

Surgical research using a range of imaging methods together with consensual collaborative clinical research shows over time that with the right carefully nuanced treatment the brain may rewire itself to compensate for lost bodily functions. Patient-empowered targeted behavioural repetition is very much a part of the story. New research shows that improvement is particularly successful and recovery more likely if the patient is involved—as well as receiving responsive and tailored therapies. *The participation of patients is important, involving observations of feeling and emotion through self-reporting comment and behavioural responses.* The research performed by Arrowsmith-Young and Pepper, patients who became experts, shows that the benefits of a particular 'treatment' or 'exercise' can 'kick start' brain plasticity and extend to 'curing' other areas of difficulty than the one being targeted.

Our feeling sensations and motivations can be the agents of actual physical recovery if the right individual therapeutic approach is found. One has only to think of the already mentioned 'placebo effect,' the expression 'mind over matter' and the 'dark tunnel or spiral of depression' to understand the possibilities. It is proved scientifically that as well as for medical interventions and recoveries, an individual's feelings and emotions are key to developing their physical neurological capabilities on many levels. The focused activity may have a beneficial effect on a surprising range of other problems. The work of clinical and neuroscience practitioners above proves conclusively human sentience is integral to brain plasticity through both 'conscious' and 'unconscious' feelings, emotions, perception and thought.

1.5 Consciousness, Memory and Regeneration in Sleep

Different concepts around the idea of 'human consciousness' have been the subject of a great deal of speculation by both psychologists, educational and philosophers like Daniel Dennett. They have looked at the subject from different perspectives and through different lenses. However, up until now and for the foreseeable future these are all generally still dependent on observation, qualitative research and analysis. Dennett is particularly dismissive of neuroscience having any effect upon our understanding of education. Matthew Cobb puts the pure science aspect of this into startling context for those of us who might be complacent in our views. He goes from Charles Darwin (1809–1882) through Camillo Golgi (1843–1926) to Santiago Ramón y Cajal (1852–1934) and Francis Crick (1916–2004) in his history of our ideas about the brain.

Cobb explains the sheer complexity of the problem and says he guesses it will be another 50 years before the human brain is mapped. For example, as he writes (2020) the maggot brain has been mapped up to 70% of its 10,000 neurons by Croatian neuroscientist Marta Zlatic's team at Cambridge University. It has so far been found to have 1.36 million synapses and two metres of neurons—all of which can be contained within a small ink dot the size of a full stop. Consciousness and memory are always actively interconnecting in any given moment. Memory supports body awareness as we apprehend and/or react to different situations. Ability nets are embedded by past experience and yet they affect and are affected by our responses and outcomes.

Neural net connections enable our ability to create, learn about and maintain 'useful' new skills. This is because both mental and physical awareness and response systems depend on a multiplicity of newly acquired and established neural memories (e.g. as in 'muscle memory'). The added capacity neural nets allow the nervous system to connect and process input from and responses to new experiences through electrochemical reaction. The process of neurogenesis is explained by neuroscientist Sandrine Thuret below. Neurogenesis is difficult to understand, but the fact that it is proven should be impacting our thoughts and affecting our work as teachers and social practitioners. Activating our brains and memorising experience and information enables us to feel ***conscious in various and complex particular ways*** at every level in every area of ability, behaviour and knowledge. See Video Endnote.[16]

Awareness 'actualises' knowledge and helps us to register 'facts' in the physical body-brain and 'mind' causing us to embed memories. Brain plasticity constantly interacts with a multiplicity of memories and 'conscious feeling responses.' Over the past 20 years we have been able to observe neuroplasticity in action in laboratories, but verbal and behavioural feedback have often been overlooked. However, they do provide vital direct evidence of neuroplasticity. Neuroplastic change can be demonstrated and explained by doctors, therapists, clinical, teachers and neuro-psychologists through qualitative narrative data. In 'scientific' social research particularly this data enhances clinical results. It can produce more useful interpretations insights, essential information, understanding and progress than quantitative data alone. I would suggest that from a social research perspective the full meaning of independent human behaviour is ultimately impossible to read in all aspects through automatic methods of testing and observation without making a qualitative analysis. This is the case no matter what system, method or device for data collection is employed.

In 2015 after publishing his second book Norman Doidge was interviewed at the Centre for Dialogue at Simon Fraser University (2015b). Doidge explains that when neurons are motivated to fire repeatedly in a pattern the signal becomes stronger and faster. Gradually the brain learns and becomes more capable with practice, getting better at whatever it chooses to do. He says that plasticity works in a similar way to when skiers choose to make "repeated tracks in the snow" and that when that activity is well established it is hard to take a different route. He explains that this process contributes to memory, which gives rise to both good and bad habits (and actions!). This ties in with the behaviourists' discovery that certain stimuli can create particular types of behaviour in some people. See Video Interview Endnote.[17]

The process of the establishment of neural nets varies and is affected by a unique and complex number of personal sensory actions. These are affected by bodily influences such as strings of peptides of amino acids, hormones, DNA, nervous response systems, habit, memories, emotional and instinctual responses. These operate in diverse ways making connections without engaging our total attention, but even so can influence social change. This justifies and explains the growth of focus group research, where group interaction uncovers emotion models, leverages memory, producing fresh thoughts and spontaneous data responses. Our responses are by turn and in combination selective, automatic and random. Throughout our lives the brain's control systems in the cerebral cortex develop and adjust to perform all kinds of modus operandi, embedded within neural nets of connection enabling our engagement with the world.

Fine discrimination in interpretation and adjustment are key at every level and variety of human endeavour. Our brain plasticity is what enables this ability to be taught, speeded up and/or refined given the 'right' stimuli. Doidge observes that understanding and researching how plasticity operates might even help us 'sculpt' our own brains! Exams are the human learning activity by which most education systems encourage humans to sculpt their intellect and demonstrate their learning success. Perhaps this is what we have already been doing in education and not always in a good way! In which case, when governments and employers complain of lack of character, literacy and numeracy skills, practical innovation, analytic ability and self-motivation perhaps they should ask teachers to research these educational goals and give them the freedom to act upon their findings.

At any given stage of life our brains are embedded with their current, habitual neural response systems. Even so, if they are sufficiently stimulated, they have potential capacity and are open to change. This ability is essential to survival, whether or not it turns out to be useful or detrimental and is to some extent involuntary. Crucially for learning, abilities depend on the existence of established and potential connections embedded, that enable and are necessary for a particular activity. The good and efficient management of our memory through our conscious efforts is an 'executive' ability we learn by practice. Education is all about building and managing our capacities, learning and potential in relation to outside events. This depends in turn on exposure to a curriculum tailored to stimulate and imprint the mind with valuable human experiences. These will never be entirely predictable.

The competitive nature of plasticity is described as 'a war of nerves' by Doidge. He says that when we do not exercise a skill then its neural map space is taken up by competing skill maps. The question of 'dormant abilities' arises here. Sometimes we remember old skills and sometimes we don't. Although there may be 'an average amount' of practice necessary to acquire particular skills, there are examples of child and adult prodigies and savants to consider and the influences of physicality, hormones and DNA. The ways neuroscientists scan the 'lit up' active brain is still in a state of relative infancy as a science, for example, electroencephalogram and event-related potentials data. *Teachers and psychotherapists are often also in a good position to observe and record how different inspirational educational stimuli work in different settings with different learners.*

Neuroscientist and sleep expert, Matthew Walker tells us that sleep plays an essential part in neuro net servicing and the storage of memories (2017). He has studied the subject for over 20 years. He says that sleep performs a fundamental service which has a profound effect upon our ability to process the events, realisations, thoughts and memories of each day. He has studied the patterns and cycles of electrical brainwaves which are active in the brain at night. There are three sources of data recorded by polysomnography. This process combines electrical brainwave activity with eye movement and muscle activity to produce a sleep graph. Currently it is thought there are two types of sleep cycles discovered by Eugene Aserinsky and Nathaniel Kleitman in 1953. They are called rapid eye movement (REM) and non-rapid eye movement (NREM).

Walker believes the evidence suggests that the 'back-and-forth interplay' between REM and NREM types of sleep cause the brain to update and remodel our neural nets at night. His research has led him to the conclusion that NREM sleep helps the brain to shed unimportant information and REM strengthens important or significant memories. REM sleep which peaks at around 2 am and 7 am is the period of dreaming when essential experiences are connected together. These peaks are called spindles and indicate cognitive activity. See Academic Paper Endnote.[18]

He tells us that the multiple functions that are performed by the body during sleep are essential to our physical and mental health. Sleep uses up one-third of our human life, but is also essential to life. It gives us many health benefits. We need at least eight hours sleep a day as adults to maintain ourselves in our best state of health. Children require much more sleep; their mental and physical growth is reduced by sleep deprivation in ways which are not always obvious. For example, Walker tells us of the politicians Prime Minister Thatcher of the United Kingdom and President Reagan of the United States boasting of only needing five hours of sleep at night. They both died of Alzheimer's disease—a degenerative brain condition now clearly linked to sleep deprivation.

A lack of sleep suppresses the body's immune system. Every major organ in the human body is regenerated by sleep and damaged by sleep deprivation. Sleep lowers blood pressure and a lack of sleep lowers resistance to cancer, heart attacks and cardiovascular disease. Sleep deprivation is associated with major psychiatric conditions including depression, anxiety and suicide. It can cause bodily cravings that lead to obesity as we search round for a substitute to compensate for lack of sleep. It can make us feel hungry as tiredness makes us look for a quick fix by eating sugar and carbohydrates. Walker says there is a correlation between the rise in obesity in some societies from 1950 onwards and the decline in regular and sufficient hours of sleep.

Melatonin is a natural hormone secreted by the body which is associated with the sleep cycle or 'body clock.' The production of this hormone may be inhibited by the state of alertness required by our current way of life, but according to Walker taking a supplementary dose of this hormone does not necessarily improve sleep. Similarly the use of 'sleeping pills' only produces a chemical coma with few of the benefits of natural sleep. Sleep on the other hand services the brain by recalibrating emotional circuits and reboots the body's metabolic state by balancing insulin and glucose. It enriches the ability to learn, memorise and make logical decisions. It modifies painful memories and processes past and present memories in a virtual reality theatre of dreams.

Modern lifestyles involve being constantly alert to the internet and interrupted by the smartphone. This has caused us to develop demanding and stressful habits which have changed our daily routines and rhythms of life. We exist in a constant state of potential distraction caused by electric light, social media, blue screens and the constant background of 'white noise' caused by electrical devices. Our modern life is often sedentary. It places demands on us to concentrate and keep alert. We may tend to compensate for sleep loss and lack of exercise by taking large amounts of caffeine, carbohydrates and chemical stimulants, but these may be damaging long-term. Lack of sleep creates a state of constant jet lag. All of this can contribute to a 'stressful' and potentially shorter life. See next section.

1.6 Ageing Successfully and Keeping an Active Mind

The sheer biological complexity with which the brain operates has the potential to continue to be neuroplastic throughout our lives. Unfortunately the old adage of 'if you don't use it you lose it' fits in *too well* with the current 'new reality' and 'convenience life-style' of modern life. Our brains use a multiplicity of neural pathways affected by conscious and unconscious stimuli, incomprehensible and inaccessible to science. *We really do have to do our own research on ourselves!* These systems have unexpected latent and resting potentials. For example, it has been found that the brain is so dynamic that there can be miswiring errors in neuronal maps and that these maps move around (Merzenich cited by Doidge, 2007).

A baffling story about the survival and recovery of one famous neuroscientist's father's brain is told by Doidge and illustrates this point. Paul Bach-Y-Rita's father Pedro had a catastrophic stroke at the age of 65 and became paralysed and helpless. Paul's brother George, a medical student, undertook his father's long-term care and physiotherapy. He came up with the idea of taking his father through similar learning experiences to those of a young child. He got him to crawl and do all kinds of infant games, using infant developmental stages as a guide. He rolled marbles and washed up, manipulated common objects and learned to feed himself all over again.

Pedro went on to 'recover' all his faculties and it was only upon his death, caused by a heart attack seven years later, that his son discovered the extraordinary amount of damage his father's brain had sustained. As an interested neuroscientist, Paul was invited to look at his father's post-mortem brain. Although it was a strange and off putting experience, he was profoundly informed by the realisation of the extensive damage sustained. He realised that 97% of the nerves that ran from the cerebral cortex to the brain stem at the base of the neck had been destroyed. It had never 'healed' in the conventional sense, but his father's recovery had shown his brain's extraordinary ability to remember and reorganise itself.

Brains develop and age in variable ways according to common patterns of electrochemical hormonal activity but also through unique biological transformations. These are now researched for very different medical and academic reasons in a variety of situations independently worldwide by a wide range of scientific and medical specialists. This new body of knowledge is based on the whole complicated multifaceted array of human life experience—in combinations of diverse cultural experiences through the personal lives of the world's teeming multitudes. The discovery of neuroplasticity is a game changer that can profoundly affect our views on human existence around the world.

We have always known about capable wise old people—many of them developing their capabilities and ideas well into extreme old age. Brain plasticity now explains why the human brain is sometimes extended throughout old age, with new skills developing. For example, Einstein's last theory was refuted but proved after his death. The competitive nature of plasticity is an interesting phenomenon that may be particularly relevant as we experience ageing. Taking language as an example, plasticity causes our native language to dominate the brain's linguistic map space, making it harder to learn new languages. However, learning a new language has recently also been found to fend off mental deterioration and even improve cognition in old age.

Doidge tells us 'neural competition' may be a reason why 'bad' habits get established and gain advantage. He suggests what this is a reason it is sometimes so hard to make a change. The established habit is hard to replace and unlearn. Perhaps we haven't wanted to change enough, but also perhaps we (or our carers) didn't believe in the possibility of change. This implies that more understanding about how our brains work, are stimulated and can be better 'self-managed' could be useful. We have not always fully understood the possibilities for 'ordinary' people to extend their brain development at every stage of life. Again there are huge implications for the way we view and deal with people in society as we come to realise that stereotyping of other people of all ages may create, exacerbate and perpetuate their problems.

Masako Wakamiya is a Japanese woman and a former banker, aged 83 is experienced in digital coding, product design and development. She has become a designer, app developer and entrepreneur in old age, continuing to learn in her areas of interest. She helped create the website 'Mellow Club' in 1999 for retired people. She developed a digital archive called 'Mellow Denshoukan' with stories told by elders about their lives both during and after the Second World War. At the age of 80, she learned how to code, developing and publishing an app for Apple store. It is called

Hianadan, a Japanese doll game designed for an iPhone. She also designs products for sale. For example, 3-D printed pendants, geometric patterned clothing, gift wrap, and handbags—one decorated with LED lights. See Article Endnote.[19]

Wakamiya has since made a speech for Apple in Tokyo telling education and technology companies not to forget about senior citizens. Education technology products overwhelmingly target young people but some companies are now starting to realise they can develop an older audience. In the past, senior citizens and their carers have tended to assume their inability to learn and been inclined to think that technology is impossible for them to manage. There is a growing industry around serving adult learners in higher education, but we are still inclined to overlook older people's own skills, abilities and experience. While I was researching online I met Joyce Williams (Grandma Williams), an 80-year-old blogger. She had been a writer and initially joined a class called 'Blogging for Beginners'—her fellow students were in their 20s and 30s. See Blog Website Endnote.[20]

Williams decided to write with a particular underlying theme—"the sad image of old age being portrayed by the media." She has been interviewed on UK national television and says that in her experience, "Old age is a happy period of life, and for most people one of the best!" Successful older people are alert and interested and able to engage with and deal with new experiences and they can relate to the world, are able to make new relationships and sustain old ones. A study at Michigan State University found that people who are optimistic help their partners to stay healthy as they grow old together (Armstrong, 2019). The choice to remain cheerful and interested in leading an 'independent' life in extreme old age is talked about by John Leland in his book *Happiness Is a Choice You Make* (2018). He spent a year interviewing and listening to seven over 80-year-olds in New York before writing about them.

All of these people had found different ways to interact socially and were philosophical, accepting and upbeat about their situation. As social animals, a feeling of value and 'place' in society is important for human happiness. Our emotional outlook and approach to the world around us is profoundly affected by relationships, education and environment. The most 'successful' old people, who live long productive lives continue to enjoy their lives and so are enabled to continue to function. They relate to their environment in a meaningful way, sometimes through learning from and teaching others. They have learned to communicate in useful ways, to be curious and interested in life.

However, there are other factors in the ageing process, which explain how we are biologically susceptible in nature. In her book (2019) *Borrowed Time: The Science of How and Why We Age*, Sue Armstrong tells us that our health also depends on chemical and biological factors in our environment. For example, sulphates, nitrates, hydrocarbons and heavy metals like nickel, mercury and lead we breathe in from the atmosphere through burning fossil fuels can cause brain damage. She also points out the difficulty of understanding the effects of DNA and gene mutation.

She tells us about Caleb Finch, neuroscientist, biochemist and molecular biologist who has made a lifetime study of ageing. He studies how human bodies are physically affected by their environment. He has carried out a long-term research programme on the Tismane people in Bolivia. He has discovered that these people have a gene labelled APOE e4—a high-risk factor in the development of Alzheimer's brain deterioration in the industrialised world. He has found unexpectedly this gene protects some Tismane people from a local parasitic infection, having the same type of long-term effect as Alzheimer's. There is still much for scientists to discover about human biology and the brain.

Older people can benefit their own brains and those of others by sharing their experiences. They can put memories and skills to use and benefit communities. Senior citizens might be much more utilised by their societies in educating young people about history and culture. Even if as adults we grow up questioning the information, human well-being and emotional intelligence is facilitated by close social relationships and an understanding of our own and other people's cultural history—*a social education*. Children also learn language and behaviour through engaging with others who teach formally, share and role model—*an external formal education*. They learn how to listen, memorise, concentrate, question, speak and behave for good or ill. They can also be taught practical skills, social responsibility, environmental awareness and spiritual values by wise elders. This is a time-honoured tradition that has tended to become lost in Western educational systems—social cohesion, civil discourse and a respect for experience have suffered as a result.

1.7 Conclusion

The brain's electrical networking can be observed, filmed and photographed, but its actions are often difficult to interpret in a useful way. New knowledge in practice comes through looking for original solutions through close and careful observation, verbal and behavioural feedback. Many current qualitative research trials in medicine are close up and personal and cannot be replicated or predicted exactly for each individual. Principles can apply and methods tried but adaptations are ultimately made by the individual patient's brain with the clinician's help. *This is about the brain transforming itself, while the insights, techniques and types of support offered inform skills and knowledge for possible interventions.*

Neuro-therapists have identified some ways plasticity enables physical, chemical, electrical biological adaptations to change the brain's structural and functional properties as it develops and 'learns.' New information is being gained about how neuroplasticity enables humans to thrive and learn to survive. Narrative data from clinicians' and patients' different points of view suggest pathways to medical recovery for specific disabilities as treatments are discovered. Patients and their physicians have learned about healing processes together. They have worked strategically, adapting interventions to improve lives in unique and groundbreaking ways. These stories can provide links and inspire new possibilities for those action researching in other social situations.

As well as looking for common effects and similarities experts can make inferences, looking for variation and difference in behaviour in order to support sometimes unpredictable and hidden developments. These often take time to emerge and become recognised. We can see energy flashes and chemical responses connecting neurons using specialist electronic devices, but we can now develop ways to support the brain-body with more confidence. For example, computer chips are now being used to help disabled people who are severely nerve damaged to walk again, read, manage and perform tasks. We still cannot explicitly observe and comprehend all of the complex forms of cognitive activity in humans, but we can do better by enlisting active and individual collaboration.

We 'observe neural activity,' but its meaningful interpretation as far as understanding cognitive intelligence in all its manifestations goes will always be a work in progress. The very existence and nature of brain plasticity, since it is always evolving in diverse ways, of itself demonstrates why this is so. Clinical psychology can contribute a great deal to our knowledge with its medical interventions and analytical reflections upon the subconscious logic behind our behaviour. However, the discovery of plasticity gives us a better existential perspective from which to investigate how our evolutionary biology has 'invented' us as specific individuals, cultural groups and societies. All of this appears to be entirely compatible with my initial theory. We are all unique and our feelings and emotions are likely to make sense in some way within our own frames of reference, even if other people do not agree.

Notes

1. Erin Wayman (2012) 'When Did the Human Mind Evolve to What It is Today? Archaeologists are finding signs of surprisingly sophisticated behavior in the ancient fossil record' 25 June 2012, Smithsonian Magazine https://www.smithsonianmag.com/science-nature/when-did-the-human-mind-evolve-to-what-it-is-today-140507905/ Accessed 31/03/2018.
2. Khan Academy (2015) *Diagnosing strokes with imaging CT, MRI, and Angiography* NCLEX-RN https://www.youtube.com/watch?v=xC55TgPmfZ0 Accessed 20/02/2020.
3. Stevan Harnad (1985) 'Donald Hebb (1904–1985)-obituary'. Princeton Harnad E-Print Archive and Psycoloquy and BBS Journal Archives. https://www.southampton.ac.uk/~harnad/Archive/hebb.html Accessed 01/04/2021.
4. Nature Science & Technology video 20 July 2016 *The ultimate brain map* https://www.youtube.com/watch?v=UHDfvfYCY0U Accessed 0/02/2020.
5. Video Linda Geddes (2016) 'Human brain mapped in unprecedented detail. Nearly 100 previously unidentified brain areas revealed by examination of the cerebral cortex' NATURE | NEWS Nature Journal: 20 July 2016 https://www.nature.com/news/human-brain-mapped-in-unprecedented-detail-1.20285 Accessed 12/02/2019.
6. Brain Scanning Methods

    - Lesion studies used in neuropsychology. Lesions may damage other systems which happen to pass through the lesion site. They occur through damage, for example, an accident or tumour.
    - Stimulation used in both neuropsychology and neuroscience involving feeding a chemical or electrical signal into part of a neural circuit and measuring its consequences.
    - Non-invasive stimulation or TMS (transcranial magnetic stimulation) used in neuroscience is a relatively recent development in the experimental neurosciences. Microelectrode recordings.
    - CT Scan; used in neuroscience. They are used to construct a structural visual image of the brain to identify problematic brain tissue, but do not reveal much about how the brain actually thinks.
    - PET (Positron emission tomography) used in neuroscience. This uses low radioactivity to identify the temporary presence of com-

pounds like glucose or oxygen in the brain to build up images of the brain showing cortical activity.

- fMRI (Functional magnetic resonance imaging) used in neuroscience. This measures changes in oxygen levels in the brain that indicate the blood flow in cortical activity. The amount of oxygen carried in the blood affects the blood's magnetic properties. (Whole brain images can typically only be collected every two seconds.)
- EEG (Electroencephalogram and event-related potentials) used in neuroscience and ERPs (Electrochemical signals) are the basis of communication between nerve cells and these were first recorded on the scalp in 1924

[School of Psychology & Neuroscience 2018, University of St. Andrews https://www.st-andrews.ac.uk/psychology/research/brainimaging/ Accessed 12/02/2019].

7. Machell, B. (2020) *Dagmar Turner played during surgery so the surgeons could identify the parts of her brain that control motor movements.* King's College Hospital, London, London: The Times (Saturday Interview 22 February 2020, 12.00 am) https://www.thetimes.co.uk/edition/news/dagma-turner-was-prepared-to-live-a-shorter-life-as-long-as-she-had-her-music-tgz2mmv37 Accessed 22/02/2020.
8. Joseph Critchlow Meet the 2019–2020 Essay Winner Joseph Critchlow, age 13, from Liverpool, United Kingdom. The theme for the contest was "Journey of Peace." Sponsored by the City of Liverpool Lions Club. International Essay Contest - Lions clubs around the world. https://youtu.be/KPpbO4Qd_Iw Accessed 12/01/2021.
9. Special Data Contribution

   'Journey of Peace' by Joseph Critchlow aged 14—St.Vincent's School for Sensory Impairment Liverpool, United Kingdom, 2019.

   I looked at a World in conflict. Who could help me answer a question, how can we solve the world's problems? There seems no way as we fight each other in our neighbourhoods, Cities and Countries for so many reasons. How can we journey to save our forests and Oceans if we don't have peace first? I looked at my City, Liverpool and wondered who could answer my peace question? I went to a City Gateway, Lime Street Train station, and in my mind asked the statue of the great entertainer Sir Ken Dodd, 'how can we find peace?' I felt he said 'share laughter, laugh together, happiness is a key.' I looked up at the Liver bird statues, a symbol of our City at a Gateway to the Oceans. I asked 'how can we find peace?' I felt they said to me 'we can see across the world and time, we can see the Statue of Liberty. Many people look at what is different, not the same, find what is the same and come together there. This is a key.' I went to the statue of John Lennon and asked 'how do we find peace?'

I felt he said 'Give peace a chance. This is a key.' I wondered, if just one person stopped and did just that, would others do the same? Could I be that 'mustard seed' from which things grew? Could my voice be heard? I went to the statue of Noel Chavasse, who sacrificed his life in the Great War winning the highest award for bravery the Victoria Cross, twice. I asked 'how do we find peace? I felt he said 'do what is right, help others. Learn from the mistakes of history we paid for in blood. Use your voice and strengths to tell others and help them learn. This is a key.' I stopped at the statue of John Hully, a founder of the modern Olympics. I asked him 'how can we find peace?' I felt him say 'sports are a way we can come together and celebrate what is the same, where we can have fun, keep healthy and make friends. This is a key.'

My Journey to find an answer was complete. I looked at all my keys. Although I am visually impaired I can lead the way, my voice can be heard, I can bring others together. I may not see you well, the colour of your face or the clothes you wear, but I can help the world find peace by my example. As Gandy said "there is no pathway to peace, peace is the pathway," let me help lead you along that path by following my voice. Let us come together in things like music and sports as better ways to see we are just the same. It is from here we can work together in making a safer, better world. We all have the keys.

10. The Woman Who Changed Her Brain: Barbara Arrowsmith-Young https://barbaraarrowsmithyoung.com/bio/ Accessed 23/02/2020.
11. Vilayanur Subramanian Ramachandran (2019) 'Embodied Brains and Disembodied Minds' TED TALKS, UCSD, 7 January 2019 https://www.youtube.com/watch?v=3PVatC2Noyo Accessed 29/04/2020.
12. Professor V.S. Ramachandran (2016) 'Releasing Phantom Limb Pain', University of California Television (UCTV), San Diego https://www.youtube.com/watch?v=jNTCFSuKEMA Accessed 29/04/2020.
13. Joshua W. Pate (2018) The fascinating science behind phantom limbs, 4 October 2018 https://youtu.be/KdihphPp1Q0 Accessed 29/04/2020.
14. Moskowitz, M. H. and DePoloGolden, M. (2013) *Neuroplastic Transformation Workbook: How To Use The Power Of Your Mind To Heal Your Brain, Your Pain, And Change Your Life* LLC: Neuroplastic Partners https://elitemanmagazine.com/heal-your-brain-your-pain-dr-michael-moskowitz-dr-marla-golden/ Accessed 24/10/2019.
15. Doidge, N (24 March 2018) Conversations That Matter - Dr Norman Doidge and the power of the brain, Vancouver Sun https://www.youtube.com/watch?time_continue=234&v=dEacWNFEprg Accessed 24/10/2019.
16. Sandrine Thuret (2015) 'You can grow new brain cells. Here's how' 30 October 2015 London: TED Talk https://www.youtube.com/watch?v=B_tjKYvEziI Accessed 05/04/2019.

17. Doidge, N. (7 April 2015) Brain's Healing Energies The Agenda with Steve Paikin https://www.youtube.com/watch?v=ifYcE4-eI_s&feature=youtu.be Accessed 05/02/2020.
18. Aserinsky E, Kleitman N (1953). 'Regularly Occurring Periods of Eye Motility, and Concomitant Phenomena, during Sleep'. Science. 118 (3062): 273–274. Bibcode:1953Sci...118..273A. doi:10.1126/science.118.3062.273. PMID 13089671 https://science.sciencemag.org/content/118/3062/273 Accessed 05/04/2019.
19. Johnson, Sydney (5 November 2018) 'Meet the 83-Year-Old App Developer Who Says Edtech Should Better Support Seniors' Edsurge News, Available at: https://www.edsurge.com/news/2018-11-05-meet-the-83-year-old-app-developer-who-says-edtech-should-better-support-seniors Accessed 13/11/2018.
20. Williams, Joyce (2020) Grandma Williams' blog 'about the fun, the enjoyment, the evolving serenity of later years' Available at: https://grandmawilliams.com/about/ Accessed 25/03/2020.

References

Armstrong, S. (2019). *Borrowed Time: The Science of How and Why We Age*. Bloomsbury Sigma.

Arrowsmith-Young, B. (2012). *The Woman Who Changed Her Brain: How We Can Shape Our Minds and Other Tales of Cognitive Transformation*. Random House.

Aserinsky, E., & Kleitman, N. (1953). Regularly Occurring Periods of Eye Motility, and Concomitant Phenomena, During Sleep. *Science, 118*(3062), 273–274. https://doi.org/10.1126/science.118.3062.273. https://science.sciencemag.org/content/118/3062/273. Accessed 05 Apr 2019.

Bach-y-Rita, P. (1979). *Recovery of Function Following Brain Injury: Theoretical Considerations*. Academic Press Inc.

Beaudet, A. (2021, February 22). *Brain and Behavior Seminar: Early Hominin Brain Evolution: What Can Fossil Endocasts Tell Us?* Bath Spa University. https://www.youtube.com/watch?v=zLJ4jnPvO6g. Accessed 03 Apr 2021.

Bliss, T. V., & Lomo, T. (1973). Long-Lasting Potentiation of Synaptic Transmission in the Dentate Area of the Anaesthetized Rabbit Following Stimulation of the Perforant Path. *The Journal of Physiology, 232*(2), 331–356. https://doi.org/10.1113/jphysiol.1973.sp010273. Accessed 03 Apr 2021.

Boldrini, M., Fulmore, C. A., Tartt, A. N., Simeon, L. R., Pavlova, I., Poposka, V., Rosoklija, G. B., Stankov, A., Arango, V., Dwork, A. J., Hen, R., & Mann, J. J. (2018, April 5). Human Hippocampal Neurogenesis Persists Throughout Aging. *Cell Stem Cell, 22*(4), 589–599.e5. https://doi.org/10.1016/j.stem.2018.03.015

Brain Scanning Methods. (2018). *School of Psychology & Neuroscience.* University of St. Andrews. https://www.st-andrews.ac.uk/psychology/research/brain-imaging/. Accessed 12 Feb 2019.

Bronfenbrenner, U. (1979). *The Ecology of Human Development.* Harvard University Press.

Cobb, M. (2020). *The Idea of the Brain: A History.* Profile Books Ltd.

Crick, F., & Koch, C. (1990). Towards a Neurobiological Theory of Consciousness. *Seminars in the Neurosciences, 2*, 263–275.

Critchlow, J. (2019). *Journey of Peace.* St.Vincent's School for Sensory Impairment.

Dennett, D. C. (1991). *Consciousness Explained.* Penguin Random House.

Dennett, D. C. (2017). *From Bacteria to Bach and Back: The Evolution of Minds.* Penguin Random House.

Doidge, N. (2007). *The Brain That Changes Itself: Stories of Personal Triumph from the Frontiers of Brain Science.* Penguin.

Doidge, N. (2015a). *The Brain's Way of Healing: Stories of Remarkable Recoveries and Discoveries.* Penguin.

Doidge, N. (2015b, April 7). *Brain's Healing Energies the Agenda with Steve Paikin.* https://www.youtube.com/watch?v=ifYcE4-eI_s&feature=youtu.be. Accessed 05 Feb 2020.

Doidge, N. (2018, March 24). *Conversations That Matter – Dr Norman Doidge and the Power of the Brain.* https://www.youtube.com/watch?time_continue=234&v=dEacWNFEprg. Accessed 24 Oct 2019.

Dweck, C. (2006). *Mindset: Changing The Way You Think to Fulfil Your Potential.* Random House.

Eaton, H. (2018). *The Brain Pioneer: The True Story of How Barbara Arrowsmith-Young Used Brain Science to Help Children with Learning Disabilities.* Barnes and Noble.

Finch, C. (2007). *The Biology of Human Longevity: Inflammation, Nutrition, and Aging in the Evolution of Lifespans.* Academic Press-Elsevier.

Geddes, L. (2016, July 20). Human Brain Mapped in Unprecedented Detail. Nearly 100 Previously Unidentified Brain Areas Revealed by Examination of the Cerebral Cortex. *NATURE | NEWS Nature Journal.* https://www.nature.com/news/human-brain-mapped-in-unprecedented-detail-1.20285. Accessed 12 Feb 2019.

Glasser, M., Coalson, T., Robinson, E., et al. (2016). Academic Paper 'A Multi-Modal Parcellation of Human Cerebral Cortex'. *Nature, 536*, 171–178. https://doi.org/10.1038/nature18933. Accessed 12 Feb 2019.

Greenfield, S. (2014). *Mind Change.* Penguin Rider.

Harnad, S. (1985). Donald Hebb (1904–1985)-Obituary. *Princeton Harnad E-Print Archive and Psycoloquy and BBS Journal Archives.* https://www.southampton.ac.uk/~harnad/Archive/hebb.html. Accessed 01 Apr 2021.

Hawkins, J. A. (2017). *Feelings and Emotion-Based Learning; A New Theory*. Palgrave.
Hebb, Donald Olding (1949). *The Organization of Behaviour*. John Wiley & Sons.
Jeewon, O., Chopik, W. J., & Kim, E. S. (2019). The Association Between Actor/Partner Optimism and Cognitive Functioning Among Older Couples. *Journal of Personality*, 2019. https://doi.org/10.1111/jopy.12529: Michigan State University, USA: “A happy partner leads to a healthier future.” *ScienceDaily*, February 11, 2020; Available at: www.sciencedaily.com/releases/2020/02/200211092555.htm. Accessed 12 Feb 2020.
Johnson, M. H., & de Haan, M. (2015). *Developmental Cognitive Neuroscience: An Introduction*. Wiley & Sons Ltd.
Johnson, S. (2018, November 5). Meet the 83-Year-Old App Developer Who Says Edtech Should Better Support Seniors. *Edsurge News*. https://www.edsurge.com/news/2018-11-05-meet-the-83-year-old-app-developer-who-says-edtech-should-better-support-seniors. Accessed 13 Nov 2018.
Joseph Critchlow Meet the 2019–2020 Essay Winner Joseph Critchlow, Age 13, from Liverpool, United Kingdom. The theme for the contest was “Journey of Peace”. Sponsored by the City of Liverpool Lions Club. International Essay Contest- Lions Clubs Around the World. https://youtu.be/KPpbO4Qd_Iw. Accessed 12 Jan 2021.
Khan Academy. (2015). *Diagnosing Strokes with Imaging CT, MRI, and Angiography*. NCLEX-RN. https://www.youtube.com/watch?v=xC55TgPmfZ0. Accessed 20 Feb 2020.
Klein, R. G., with Edgar, B. (2002). *The Dawn of Human Culture*. Wiley & Sons Ltd..
Leland, J. (2018). *Happiness Is A Choice You Make: Lessons from a Year Amongst the Oldest Old*. Sarah Crichton Books.
Loria, K. (2018, August 22), 7:42 AM Sleep Deprivation Can Kill You – Here’s What Sleeping Less Than 7 Hours Per Night Does to Your Body and Brain. *Business Insider Journal*. https://www.businessinsider.com/sleep-deprivation-effects-on-your-body-brain-2018-8?utm_content=buffer08631&utm_medium=social&utm_source=facebook.com&utm_campaign=buffer-ti&r=US&IR=T. Accessed 21 Mar 2019.
Luria, A. R. (1973). *The Working Brain: An Introduction to Neuropsychology*. Penguin Books.
Luria, A. R. (1976). *Cognitive Development: Its Cultural and Social Foundations*. Harvard University Press.
Machell, B. (2020, February 22). *Dagmar Turner Played During Surgery So the Surgeons Could Identify the Parts of Her Brain That Control Motor Movements*. King’s College Hospital, London: The Times (Saturday Interview, 12.00am). https://www.thetimes.co.uk/edition/news/dagma-turner-was-prepared-to-live-a-shorter-life-as-long-as-she-had-her-music-tgz2mmv37. Accessed 22 Feb 2020.

Merzenich, M. (2013). *Soft-Wired: How the New Science of Brain Plasticity Can Change Your Life* (2nd ed.). Parnassus Pub. isbn:978-0989432825.

Moskowitz, M. A. (2010). *Reading Minds: A Guide to the Cognitive Neuroscience Revolution*. Routledge.

Moskowitz, M. H., & DePolo Golden, M. (2013). *Neuroplastic Transformation Workbook*. The Bay Area Medical Associates, LLC: Neuroplastic Partners.

Nature Science & Technology Journal. The Ultimate Brain Map, July 20, 2016. https://www.youtube.com/watch?v=UHDfvfYCY0U. Accessed 20 Dec 2020.

Pate, J. W. (2018, October 4). *The Fascinating Science Behind Phantom Limbs*. https://youtu.be/KdihphPp1Q0. Accessed 29 Apr 2020.

Pepper, J. D. (2012). *Reverse Parkinson's Disease*. John Douglas Pepper Website. www.reverseparkinsons.net

Perry, B. D., & Szalavitz, M. (2006). *The Boy Who Was Raised as a Dog: And Other Stories from a Child Psychiatrist's Notebook*. Hatchette Book Group.

Plomin, R. (2018). *Blueprint: How DNA Makes Us Who We Are*. Penguin Random House.

Ramachandran, V. S. (2002). *Encyclopedia of the Human Brain* (Vol. 1–4). Elsevier Science Ltd.

Ramachandran, V. S. (2003). *The Emerging Mind: The BBC Reith Lectures*. BBC/Profile Books Ltd.

Ramachandran, V. S. (2016). *Releasing Phantom Limb, Pain*. University of California Television (UCTV). https://www.youtube.com/watch?v=jNTCFSuKEMA. Accessed 29 Apr 2020.

Ramachandran, V. S. (2019, January 7). *Embodied Brains and Disembodied Minds*. TED TALKS, UCSD. https://www.youtube.com/watch?v=3PVatC2Noyo. Accessed 29 Apr 2020.

Renner, M. J., & Rosenzweig, M. R. (1987). *Enriched and Impoverished Environments: Effects on Brain and Behavior*. Springer. ISBN978-3-540-96523-7.

Rose, H., & Rose, S. (2012). *Genes, Cells and Brains*. Verso.

Rosenzweig, M. R., Krech, D., Bennett, E. L., Diamond, M. C. (1962). "Effects of environmental complexity and training on brain chemistry and anatomy: A replication and extension". *Journal of Comparative and Physiological Psychology*. 55 (4): 429–437. https://doi.org/10.1037/h0041137.PMID14494091

Skinner, B. F., Ferster, C. B., Cheney, C. D., & Morse, W. H. (1957). *Schedules of Reinforcement*. The Skinner Foundation.

Stenstrom, J. (2018, February 13). How to Use the Power of Your Mind to Heal Your Brain, Your Pain, and Change Your Life – Dr. Michael Moskowitz & Dr. Marla Golden. *Elite Man Magazine* (Ep. 158). https://elitemanmagazine.com/heal-your-brain-your-pain-dr-michael-moskowitz-dr-marla-golden/. Accessed 23 Feb 2020.

Taub, E., Burgio, L., Miller, N. E., Cook, E. W., III, Groomes, T., DeLuca, S., & Crago, J. (1994). An Operant Approach to Overcoming Learned Nonuse After CNS Damage in Monkeys and Man: The Role of Shaping. *Journal of the Experimental Analysis of Behavior, 61*, 281–293.

The Academic Press Profiles in Science. U.S. National Library of Medicine. https://profiles.nlm.nih.gov/spotlight/sc/catalog?utf8=%E2%9C%93&exhibit_id=sc&search_field=all_fields&q=Koch. Accessed 05 Apr 2021.

The Woman Who Changed Her Brain: Barbara Arrowsmith-Young. https://barbaraarrowsmithyoung.com/bio/. Accessed 23 Feb 2020.

Thuret, S. (2015, October 30). *You Can Grow New Brain Cells. Here's How*. TED Talk. https://www.youtube.com/watch?v=B_tjKYvEziI. Accessed 05 Apr 2019.

Walker, M. (2017). *Why We Sleep: The New Science of Sleep and Dreams*. Allen Lane.

Wayman, E. (2012, June 25). When Did the Human Mind Evolve to What It Is Today? Archaeologists Are Finding Signs of Surprisingly Sophisticated Behavior in the Ancient Fossil Record. *Smithsonian Magazine*. https://www.smithsonianmag.com/science-nature/when-did-the-human-mind-evolve-to-what-it-is-today-140507905/. Accessed 31 Mar 2018.

Williams, J. (2020). Grandma Williams' Blog 'About the Fun, the Enjoyment, the Evolving Serenity of Later Years'. Available at: https://grandmawilliams.com/about/. Accessed 25 Mar 2020.

Zadina, J. N. (2015). *The Emerging Role of Educational Neuroscience in Education Reform*. Tulane University School of Medicine: Published by Elsevier España, S.L.U. on behalf of Colegio Oficial de Psicólogos de Madrid. This is an open access article under the CC BY-NC-ND license.

Zlatic, M. (2004). *Establishment of Connectivity in the Embryonic Central Nervous System of Drosophila*. cam.ac.uk, PhD thesis, University of Cambridge. OCLC 890159518. British Library EThOS uk.bl.ethos.615935. https://ethos.bl.uk/Home.do;jsessionid=687EDE6B3BFA4FEC9CD134C7FF1AD231. Accessed 05 Apr 2021.

CHAPTER 2

The Importance of Feelings and Emotions

2.1 Introduction

In the last chapter I looked at Doidge's stories about some of those groundbreaking neuroscientist researchers and their patients, who contributed to the discovery of brain plasticity. There were key patients who contributed ideas and helped reveal more about how neuroplasticity works. These people's feelings and emotions were previously traditionally ignored in the analysis and treatment of their conditions. Yet their motivations and understanding and those of their doctors and therapists as they researched solutions made a huge impact upon their success or failure and were a driving force behind their recovery.

According to Lisa Feldman Barrett in her book *How Emotions Are Made* (2017), a vast body of historical emotional research data failed to prove anything about human emotional intelligence using traditional methods. This was because of the assumptions of the researchers that all humans interpret emotions in the same way. They even performed widespread studies in which they presented emotions portrayed by Western faces in photographs and expected them to be recognised similarly across cultures. 'Wrong' answers were given and conclusions were drawn. She found, after meta-analyses of a huge number of studies on

Special thanks to Dr Anna Maria Rostomyan for 'Background Knowledge' (2020) – Corporate Communications Specialist, Yerevan State University.

J. A. Hawkins, *Brain Plasticity and Learning*,
https://doi.org/10.1007/978-3-030-83530-9_2

populations around the world, that they *actually* prove the importance of emotions by default. See Video Presentation Endnote.[1]

In the light of physical neuroplasticity research in the laboratory, these same studies evidence the importance and the central influence of the emotions in cognition. Participants just didn't give the answers expected by the researchers! The conscious thought aspect of rationalising events involves connecting to emotional insights and intuitions about external happenings and our own behaviour. This is not *just* immediate but happens through the variable processes of retrospective reflection. As Malcolm Gladwell describes in his book *Blink: The Power of Thinking Without Thinking*, sometimes human beings make astonishing judgements in life-saving situations, but can at other times make life-threatening mistakes (Gladwell, 2005). This behaviour occurs 'in the moment' but can often only be explained (if at all) in retrospect.

In this chapter I am going to look at some of the published research around metaphysical and emotional aspects of thinking. In doing so, I have created my own explanations around the subject and made some points from my personal and teacher's point of view in the hope of gaining a better-informed perspective.

2.2 Historical Difficulties in Understanding Our Emotions

Suzanne Langer (1895–1985) was a philosopher and psychologist in a traditional scientific era which was generally dismissive of women in science. She was convinced that feelings and emotions were fundamental to intelligence, but she had some difficulty in substantiating her argument as a woman. There was no scientific evidence of neuroplasticity and the subject was largely ignored. However, she was incredibly perceptive and her descriptions were surprisingly accurate and enlightening. Some of them are appropriate and apt to describe modern clinical neuroscience research observations today. Like many groundbreaking neuroscientists and psychologists in this field she was informed by personal experience. She was neurodivergent herself and had synaesthesia. She experienced different sensations of colour when she heard particular sounds.

The philosopher Alfred North Whitehead (1861–1947), a leading mathematician, became interested in a metaphysical 'philosophy of science' in which he claimed that reality consists of processes relating to other processes. Metaphysics is an aspect of philosophy that explores the fundamental nature of reality, the relationship between mind and matter, solid reality and attribution—potential facts and actual facts. He disagreed with Descartes' 'scientific' view that science is the study of physical materials or objects. He said that there are many contributory singular causes for an event. He believed that *processes* were more important than things. Apparently he stated he did not believe reality is made of separate or isolated bits of matter. This seems rather a sweeping conclusion, although it may be seen to be true in some ways—in the sense that even mountains move over time. From a realistic human perspective, we should definitely consider objects and processes in the world over which we have no control when weighing up human processes and effects.

Whitehead wrote a book about human thinking or cognition called *Process and Reality* (1929) stating that reality is made up of many interrelated processes. In his theory different human perceptions are problematic to making fixed determinations. He pointed out that human ideas, choices and behaviours have consequences that affect 'reality' in the world. His philosophical ideas have lately become more popular because of concerns about how our world ecological environment has been affected by human activities. This is seen by some philosophers and pundits as creating a shift away from materialism to organic realism. *Whitehead was interested in abstract thought and like Einstein believed creativity to be important.* He was also a mathematical physicist and produced a Theory of Gravity as an alternative theory to Einstein's 'Theory of Relativity.'

It seems to me that *the fact of neuroplasticity* can now provide us with an overview and lighten our darkness in this area of human understanding. Human ideas affect the world because we have a built in biological propensity or natural tendency to build 'logical' neural nets and to behave according to this 'knowledge.' Many are embedded physically in memory in particular patterns of thought by 'supporting evidence' selected from limited, mixed or insightful experiences—for example, just plain ignorance and self-interest, unexpected incidents or informed perceptions. This is how our prejudices and self-interests turn into *habits of thought and character*. Such processes can lead to war but can also lead to an obsessional interest in an idea leading to great invention or work of art. Therefore neural nets can and do connect 'logically' and produce material consequences in the world and are potentially adaptable and variable! As an artist friend of mine often says, "If I can imagine it I can make it." *Unfortunately some ideas have been imagined with negative effects!*

We have a human ability to create our own virtual reality with imagination and then to actually not only produce a material outcome but also a social effect. For example, computer technology and the 'world wide web' or internet had to be imagined before being invented. John Shotter (1927–2016) was a psychologist who realised this social effect and questioned mainstream ideas in psychology. He helped change the direction of academic psychology and psychotherapy practices. John Shotter wrote books calling for new forms of psychological inquiry. He pointed out that psychologists needed to be able to take more account and be more aware of their 'subjects' as individuals and not become too involved with and 'blinded' by their own supposedly 'objective' point of view and their own *need* to explain in 'psychological' terms and language. He spoke about getting a 'feel for the you' in the social interaction/discussion dynamic going on around him (1985). See Article Endnote.[2]

Shotter proposed that if one could focus attention away from the 'I' one could challenge the traditional scientific view that natural psychological and sociological 'diagnosis' can be prescribed according to 'received' professional practice rather than looking beneath appearances. He claimed that 'reality' is socially constructed and supported by disorganised everyday conversational activities. His interdisciplinary analysis highlights linguistic relativity in the moment and *the imaginary nature* of many of the socially constructed 'things' we talk about. He analysed these by using examples from psychotherapy and everyday life (1993). As a psychologist he demonstrated and taught a more responsive, engaged, and sensitive way to understand and analyse our social interactions.

Professor Annette Karmiloff-Smith (1938–2016) studied developmental disorders (such as Autism and especially 'Williams Syndrome') and studied with Piaget. She had a long distinguished career and was a research fellow at the Developmental Neurocognition Lab at the University of London. She became an expert in developmental disorders and supported the theory of Neuroconstructivism. This is a theory that suggests that each person's individual developmental history (ontogeny) is due to interactions between their genetic inheritance and the varied and changing environments they encounter. This is how the brain sculpts itself progressively, becoming more differentiated and specialised over time throughout each lifetime. See Video on Endnote.[3]

Her work is sometimes positioned between the theory of Psychological Nativism (human abilities are hard wired at birth) (Fodor 1935–2017) and the traditional Tabula Rasa (Blank Slate) view (learning development is solely due to environment) (Pinker, 2002). However, it may just simply be a different theory altogether. She argued against the innate modularity of mind theory supported by Jerry Fodor. This is the idea that a brain has neural modules which have clearly defined functions. She said such approaches take a "mosaic-like" approach to cognitive development as though different systems within the brain develop separately from each other based on genetic code. Instead, she put the emphasis on innate biases (or perhaps ability tendencies) that aid learning and direct attention. She thought perhaps the autistic brain might be better understood in a variety of interrelated areas such as motion perception, visual search and multitasking (Karmiloff-Smith, 1996). See Interview Endnote.[4]

Mark Johnson (1960–) is a developmental cognitive neuroscientist. He has developed an Interactive Specialisation approach to how the brain works. He views brain development as resulting from a series of back-propagated interactions arising from genetics, brain, body and environment. He believes cognitive development to be stochastic (random), network based (neural) and interactive. His ideas appear to be compatible with more recent research into epigenetics and gene-regulated networks (Rose & Rose, 2012). Apparently genes can to some extent alter and adapt throughout life. This may explain why some families can trace similar talents and abilities down the generations in some children, but not in others—*nature and nurture are certainly interactive in complex and unpredictable ways.*

In consideration of all of the above, it is possible to conclude that, in the light of brain plasticity, the old binary argument, the 'nature versus nurture' dispute is dead because both appear to overlap and interact and to be essential contributors to intelligence. The idea that subconscious thought is 'logical' in its own terms appears to be compatible with the individual work of both Karmiloff-Smith and Johnson. The 'neuroconstructivism' idea and the 'interactive specialisation' idea are not incompatible either, given the capacity and versatility of the human brain. Currently Johnson's ideas about 'back-propagated interactions' that are random, neural and interactive arising from genetics, brain, body and environment appear to give a fairly adequate description and overview.

Neuroplasticity shows (through neural imaging) that the brain definitely does make complex and variable connections even when we are unaware of it. Johnson co-authored an academic book that discusses the relationship between cognition, the brain and the environment (Mareschal et al., 2007). It argues that the brain apprehends and uses a multitude of fragmented representations that are just enough to make quick-fire judgements and predictions. The authors produced a second volume containing a number of neural network models that show how these representations appear to change during learning. They cause behaviour and do not exactly mirror the environment.

The brain responds to fragmented messages and responds when they connect and make meaning. For example, in social situations towards a popular fashion, idea or mood. This might explain how politicians and advertisers have been able to cause populations to respond to emotional slogans and appealing memes in history and currently on social media posts: for example, "for the many not the few," "you've never had it so good," "get brexit done," "the people united can never be defeated," "make America great again," "the old people have stolen our future," "keep the red flag flying here." Johnson has studied the development of brain networks which subserve social cognition. He has written a textbook with Michelle de Haan called *Developmental Cognitive Neuroscience* (2015).

Feldman Barrett says the traditional view of intelligence was generally accepted by Hippocrates, Aristotle, Descartes, Freud, Darwin and other great thinkers up to the present day. These great men were of their time and obviously cannot now speak on the subject, but as the human brain has had the same capability over millennia. It is possible they were engaged in dialectic discussions on the subject that have not survived. Interpretations of their work are sometimes seen through a biased lens. Texts are taken up out of context and used to confirm other people's ideas and assumptions. In education, I have found this has been the case with works of William James and John Dewey used selectively to support or deny particular schools of educational thought, for example, traditional versus progressive arguments.

Towards the beginning of her career Feldman Barrett made four meta-analyses (re-examinations) of results of a large number of previous research programmes with data on emotion (involving 220 physiology studies and 22,000 people). She found the findings were misinterpreted or unproved. As already stated, some asked subjects from different cultural and ethnic backgrounds, to give emotion labels to images of white people's facial expressions. The studies failed because humans view emotional expressions differently, labelling them with different words, meanings, nuances and concepts. A paper by a group of nine international authors from North America, New Zealand, Germany, the United Kingdom and Australia recently published in the *American Association for the Advancement of Science* journal supports this conclusion (2019). See Academic Endnote.[5]

During our history our inability to appreciate our own brain complexity has led to fixed and unconsciously inherited attitudes towards our brain's actual capabilities. Some of these self-fulfilling beliefs and prophesies about ourselves have held 'progress' back in a range of personal and social situations. They have informed the 'post truth' era producing 'virtual truth' slogans and memes affecting real life situations of populations and produced 'self-fulfilling prophesies' of lifestyle. They have even affected brain science and psychology when the tendency to seize upon partial truths as representing whole truth has restricted different levels of human development. Knowledge from neuroscience gives pause for reconsideration. It can encourage us to collect more knowledge and question current information to research and collaborate in cross-checking our assumptions.

2.3 Neuroscience and the Basis of Emotional Intelligence and Rationalisation

As previously mentioned, Lisa Feldman Barrett, an American neuroscientist, has studied and researched human emotion for 25 years (2017). She says most previous 'scientific' research in her subject was based on the idea that emotions are a quantifiable phenomenon. Traditional psychologists researched assuming emotions to be 'fixed components' in biology that cause common responses—behaviour which can be categorised. The idea that everyone feels named emotions in the same way was generally accepted and emotional conformity and categorisation were even assumed to be 'rubber stamped' on our brains. In her book *How Emotions Are Made* she explains how neuroscience now has biological evidence to show emotions being created. She has produced a plausible causation theory, which appears to be useful when we are analysing and justifying complex thought processes for various purposes.

At first sight Feldman Barrett's work may appear disingenuous to teachers, creative practitioners, therapists, engineers and scientists who work in and enjoy the arts, humanities, arts, social and sports sciences. There are after all great and small works of literature, history and biography that unpick human nature and psychological difference in endless variety, while exploring and describing iconic human themes. However, when we are struggling with those who do not understand or deny our creative insights and frames of reference, her pure neuroscience analyses back us up with concrete evidence in scientific terms. She shows how the traditional scientific approach to feelings and emotions has failed humanity. In the long term, it has not only caused a failure to understand these vital brain functions, but has encouraged the dismissal and denial of their influence on many important and even potentially dangerous aspects of scientific research.

In her Constructed Emotion Theory, the brain serves the body and is constantly responding to the resources in the body. It helps those resources to be kept in a healthy range. Feldman Barrett calls this phenomenon the Body Budget. A change in the body budget causes feelings—she has named these feelings the state of Affect. Body budget and affect are liable to change and vary at any given moment as body resources change. Affect is different to emotion. It has two dimensions—valence and arousal (pleasant or unpleasant)—aroused or unaroused (agitated or calm). Affect is a part of consciousness informed by both the body and brain 'in concert' and is currently a mystery. Further research into interoception (the senses informing the internal state in one's body) and proprioception (the senses informing self-movement and body positioning awareness) is ongoing. Annie Murphy Paul's recent book *The Extended Mind* (2021) gives us much to think about on this subject.

Good and bad things affect our body budget, but we are not born knowing how to control it and it is a process we learn through experience. Our brains construct emotion models of the world that help us to regulate and control ourselves. The models we construct to help us are predictive. They suggest what is or may be happening.

As we proceed in living, we constantly predict and correct. Our brains are governed by this prediction and correction—not *just* by stimulus response (reaction). The brain is (to some extent) a custodian as it predicts and chooses how to respond from information inside and outside your body. For example, it tells you when you are going to need food or drink and when to stop before you have an accident. The models you create in your brain are based on past experience and tell you what is going to happen. If your prediction is wrong, the brain updates its model in response to challenge and it will modify your body budget and make you feel some affect. These emotion models of the world are made up of concepts and categories. They are flexible and can combine old concepts to make new concepts. They come together to combine and are based on our individual goals. It is hard to grasp a new theory that currently seems so non-intuitive but it has important implications for how we understand and view ourselves in the world. See Endnotes.[6,7]

As mentioned before, we now know that our electrical impulses travel at variable rates between 2 mph and 200 mph throughout the body's nervous system in order to connect as we are variously stimulated. Electrical connections interact both mentally (e.g. memory and thought) and at the same time physically (e.g. heart rate, blood flow, breathing, sick stomach, adrenalin rush, shivering). These electrical 'messages' or 'feedback loops' fire back and forth several times in milliseconds creating networks which can be seen 'live' in real time on MRI scans. Apparently, they are not consistently identical either in pattern or in position. Human brains operate atypically and this may be because this enables us to cope with unexpected or random biological and external events in our everyday lives.

This does not seem so surprising when we consider we have billions of neurons stimulated by the body's changing physical state, combinations of hormones, activated chemicals and central nervous system reactions. This is how we have learned to cope in a world of changing external inputs over which we have limited control. As 'situations' develop, in any given moment we are bombarded with stimuli causing physical and mental responses (e.g. thought, behaviour and memory). It is not surprising therefore that, caught within this multitasking dynamic, we are not always able to be fully aware and in command of all of the connections behind our thoughts and actions!

Suzanne Langer (1895–1985), as previously mentioned, was a woman ahead of her time who deserves recognition as a neuroscientist. Her descriptions were perceptively accurate and are worth remembering. "She talked about symbolism, feelings, sensations, perceptions, reasoning and language limitations. Langer explained that we experience the world through our bodily biology, our capacities and sensibilities. She said the body's many regulating processes include neural, vascular, hormonal, physiological, autonomic, sensory, gustatory, kinaesthetic and more. She described the brain as organising its own world view being in a state of constant change as it adapts to sensory information. It works with the body with rhythm, flow, feedback, feedforward, loops, circuits and entrainments. These processes, affected by the serendipity of external events, result in human learning experiences" (Hawkins, 2017).

Connections between neurons are activated by both physical activity and sensations in our minds (memories, moods and thoughts). Each of these processes involves different feeling sensations even when not 'emotional.' These pulses of power or 'packets' of energy are fired up and driven by electrochemical impulses, which follow habitual, reactive and random connections to form neural circuits in the brain. We experience all of these in our everyday lives and we make choices. The fact that random thoughts arise unbidden in our minds does not necessarily mean that our behaviour will follow them automatically. We learn the ability to exercise choice. We constantly learn to accept and use or dismiss thoughts. Intelligence is involuntary, but also to some extent a variable skill that is acquired, deployed and sometimes forgotten. See Videos on Website Endnote.[8]

Our imagination allows us to consider and play around with a range of optional ideas. It also allows us to understand and empathise with other people's experiences and feelings. Logic is used in many subject disciplines and is considered to be a branch of philosophy. The term is used to describe a range of different types or methods for reasoning or rationalising. At its most basic it can be understood as simply making meaning by connecting ideas. It is probably not coincidental that this is also the way our billions of neurons create intelligence! The brain multitasks and is multidimensional—constantly rationalising—weighing up and assessing the world. In order to do this it has to make random as well as logical connections. These are subsequently at some point embedded, shelved (dormant), accepted or discarded.

The deficit model or framework of logical diagnosis is based on scientific medical models used historically, but it has to be used appropriately. This is also true of deconstruction analysis in academia. Such approaches tend to look for and focus on what is wrong and depend heavily on the judgements of the researcher or therapist. This can be useful medically and socially in a range of disciplines, provided alternative opinions are considered. There are four notable difficulties in the research and study of psychology.

- The psychologist's brain and the client or 'subject's' brain are often effectively studying themselves.
- Brains are subconsciously influenced and liable to spontaneous change.
- Contexts and situations change.
- Positive outcomes are often neglected and overlooked.

Psychoanalysis can be seen as a way of tracking back logically through 'logical' brain connections, but is often inaccurate being dependent on the *logic* and listening skills of the psychiatrist and the *logical* independent choices of the patient (Freud 1856–1939). These problems are mitigated in a phenomenological research approach where person, place and time are recorded in the process and rational analysis based on looking at data from alternative perspectives. Sometimes each individual's own point of view or own embedded neural frameworks cause a fixed mindset in viewing the world (Dweck, 2006). Their logic may become inappropriate when situations change. Beliefs sometimes persist as a default mode of thought, although we are proved wrong time and again every day.

Fear is one strong emotion, which can cause us to create these types of habitual thought patterns and become trapped in a 'fixed,' almost 'algorithmic' mindset! This is thought to be a possible basis of some personality disorders. The fact of neuroplasticity tells us we have the potential to adjust our mindsets and develop at any point in life if we encounter suitable stimuli (Christian & Griffiths, 2016). As teachers we often use this thinking ability positively in our lesson design to plan engaging, impactful events and environmental activities. The Constructed Emotion Theory now confirms and informs us that we are correct to do so and not to treat our learners as passive receivers of information. It is confirmed that learners transform knowledge as they absorb it adapting it to fit in with their own models of the world. *Feelings and emotions enable this process.*

Carol Dweck (1946–) is an American psychologist who has developed the idea of empowering people by informing them that a positive hopeful growth mindset improves their personal ability to develop their intelligence. Bad habits vary in importance, but can become a serious problem when ideas and opinions are fixed and new information is restricted—for example, when our ability to consider other people's legitimate points of view becomes restricted to a particular set of opinions on social media. Constantly liking, reposting and being 'liked' within an 'opinion bubble' stream on social media imposed and regulated by algorithm can encourage a fixed mindset, for example, conspiracy theories. This is how media platforms based on traditional behaviourist psychology have been used to deliberately 'reinforce' original bias. Fortunately, *we are all capable of behaving differently given knowledge and crucially the ability and opportunity to make an educated choice.*[9,10]

Problems and conflicts arise when we do not have enough information on which to base our judgements and when other people disagree with us! Logic can be a way of navigating around a set of predetermined rules, principles or declared 'facts' but these are not always suited to or sufficiently challenged in every situation to which they are applied. Humans who are open-minded and life experienced are best equipped to perform the task of understanding other humans—computers are not. Feelings and emotions are logical to that person even if they are wrong. Listening to 'mistaken logic' tells us a great deal about what people know and how they think. Sometimes they turn out to be correct! I have found this to be a useful approach to adopt researching, counselling and teaching. (Hawkins, 2017).

We make assumptions all the time and we question ourselves and each other. We learn wisdom by experience, acquiring and reconfiguring new knowledge. We find ourselves having to negotiate a meaningful way forward and adjust our thoughts as different situations arise. Of course, we do *intend* to think efficiently, but we may just be ignorant of some facts. We often miss the *logical* unconscious connections contributing until we ask ourselves, "Why did I think or do that?" Reflection is an ability we need to learn, build on and develop every day. We are inclined to think that our thoughts are based on external logical facts and sometimes they are.

We are creatures of habit. Our adult minds are embedded with established assumptions and those of our teachers, peers and parents. For example, according to Greg Lukianoff and Jonathan Haidt, there are ideas, frameworks and memories, which we should be taught to confront and challenge (Lukianoff & Haidt, 2018). In their book *The Coddling of the American Mind* they tell us that good intentions and bad ideas are setting up current college students for failure. This commercially driven trend has led to the 'no platforming' idea of not exposing oneself to other people's ideas. In fact, our internal logic systems have the potential to adapt to changes when novel stimuli in new situations produce new insights and attitudes. We need to be capable of making decisions, informing our lives, predicting outcomes and solving problems as situations and environments change.

Brain plasticity informs us that we can be adaptable animals throughout our lives, accumulating experiences, building up a 'working' and long-term memory. Logic can also be considered as a particular 'method' of reasoning and argument for making meaning and rationalising. We do this by predetermining the relationship and interdependence of a series of events and/or facts and judging by this framework. These frameworks are affected by any human value judgement error within the system. Even so algorithms are inferior to human logic because they are inflexible when conditions change. They are at best still just a modern technology engineers' tool of the technical communications industry. Highly complex computer networks are built, but essentially they are binary and founded on the same linear trajectory of scientific reasoning that goes back to Descartes' theories. ***It is their underlying purpose set by unregulated human logic that poses a threat to human societies.***

The public are learning to deal with their limitations as well as harnessing them creatively for human needs. They are a manifestation of human nature and a driving force for both good and evil in the fourth industrial revolution. They will inevitably require better regulation. A change in neural behaviour is indeed a modification, but the human brain has massive capacity and billions of neurons each making multiple connections. It is important to know as a teacher that externally targeted 'modifications' are liable to self-modify and are therefore extremely hard, if not impossible to track. Fortunately this is also a problem for behaviour modification experts! This may explain the continuing failure of technology companies to 'reverse engineer' the brain in order to create AI systems (Skinner et al., 1957). There are many subconscious operations in the incredibly diverse, random and potentially dynamic human brain and they are not necessarily trackable!

2.4 Feelings and Emotion-Based Learning: Awareness, Meaning and Inference

Feldman Barrett found, on reanalysis, the vast majority of historical research programmes that believed in the categorisation of emotions proved the opposite. They failed because they found that ***emotions are consistently inconsistent.*** She concludes that in the light of current neuroscience research emotions are never exactly replicated. Previous researchers were mistaken when they tried to categorise each emotion type by its scientific properties or fingerprint in scientific jargon. She tells us experiences of emotion are biologically unique because they exist in each moment through neuroplastic, neurological networks that have a propensity for variation and readjustment.

The human tendency towards variable and apparently contradictory behaviour is due to the way the body and brain operate and adjust to risk. This conclusion is supported by the evidence from my own educational research with school refusers, whose behaviour eventually became with patience, to some extent understandable over time. Emotions are goal-based concepts that we construct based on our past experiences. The emotion we construct depends on our perception. It arises as we face situations and depends on three aspects—what our goal is; what is happening inside;

and what is happening outside of us. If we predict someone or something has or is about to hurt us or doesn't value our lives, our bodies change. We will feel an agitated affect and we react. The body-brain may construct the emotion of anger or fear. However, if one understands the reason for this action one may calm down, slow down and stop an overreaction.

Emotions are predictions for action about how to achieve a goal using concepts based on past experiences. Sometimes if we understand why another person has upset or hurt us, we construct the emotion of sympathy or empathy. Justin Deol (2019b) explains how emotions prescribe our actions and that there are lots of internal variations in the types of emotion we can feel and how they are expressed. In his video he explains hypothetically how two people might acquire quite opposite but fixed mindsets or conceptual models of the world. Their emotion-driven attitudes (models) they carry around in their everyday lives are based on opposite goals that make them take two opposite paths. Justin explains each person's logical reasoning based on childhood experience giving rise to their goals and models. Neither of them are totally right and neither are totally wrong—there are elements of truth on both sides, but they both need to make adjustments.

In order to be in control of themselves in any situation, the two people in the example need to listen. They need to be more aware and adjust their ideas in order to determine the truth and decide how to act. We are ultimately responsible for how we construct our model of the world because this contributes to our predictions. Many of us have not known this, but now that we are aware of brain plasticity, we can improve our model of ourselves and our actions in the world. We are surrounded and affected by an invisible cocoon of culture, the worldview structure or system we learned from birth and carry around with us. If we can challenge our personal systems, gain ideas and adapt it, we can benefit ourselves and others. If we learn to adapt the invisible structure around us, we can better decide on our future actions and realities.

See Animated Video Endnote.[11]

Different concepts and frames of reference gained by listening to others help us see the same scenario in different ways. They help us to adapt to different roles appropriately for different purposes. We need to challenge ourselves to go into new situations, pay more attention and listen to improve these skills. It is ironic that emotional aspects of learning are the least understood area of cognition in traditional psychology. This is extraordinary because what we are 'aware' of and 'how we feel' is arguably the most important area of personal consciousness and concern in everyday life. Feelings and emotions are major communication channels for conscious thoughts. They contribute to and are intrinsic to intelligence. We puzzle about our own and other people's behaviour and motivations. We look for reasons and justifications. All the time, our views of the world, the incidents, the people and the situations we face are connected to our emotions as they are rationalised and reconfigured.

Tacit Knowledge is 'embedded' neuroplastic knowledge which depends on subconscious neurological connection—in other words a kind of logic. Anna Maria Rostomyan, a corporate communications specialist says that this background knowledge is very important in modern business because a person's emotions both positive and negative are part of his/her background knowledge stored in the rational mind and, together with the emotional mind, shape human cognition. Human beings base predictions, conclusions and feelings on past experience, their own and others' opinions, absorbed values, hearsay and role model behaviours. We look to find a way to behave that satisfies our needs as we make our way forward through life. These decisions are sometimes so rapid that we can only explain when we look back and gain insight after the event. People predict and 'build' their concept system models according to their experience and situation in life. See Special Data Contribution.[12]

The Feelings and Emotion-based Learning Theory says that these are logical to people in the moment, even if we don't agree. What is more, if made aware they can adapt and extend their concept model as they learn—as teacher practitioners maybe we can inspire them to do this. Knowing this may help to enable our empathy as we look for ways to help ourselves and each other, because we will develop in some similar but not identical ways (Hawkins, 2017). Academics and philosophers often become constrained by the mental models, ideas and historical ramifications of their subject. For example, consciousness (awareness) has held a particular fascination for many of them. Descartes had a principle known as Cartesian dualism (1641). His idea was that the mind and body are not identical (incidentally 'Cartesian' is a different form of his name). He believed inputs from the sensory organs travel to the brain and from there to a person's mysterious non-material spirit.

The mind-body problem is an old paradoxical debate about thought and consciousness in the mind being somehow separate to the brain—a separate physical organ. Even though the mind and body were known to be both physical, they were considered fundamentally different in nature.

Neuroscience appears to be resolving this problem in explaining, since we all have different minds, that the 'mystery of the mind' is too complex to be understood. Descartes thought the mind, mysteriously self-aware and conscious was the seat of intelligence and separate from the 'physical' brain (Dennett, 2017). The idea of a 'soul' or intellectual element that lives on after death was handed down through the teachings of Aristotle (384 BC–322 BC) and Plato (427 BC–347 BC), amongst others and is still an open debate. Many different forms of religion impose models that include this idea, which may be connected to the mystery of life and the understandable wish to avoid death. After all we do have a *necessarily* very strong survival instinct.

Powerful people particularly rulers and the inordinately wealthy have and still do nurture the delusion of entitlement to immortality, for example, mummification or cryogenics. Some of these beliefs affect current lives in society today. Dennett points out that unlike animals we are able to think about ourselves in a consciously personal, observational and deliberate way. We are aware of ourselves in the world. Humans have an elaborate set of thinking tools. These mental tools or cognitive gadgets have widened our awareness and allowed us to manipulate the world on many levels, for example, in social practices and physical environments. Language, in its many forms, is one of them. It enables complicated written and spoken ideas to be thought about and communicated across generations boosting awareness in many areas. Mathematical formulas capable of adaptation and development are another. Celia Hayes tells us these tools or cognitive gadgets contribute to our conscious awareness. See Video Endnote.[13]

New theories of mind have been developed by psychologists and philosophers up to the present day. A central theme emerging is around an Extended Theory of Mind. Andy Clark and David Chalmers were amongst the first to suggest it (Clark, 2008). They wonder where the mind stops and the rest of the world begins. Cognition is largely stimulated externally by complicated processes involving neural, bodily and environmental stimuli. They conclude that mindful cognition is not just in the brain and defended their views in a book with other experts contributing evidence (Menary & Clark et al., 2012). Some scientists have assumed perception to be just about our senses receiving information in a passive way, but there is an emerging consensus that our true relationship to the external world is much more complex. Empirical evidence can now support that position. This gives us potential for a better understanding of consciousness in its many forms.

Neuroscientists, philosophers, doctors and psychologists have explored different aspects of 'mind extension.' Further developments have come along that emphasise that embedded (internal) and environmental (external) stimuli have complementary cognitive processes and interactive elements. Clark and others have gone on to explore the idea of a predictive mind. Jakob Hohwy explains the Predictive Theory of Mind from the perspective of philosophy and cognitive science (2013). It is supported by powerful theoretical arguments across several disciplines and is based upon a simple premise. The brain is a 'hypotheses testing' organ performing the job of predicting and testing likely outcomes and so the *purpose* of the complex, deep, and multiple nature of human consciousness is explained. This idea can feed into teachers' research into their own and their pupils' cognitive activity. In the learning situation our opinions on both sides deserve respect.

The fact the environment actively drives cognition is a justification for teaching research using the Action Research Cycle. Such research show curriculum delivery needs to be responsive to enhance learning outcomes (Hawkins, 2017, Chapter 5). See Academic Endnotes.[14] This holistic approach to teaching is explained by neuroplasticity research. It shows our minds and bodies are sculpted by attention through stimulation as they interact with the world. However, in practice the 'mind' *has* to take an 'objective' and to some extent disassociated position in order to judge the world. The 'paradox' may simply be that at the same time our senses keep us deeply tuned in to our world on many conscious and subconscious levels. *This is no surprise if one accepts our complex brains tend to be always busily multitasking on some level.* See Academic Article Endnote.[15]

Koster-Hale and Saxe say that our predictions rely on making inferences about other people's personalities, thoughts and goals using predictive 'coding' which looks for differences or errors between what we see and what we predict (2013). This fits in nicely with feeling and emotion-based learning theory when, thanks to Feldman Barrett, we are able to acknowledge that emotions are predictive system concept models (attitudes) that prescribe our actions. Our process of reasoning, which is often but not always driven by emotion uses inference. The list of associated words or similes for inference is revealing. These are words that were invented to explain many of the ways we think. We surmise and conjecture, extrapolate and construct, interpret and read meaning into situations and events. As well as objective reactive logic we can now include inferential emotional logic—both are inferential, but the second is ***intuitive***, not overtly objective. See Academic Endnote.[16]

If we can understand mental perception as predicting, testing and correcting we can bring together a range of psychological (mind) and neuroscience (physical) research under the umbrella of neuroplasticity. This should tell us more about how we predict and how we act as we experience the learning process. For example, our senses are not just passive receivers of information because neuroscience proves awareness and consciousness to be highly interactive. Within that process we create our own personal awareness from our own unique perceptual illusions and together with emotions, feelings interact with and generate conscious thoughts, memories and behaviours. Those illusions are cognitive creations which may or may not be 'correct.' They might be thought of as our personal virtual reality. For example, people who think they agree may find out later they ***actually*** have quite different concepts (or takeaways) about the same idea.

Humanity's survival, acquisition of new ideas, inventions and knowledge depend on these complicated physical and mental connections. It is their very complexity of operation, which makes the important function feelings and emotions play in our intelligence, difficult to pin down and appreciate. Yet, it is generally acknowledged that many operate subconsciously. They can make connections which facilitate or hinder learning in important ways. When people express and realise their influence, they develop conscious and subconscious ways to work with them. These are sometimes surprisingly useful to us even when the reasons why they sometimes work can't be fully explained in words. For, example virtuoso artistic performances in music, dance, drama and the visual arts, acts of great charity, acts of bravery and scientific achievement.

2.5 The Senses in Proprioception Create Affect, Inform Emotion and Aid Learning

The input we gain through our senses involves the triggering of mental adjustments such as noticing, awareness, remembering, connecting and realising thoughts, imagining consequences, responding to and adjusting with physical movement. All of these stimuli inform the brain and central nervous system and change their physical structure by changing neural connections between cells in the body. For example, new cells and connections can grow triggered by physical exercise and thought. This is still a relatively new scientific discovery for ordinary people to appreciate, but it is providing evidence which will eventually cause a huge paradigm shift away from the machine model. This new understanding has enormous cultural and medical implications for human kind and makes our value systems even more important to consider.

Matthew Walker is a neuroscientist, who is an expert on sleep has also studied the 'waking brain.' He tells us the wide-awake brain constantly 'hums' with brain wave rhythms. The wave cycles go up and down at a frequency of 30–40 times a second (2017). The pulsating beat is erratic as the waves operate in an inconsistent pattern. Apparently you cannot 'tap along' to these brain waves because they are asynchronous. Walker likens the brain to a crowded stadium full of noisy crowd activity. The wide-awake brain is superfast, frenetic and chaotic. It has to process many consciousness streams dealing with smells, sights, sounds, thoughts, memories and of course, feelings and emotions. During the day, brain activity is mainly concerned with receiving, processing and responding to information from the outside sensory world.

It is likely that emotions and emotional feelings help us to distinguish and remember those things we deem important and make adjustments that help us deal with events in coping with our lives. Walker says the thalamus in the middle of the brain is a sensory gate, which blocks the senses when we need to sleep. This is when brain waves synchronise in deep NREM sleep and the body rests. REM or dream sleep is where according to Walker we replay our visual, auditory and kinaesthetic experiences upon the 'big screens' of the brain's sensory cortices. Of course, I would want to add emotional events into the mix! If asked about it, we tend to think our underlying brain *input* must follow an ordered sequence of logical thought, but our brains have adapted to be even cleverer.

In order to cope with a changeable environment we need to constantly seek out and make meanings and this is what we do. Apparently dreaming is a part of a brain sorting, memory filing and resetting adjustment system, ready for when the need for alert wakeful brain activity returns. Different feelings interact together within a general process of embodied cognition. It is difficult and probably impossible to deconstruct all of the processes involved. We are still learning about many of them. As well as the five sensory systems of taste, sight, hearing, touch and smell, we can now include extended physiological proprioception (EPP, suggested by D.C. Simpson in 1972), which involves many more subtle senses.

Proprioception is a general body awareness sense associated with touch or haptic perception, which involves the active exploration of surfaces and objects through tactile perceptions through sensory organs.

It encompasses one's physical body image, functional awareness in action, cognitive projections and nonverbal haptic communication. Proprioception was discovered and largely researched through medical disability studies of people with different types of unusual sensory experiences caused either by their natural physical constitution or by incidental injuries. Pietro Morasso's explanation explains how haptic perception constantly interacts, impacts and creates adaptions in the brain through the nervous system, body and environment. There is an exchange of signals/energy between the nervous system, the body and the environment in a continuous stream. The biophysics of the sensory organs shapes the motor neuronal input, which is transformed by the biomechanics of the body. He uses a musical metaphor and says the brain is not the conductor of an orchestra, but rather a player in a jazz jam session. This would be impossible without neural plasticity. See Endnote.[17]

There is an interplay of body, brain and environment. Consciousness may be an emergent property of this sensory-motor-cognitive process. The biophysics of neurons, synapses and glia shape and constrain the ways in which conscious experience emerges (as in learning) and expresses itself. Haptic perception is related to touch but it is not only touch, it is not passive but active. Most activities of daily life require and involve haptic experiences, for example, hitting, grasping, weighing, palpating, pushing, caressing and cutting. Through haptic experiences we get information about the physical properties of objects—their texture, compliance, fragility, weight, mass and stiffness. Sensorimotor learning is done through haptic experience and may be responsible for the formation of self through body consciousness. See Video Endnote.[18]

Severely impaired hemiparetic patients, who have suffered a stroke, are given haptic experience and robot therapy with the strategy of minimal assistance. In addition to human physical therapy robot therapy gives a deep haptic experience to the patient and therapist and has proved quite successful in rehabilitation processes. The patient may regain a body consciousness of the damaged part with brain-driven motor capability. Outside such situations, people in the general population with different sensory experiences have probably always existed without needing to be labelled as 'disabled.' Their neural plasticity enables them to develop in unique ways. If they are unidentified, they may have compensated for difficulties by adapting mentally and physically to fulfil personal needs and social situations. It is likely that their central nervous systems have adapted and operate 'differently' to the general population. See Article Endnote.[19]

Empathising with another person's emotional feelings involves tacit knowledge enabling a shift in understanding. Apart from facilitating social interactions, empathy provides us with a means of gaining useful information. For example, how do others see us, how can social interactions operate for mutual advantage and what can usefully be gained by seeing the world from other personal and cultural perspectives? Michael Polanyi tells us that tacit or implicit knowledge is the kind of knowledge that (unlike facts about the world) is difficult to transfer to another person indirectly by means of the written word or by listening to instruction. For example, speaking a language, riding a bicycle, making a cake, playing a musical instrument, designing, making and using machinery. All of these things, although they may include written instruction, have to be taught in large degree by social interaction, example, copying and trial and error experience.

He says it is possible to know a great deal more than we can tell and uses the term tacit knowing because although all knowledge is based in tacit knowledge, we tend not to fully appreciate this. This view is increasingly accepted in practical areas of knowledge management. Animals also have tacit knowledge, but humans have the ability to transmit and interpret it through complex behaviours with facial expression, talking and writing. People aren't always fully aware of or able to explain these skills (Chugh, 2015). In the case of children who have been abused, the transferring of tacit knowledge can be beneficial or harmful. Young children especially do not have the experience to understand what is happening and cannot explain their situation even though they may know something is wrong. It requires a great deal of empathy and open-mindedness in order to help them. Communities of practice can leverage these skills through shared experience.

Etienne Wenger's work on communities of practice seems to be in accord with Gilbert Ryle's (1900–1976) description that knowledge is not just 'know what' but also 'know how.' Wenger recommends that communities that collaborate and share knowledge operate more efficiently to everyone's benefit. Embodied knowledge or tacit knowing can often be better imparted by social interaction and modelling. Crafts people and professional practitioners such as surgeons have to make swift judgements without explicitly reflecting on the procedures or principles involved. Some of their tasks have to be performed skilfully without deliberation or consciously focused attention. The nervous system enables embodied knowledge which is different to explicit knowledge, but they cannot be separated in practice and are both essential. Historically indigenous peoples seem to have realised this very effectively in their education of the young. The sensory aspect of memory appears to have been much more utilised in historic educational systems than today.

Professor Lynne Kelly (2017), herself of indigenous descent, explains how perhaps we are missing out on some useful and intelligent approaches to teaching, memorising and learning. For example, the original peoples of Australia recorded their history, medicine, engineering and survival technologies by walking around the landscape using enduring features of the landscape as aids to memory. They 'recorded' knowledge through their art and their landscapes—memorising 'scientific facts' within stories in a long oral tradition passed down from parent to child. Many of the facts thus recorded are more accurate than written historical records when validated by science—for example, planetary eclipses, comets, earthquakes, droughts and floods to the year, day and hour. The connection between lived awareness and emotionally embodied memories deserves a great deal more analysis in all aspects of human learning to increase our understanding in many areas of human knowledge.

2.6 Empathy as a Restorative Therapy and an Essential Thinking Strategy

An empathic approach is needed across the whole spectrum of 'types of children' teachers may encounter including those who are labelled gifted and talented. Those with 'natural' talents for particular sports, dance, music, acting, mathematics, languages, mechanics and engineering are often envied and admired. However, we all need a loving social environment and help to develop our natural talents and we all need support to overcome attendant difficulties. These learners often feel different and need good empathic professional teaching support. For example, Michael Clay Thompson explains that this can put some learners at an emotional disadvantage. Unless supported, having exceptional physical, artistic, scientific curiosity, mathematical and three-dimensional perception can also set children at a social disadvantage. See Website Article Endnote.[20]

If learners are neglected and/or educated without empathy, problems may develop. We are neuroplastic in diverse ways. We need understanding and an encouraging and adaptable education programme with which to engage and develop. There are questions to be asked by teachers and important research to be carried out. This can be informed by a range of observational reporting by all involved contributing different points of view. This should include learners and parents. Teachers commonly come across behaviours that are inexplicable to them, because the pupil's back story is not known or understood. Although always possible, it need not *necessarily* have been caused by any type of extreme trauma, but simply occurred through accidental happenstance. Narratives which explain a child's unique experience in different cases help a teacher to develop empathy and practice, changing their responses appropriately even to 'apparently ordinary' children.

Bruce Perry's comprehensive clinical work with children badly affected by adverse childhood experiences (ACEs) evidences and explains neuroplasticity as he researches the complex, unique and nuanced effects on children of trauma, deprivation and 'therapeutic' intervention (2006). His narrative data evidences the view that our body-brains are continuously developing and learning from birth through childhood and adult life, expanding, pruning and refining our neural net capacity. He shows us how our network of awareness and memories becomes established and can reconfigure. This idiosyncratic adjustment process continues throughout life according to situation, memories and experience and it can have a negative or positive effect depending upon individual responses to neglectful and remedial experiences.

Perry's work has led to some enlightening discoveries about babyhood. He has seen how relationships influence children's cognitive development. He has made it his life's work to understand how trauma affects children (Perry & Szalavitz, 2006). He has found some original ways to help them. When he trained as a psychiatrist and neuroscientist in the 1980s such children were being treated for mental health difficulties and problem behaviour. Their personal histories of trauma were traditionally deemed coincidental and even irrelevant. His work has helped him understand the value of human relationships. Perry finds the developmental process involves two kinds of interacting neural response systems. A complex social set of specifically 'human' relational sensory perceptions about others, for example, significant carers and parents and a system of reward, gratification, comfort and safety (Szalavitz & Perry, 2010).

For example, a mother he calls 'Virginia' had a child who failed to thrive and grow called 'Laura' in a key case example. At four years old, the baby was baffling (and fascinating) the medical profession by failing to grow since her birth. A perceptive social worker suggested a psychologist investigate. The first attempt by an anorexia specialist failed. He thought Laura might represent an example of infant anorexia. He thought refusing to eat was a plea for attention. Virginia was told not to interact with Laura at meal times. Since Victoria was already not engaging with Laura in a 'normal' manner, her unintentional neglect of Laura was mistakenly encouraged. Perry inherited a pile of medical files written by medical specialists about Laura's condition. She had gone through extensive gastrointestinal examination and although she had a highly calorific diet, she still failed to grow.

The unravelling of Laura's problem did turn out to be 'psychological' but in a different way. Perry made an investigation, not only of her history, but also of Virginia's own story. After questioning, Virginia revealed she had been 'deprived' of adult nurturing relationships as a child in the government care system. She had not known her biological parents and been moved around every six months in a system which favoured discouraging attachment to any particular family. She did love her child but she did not know how to be a mother. Victoria's own childhood history turned out to be key to unlocking Laura's recovery therapy. Perry arranged for Victoria to stay in a professional foster carer's home with Laura for some months so that she could be taught how to care lovingly for Laura.

The carer, an experienced professional foster mother had previously shown Perry the value of 'maternal love' by persisting in hugging and cuddling a seven-year-old boy when she felt he needed it. 'Experts' thought her behaviour was inappropriate, but Perry listened to her and realised she was right. She demonstrated her skills by providing appropriate attachment therapy as a compensation for deprivation and it worked. When the child recovered, he no longer needed such close attention. Victoria learned from this woman how to love Laura and she started to thrive and grow. This story shows how the well-meaning 'logic,' narrow professional training, self-interest and enthusiasms of medical specialists can cause them to miss the obvious—in this case the importance of love, intimacy and good parenting. Suzanne Zeedyk says imitation is important in learning about intimacy and attunement with others and Victoria and Laura's case clearly demonstrates this (2006).

Perry believes the pleasure of human interaction is the 'glue' that creates and holds human relationships together. In his career he has proved nurturing with responsive parenting encourages neural connection, cognitive development and mental well-being. Laura's case illustrated how the nurturing of the mind through a loving relationship is essential for physical growth. Key stories by Perry explain how being left without human interaction affected different babies. Although physically fed and changed, they were left alone without stimulation or loving interaction for long periods. Some had carers who were trained not to interact with babies in orphanages to make them passive and others had inadequate parents with learning difficulties of various kinds. Such individuals very rarely, become sociopathic developing unsocial, self-destructive and violent behaviours.

Perry has had a distinguished career both in uncovering developmental neuroplastic processes and the effects of trauma on children. He has written about, considered and analysed key factual histories that document developmental deprivation. Perry's extensive phenomenological research reveals much about neural cognitive development and possible remedy. He has treated children who suffered severe traumas and mental abuse. These include rape, physical violence, neglect and emotional abuse and witnessing murder. He has researched the effects and causes of abnormal behaviour and failure to develop and shown how restorative treatments may or may not work. One 16-year-old had already become a murderer and appeared set in his ways. Perry uncovered some of the reasons, but was unable to help him. Brain plasticity developed to give us the powers we needed to survive, but sometimes deprivation causes severe embedded degeneracy and distortion.

When deprivation is diagnosed at an early stage, deliberate empathic therapy interventions prove successful. These include infant massage taught to parents, musical therapy and enlisting the support of children in an early years' class to befriend in specific ways. Perry shows us how children's plastic brains are embedded with neural nets due to carer treatment from babyhood caused by family dynamics, uncontrollable events and situations. These may also be affected by one or other of the following in various combinations: physical biology both pre and after-birth such as the effects of birth injury, illness, medicines, mother's and father's physical and mental health, especially if affected by drugs, alcohol and environmental pollution. See video Endnote.[21]

Parents and children themselves may be able to explain exactly what has caused a problem, but they *should not* necessarily be expected to do so. Too much emphasis and repetition on what has gone wrong reinforces neural embedding. Although sometimes useful to the patient who feels empowered, this may be why traditional psychoanalysis methods have become so much less popular. However, as Perry demonstrates using empathy, listening and observational reflection is a method that can work. Collaborating professionals can work out ways to support the nervous system to recover as it reveals how it can help itself in a loving environment. Neuroplasticity explains the usefulness of positive regard in person-centred counselling (Rogers, 1951). This is an interactive human communication therapy that involves developing tacit knowledge with clients in action. Empathy is a natural healing interactive process. It is also an important skill, which can be learned from modelling and experience.

Pam Jarvis is a chartered psychologist and a qualified teacher who has researched and lectured about the history and education of children and young people at Leeds Trinity University in the United Kingdom. In her blog article 'Birth to Three Really Does Matter: In Search of Security' she gives a summary of the history of human attachment theory. She tells us British psychologist John Bowlby (1907–1990) proposed the theory based on his research during the Second World War (1939–1945) (Bowlby, 1956, 2018). He had worked with children evacuated from cities who were separated from their families as babies and toddlers. He proposed that early on in life, infants construct an Internal Working Model (IWM) of their own lovability. He found this created a core model for future emotional interactions with other people. Bowlby emphasised the mother as of primary importance. See Blog Article Endnote.[22]

Jarvis tells us Schaffer and Emerson (1964) later found that babies aged 0–2 can form bonds with other adults in the first months. They found babies tended to form one primary attachment and several secondary attachments and were happy to accept 'substitute' adults. About 50% of babies were primarily attached to the mother and about 50% to the father or a family member. The person they favoured was the one who showed empathy and was most 'sensitively' responsive to them. Many powerful emotional models are based in self-preservation and the primeval instinct to survive. These have arisen from instincts based on avoiding starvation and keeping safe. Without the unnatural conveniences of modern life we are after all vulnerable animals living in a potentially hostile natural environment. As babies, we are helpless for a very long period of time and as adults we rely on our wits and brains for our survival.

Social networks have evolved to nurture us. Our brains are powerfully and acutely wired to respond strongly to other human beings. Our early life experiences create logical connections which enable thinking. Human beings need nurturing relationships in order to lay the basic brain connections and embedded systems that establish and develop neural nets. Brain plasticity and neural net capability is affected by these experiences, particularly in early childhood. As well as food, children need an emotionally engaging relationship with at least one significant adult early on. It is even better if there are several family members, who can help them to cope with positive and negative events, thoughts and behaviour. In human infants the establishment of neural networks caused by meaningful emotional memories is crucial for development.

Future awareness potentials are embedded early. Success in learning 'appropriate' and 'useful' behaviours depends on pleasurable social interactions and the nurturing behaviour of responsive others. This process continues throughout our lives through peer group pressure, encouragement and/or disengagement sometimes causing disaffection and degeneration. When we encounter neglectful or adverse situations, our brains adapt to them and may fail to learn and engage in useful social skills perhaps going on to develop 'inappropriate' habits of thought. This may lead to an inability to connect to and 'read' the possible consequences of a particular situation or course of action because we have no experience of likely outcomes. Other studies evidence that nurturing continues to play a role in psychological remediation processes such as counselling to support positive social interactions through adulthood into old age. The brain continues to have plasticity potential. See Article Endnote.[23]

2.7 Conclusion

In summary based on clinical, neuro-cognitive, psychotherapeutic and educational research the following aspects of our thinking inform our knowledge as learners and teachers. Our emotions are actively ***constructed as predictions*** or guesses as we refer, adjust and construct concepts in the moment. Instances of emotion need not look or feel the same and are not caused identically. Even the same actual experience and event affects people both similarly in some ways and differently in others. Rapid variations of feelings and emotions are normal for humans—both within and in comparison to others and are intrinsic to our thinking. There will be variations in emotional models of understanding between people involving past experience, ability, empathy and inhibition. There is also misattribution when the brain guesses wrongly—perhaps due to faulty operation through distraction, conflicting emotions, misinformation, ignorance or strong prior overriding belief.

The following points are useful to consider in improving our learning and teaching skills.

- Children are unable to think constructively, memorise and make meaning without making emotional and personal logical connections.
- Other people may have similar, but do not have identical emotional, feelings and thoughts to ourselves.
- If our own emotional, physical and intellectual responses are given without listening and reflection, we may miss out on important information.
- Assumptions are useful in new situations to make a prediction of other people's intentions, but may need to be reconsidered.
- We can gain insight by learning to change frames of reference. We are all capable of adopting different models by adjusting our perceptions.
- We all have unique perspectives, which make some kind of sense to us. These may be based on 'forgotten' events that we cannot recall.
- Learners learn best in situations which serve their immediate social and personal interests.

Teaching is one of the professions that demands high emotional intelligence in practice—in all aspects of the job, particularly in dealing with specialist subject sets of information in relation to learners, cultural conditions and settings. As teachers we may understand our students' behaviour, but sometimes we do not, especially if we are missing important information. We are not experienced or trained to consider learners' emotional feedback and interpret their behaviour, but we often learn to do it anyway. It would be better if we were trained in this type of research. It would assist us in our work. Teachers' and learners' brains have plasticity and can develop individual talents and experiences in unique ways. Each of us has our own unique potential and our abilities need to be matched to relevant learning experiences.

We all experience the world slightly differently. Everybody's sensory neural systems are operating deliberately, involuntarily and rapidly beneath conscious awareness at any given moment. At times we get confused, mix and conflate the meanings of different memories of events and feelings. This causes errors in memory association and behaviour. There are groups of people whose sensory cognition, natural physical abilities and motivations work in particular ways and perhaps not in others. It is likely that every human being has unique, networking quirks, neural caveats (specific stipulations, conditions or limitations) due to wiring differences within their biological constitutions. We are all on a continuum and a spectrum of ability in many different skill areas and aspects of cognition.

The body-brain has turned out to be much more complex than previously understood. It acts as a hyper alert control system and operates alongside and quite often independently of the stream of our conscious worded dialogue we call 'thinking.' The brain is highly interactive—it is constantly reacting and making micro-adaptations and tiny physical transformations. Feelings and emotions are an efficient and essential part of this process as our bodies react according to our perceived self-interest. Many human behaviours can now be better understood and research undertaken. The discovery of neuroplasticity also has widespread cultural implications. Some of these are explored in the following chapters.

Notes

1. Lisa Feldman Barrett (2019) 'How Emotions are Made' Providence Book Festival 25 May 2019 Interdisciplinary Affective Science Lab https://www.youtube.com/watch?v=DtIUQIHLnx0 Accessed 06/04/2021.
2. John Shotter-Guardian obituary by Michael Billig Published Thu 29 December 2016 https://www.theguardian.com/education/2016/dec/29/john-shotter-obituary Accessed 22/05/2020.
3. What is NEUROCONSTRUCTIVISM? What does NEUROCONSTRUCTIVISM mean? (Jan 3, 2017) The Audiopedia https://www.youtube.com/watch?v=EPkKpWX-ZgA Accessed 14/05/2019.
4. Inside the Psychologist's Studio: Annette Karmiloff-Smith (Nov 7, 2014) The Association for Psychological Science https://www.youtube.com/watch?v=qXmEe_m2Z9M&feature=youtu.be Accessed 21/05/2020.
5. Article Joshua Conrad Jackson et al. 'Emotion semantics show both cultural variation and universal structure' Science 20 December 2019: Vol. 366, Issue 6472, pp. 1517–1522.

   DOI: https://doi.org/10.1126/science.aaw8160 Science (print ISSN 0036-8075; online ISSN 1095–9203) is published by the American Association for the Advancement of Science, 1200 New York Avenue NW, Washington, DC 20005. https://science.sciencemag.org/content/366/6472/1517 on 30 December 2019.
6. Justin Deol (2019a) Dr. Barrett's theory-an animated vide explanation 'This is why you feel the way you feel', 1 January 2019 Skillsware Freedom

in Thought https://www.youtube.com/watch?v=M1OdhdI_3eI Accessed 06/04/2021.

7. Custodian definition—A custodian is responsible for protecting, caring for or maintain something or someone. Cambridge English Dictionary https://dictionary.cambridge.org/dictionary/english/custodian Accessed07/04/2021.
8. Feldman Barrett, L. Feldman Barrett-Website- https://lisafeldmanbarrett.com/multimedia/?t=Uzaan1HyJZ Accessed 28/05/2020.
9. Carol Dweck (2018) 'Core message-Adults and Children-Everyone can grow their cognitive abilities—rewiring' https://youtu.be/2nF90sAW-Yg Accessed 07/04/2020.
10. Carol Dweck (2019) 'Growth Mindset—Developing a Growth Mindset with Children and Adults' https://youtu.be/xuWUj9TtrOc Accessed 07/04/2020.
11. Justin Deol (2019b) Animated Video 'How to master your emotions | emotional intelligence' 15 January 2019 Canada: Freedom in Thought @ justintht @freedomintht https://www.youtube.com/watch?v=QGQQ7pJQqHk Accessed 08/04/2020.
12. Special Data Contribution

Dr Anna Maria Rostomyan (2020) 'Background Knowledge' Corporate Communications Specialist, Yerevan State University, Armenia Independent Research Group, IRG University of Fribourg, Switzerland Email annarostom@yahoo.com linkedin.com/in/dr-anna-maria-rostomyan-95b16783

Blog: https://donnannarossa.wordpress.com/

Background knowledge, also called *mutual knowledge* or *shared knowledge*, is a complex phenomenon, which is comprised of our physical, social, and mental worlds. It is the shared knowledge of the interactants in the processes of verbal and non-verbal interaction. Actually, our mental world is basically based on our background knowledge, which is being activated through verbal and non-verbal tokens and comprises both rational and emotive elements, which simultaneously partake in the higher cognitive processes (Rostomyan, 2013b).

Our brief theoretical survey here enables us to conclude that a person's emotions are evaluated as *positive* and *negative* and become part of his/her background knowledge stored in the rational mind, which, together with the emotional mind, shapes the cognition of the human. As a matter of fact, we do not view emotions as mere sensations, but rather chunks interplaying with rationality. As a result, any information, which is able to be processed in the brain, evokes certain emotions, which may, accordingly, have a great impact on one's behaviour and interpersonal relations. Consequently, in certain cases one's verbal behaviour depends on the evaluation made on the spot, due to certain parameters of the communicative situation.

Furthermore, sometimes emotions can be guided by our rational mind due to background knowledge—cultural, ethnic, social, educational, historical information stored in one's memory. Moreover, when making judgements on situations, evaluating events and individuals surrounding them, human beings are often guided by previous emotional experience, which is stored in background knowledge, which has a great impact on interpersonal relations in the process of interaction. Therefore, it follows that a person's previous emotional experience can also determine his/her present emotional state and, accordingly, become verbalized in the process of communication (Paronyan, Rostomyan, 2011).

Actually, our background emotional knowledge is no less important in processing, interpreting and comprehending this or that piece of information at hand, since far too much of what happens in the process of communicative interaction occurs on the emotional level, which is highly linked to our previous emotional experiences that are stored in this or that person's background knowledge.

Hence, in case parents and teachers take into account their children's previous emotional experiences, their feelings, beliefs, desires and yearnings, it will become easier for them to interact peacefully and more efficiently. Moreover, if teachers also take into consideration the emotional parent-child relations as a background knowledge, they will manage to build stronger relations with their students, which will undoubtedly ensure better labour output and better educative results.

References:

1. Paronyan, Sh., Rostomyan, A. (2011). The Pragmatic Impact of Background Emotional Memory on Interpersonal Relations. Armenian Folia Anglistika, *International Journal of English Studies*, 2(9), Yerevan: Lezvakan Horizon, pp. 7–14.
2. Rostomyan, Anna (2012). *The Vitality of Emotional Background Knowledge in Court*, Polemos, 6(2), De Grutyer, pp. 281–292.
3. Rostomyan, Anna (2013a) Management Techniques of Emotions in Communicative Conflict Reduction, in part 3: Communication and Management, *Communication: Breakdowns and Breakthroughs*, Probing the Boundaries, eds. Anabel Ternès, Inter-disciplinary Press, UK: Oxford, pp.141–151.
4. Rostomyan, Anna (2013b). *Huyzeri khoskayin ev voch khoskayin drsevorumnere vorpes lezvachanachoghakan qnnutyan ararka (angleren nyuti himan vra)/* A Linguo-cognitive Analysis on the Verbal and Non-verbal Expressions of Emotions (on the material of English fiction and films), dissertation, 160 pages, summary 27 pages, Yerevan, Armenia.

13. The Chandaria Lectures 2017 with Prof Cecilia Heyes – Lecture 1 'Cognitive Gadgets' Fri 8th December; Institute of Philosophy, School of Advanced Study: University of London https://www.youtube.com/watch?v=xRbyMehnUfA Accessed 15/06/2020.
14. Action research for educational change-Images Milton Keynes: The Open University. https://www.google.com/search?client=firefox-b-d&q=action+research+cycle+in+education Accessed 20/06/2020.
15. University of Rochester 'Memories Create 'Fingerprints' That Reveal How the Brain Is Organized' Neuroscience News 20 November 2020 https://neurosciencenews.com/memory-brain-organization-17297/ accessed 23/11/2020.
16. Jorie Koster-Hale & Rebecca Saxe *Theory of Mind: A Neural Prediction Problem*. Perspective Journal Volume 79, Issue 5, P836–848, (4 September 2013) Open Archive: Elsevier Inc. https://doi.org/10.1016/j.neuron.2013.08.020 Accessed 18/06/2020.
17. Pietro Morasso"(2005) "Consciousness as the emergent property of the interaction between brain body & environment: the crucial role of haptic perception"(2005) Università di Genova Istituto Italiano di Tecnologia | IIT · Department of Robotics, Brain and Cognitive Sciences www.consciousness.it/iwac2005/Material/Morasso.pdf Accessed 30/07/2020.
18. Paul Andersen (2012) The Sensory System Bozeman Science, 17 March 2012 https://youtu.be/TAzTFgPSPiU Accessed 09/04/2019.
19. "What is the Proprioceptive System?" By Griffin Occupational Therapy News (|24 March 2018) The Senses- King George V Lodge HP6 5FB; Company Number: 07140862; HCPC Registration: OT45274 https://www.griffinot.com/what-is-proprioception/ Accessed 04/12/2018.
20. Michael Clay Thompson (1998) "Is Everyone Gifted in their Own Way?" by Presentation, Indiana Association for the Gifted 1998 Annual Conference: Royal Fireworks Press Website; publishers for gifted and talented children https://www.rfwp.com/pages/is-everyone-gifted-in-their-own-way/ Accessed 08/10/2019.
21. **Dr Bruce Perry** Aug 25, 2020—Stress, Trauma, and the Brain: Insights for Educators—
    1. The Neurosequential Model https://www.youtube.com/watch?v=_3is_3XHKKs
    2. How Stress Impacts the Brain https://www.youtube.com/watch?v=COMwI2akgqM&feature=emb_rel_end
    3. The Power of Connection https://www.youtube.com/watch?v=oElS6AGwuxU Accessed 02/03/2021
22. Dr Pam Jarvis Friday, 5 April 2019 *Birth to three really does matter: in search of security* On Blogger since April 2013 http://histpsych.blogspot.

com/2019/04/birth-to-three-really-does-matter-in.html?m=1 Accessed 24/10/2019.

23. Mark Manson (2015) The Four Stages of Life: As we transition through life, we experience traumas, reshuffle priorities, and lose relationships. But this all allows us to reach our true potential. https://getpocket.com/explore/item/the-four-stages-of-life?utm_source=pocket-newtab-global-en-GB Accessed 19/07/2020.

References

Action Research Cycle Images Action Research for Educational Change-Images. The Open University. https://www.google.com/search?client=firefox-b-d&q=action+research+cycle+in+education. Accessed 20 June 2020.

Andersen, P. (2012, March 17). *The Sensory System Bozeman Science*. https://youtu.be/TAzTFgPSPiU. Accessed 09 Apr 2019.

Audiopedia. (2017, January 3). What is NEUROCONSTRUCTIVISM? What does NEUROCONSTRUCTIVISM mean? *The Audiopedia*. https://www.youtube.com/watch?v=EPkKpWX-ZgA. Accessed 14 May 2019.

Ballantyne, et al. (2005). Measuring Environmental Education Program Impacts. *Australian Journal of Environmental Education, 21*. University of Queensland.

Barrett, L. F. (2019). 'How Emotions are Made' Providence Book Festival, May 25, 2019. Interdisciplinary Affective Science Lab. https://www.youtube.com/watch?v=DtIUQIHLnx0. Accessed 28 May 2020.

Bickle, J., Mandik, P., & Landreth, A. (1999). The Philosophy of Neuroscience. In *The Stanford Encyclopedia of Philosophy* (First published Mon June 7, 1999; substantive revision Fall 2019 Edition), E. N. Zalta (Ed.). https://plato.stanford.edu/entries/neuroscience/. Accessed 07 May 2020.

Billig, M. (2016). John Shotter-Guardian Obituary. Thu December 29, 2016. https://www.theguardian.com/education/2016/dec/29/john-shotter-obituary. Accessed 22 May 2020.

Bowlby, E. J. M. (1956, 2018). *Attachment: The Attachment and Loss Trilogy*. Pimlico.

Chivers, T. (2019). *The AI Does Not Hate You: Superintelligence, Rationality and the Race to Save the World*. Orion.

Christian, B., & Griffiths, T. (2016). *Algorithms to Live By: The Computer Science of Human Decisions*. William Collins.

Chugh, R. (2015). Do Australian Universities Encourage Tacit Knowledge Transfer? In *Proceedings of the 7th International Joint Conference on Knowledge Discovery, Knowledge Engineering and Knowledge Management* (pp. 128–135). https://doi.org/10.5220/0005585901280135. isbn:978-989-758-158-8.

Clark, A. (2008). *Supersizing the Mind: Embodiment, Action, and Cognitive Extension*. Oxford University Press.

Claxton, G. (1997). *Hare Brain Tortoise Mind*. Fourth Estate Limited.

Collard, P. (2010). *The Little Book of Mindfulness:10 Minutes a Day to Less Stress More Peace*. Octopus Publishing.

Collard, P. (2014). *The Little Book of Mindfulness: 10 Minutes a Day to Less Stress, More Peace*. Gaia Books.

Cox, T. (2018). *Now You're Talking: The Story of Human Conversation from Neanderthals to Artificial Intelligence*.

Custodian Definition—A Custodian Is Responsible for Protecting, Caring For or Maintain Something or Someone. *Cambridge English Dictionary. https://dictionary.cambridge.org/dictionary/english/custodian*. Accessed 07 Apr 2021.

Dennett, D. C. (1991). *Consciousness Explained*. Penguin Random House.

Dennett, D. C. (2017). *From Bacteria to Bach and Back: The Evolution of Minds*. Penguin Random House.

Deol, J. (2019a, January 1). *Dr. Barrett's Theory 'This Is Why You Feel the Way You Feel'*. Freedom in Thought @justintht @freedomintht Skillsware Freedom in Thought. https://www.youtube.com/watch?v=M1OdhdI_3eI. Accessed 06 Apr 2021.

Deol, J. (2019b, January 15). *How to Master Your Emotions | Emotional Intelligence*. Freedom in Thought @justintht @freedomintht https://www.youtube.com/watch?v=QGQQ7pJQqHk. Accessed 08 Apr 2020.

Dixon, N. (2014, January 20). Effective Means to Capture Tacit Knowledge, Rudolf D'Souza, CEO of InKnowin consulting and co-chair of KM Asia 2013, asks Nancy Dixon eClerxService. https://youtu.be/iba44fxq1hQ. Accessed 11 July 2018.

Dweck, C. (2006). *Mindset: Changing The Way You Think to Fulfil Your Potential*. Random House.

Dweck, C. (2018). *Core Message-Adults and Children-Everyone Can Grow Their Cognitive Abilities—Rewiring*. https://youtu.be/2nF90sAW-Yg. Accessed 07 Apr 2020.

Dweck, C. (2019). *Growth Mindset—Developing a Growth Mindset with Children and Adults*. https://youtu.be/xuWUj9TtrOc. Accessed 07 Apr 2020.

Euler, M. J. (2018). '1.The idea that experience and expectations are central to cognition is typically attributed most directly to Helmholtz' (von Helmholtz and Southall, 1924) ideas on unconscious perceptual inference (Clark, 2015; Friston, 2005; Hohwy, 2013). 2. Here, predictive processing is meant to subsume an active engagement with—or sampling of—the environment. More generally, this is known as "active inference" (a corollary of the free energy principle, as outlined here' Neuroscience and Biobehavioral Reviews 94 (2018) 93–112 https://doi.org/10.1016/j.neubiorev E-mail address: matt.euler@psych.utah.edu © 2018 The Author. Published by Elsevier Ltd. This is an open access article under the CC BY-NC-ND license. http://creativecommons.org/licenses/BY-NC-ND/4.0/. Accessed 21/03/202.

Feldman Barrett, L. (2017). *How Emotions Are Made; The Secret Life of the Brain*. Pan Macmillan.

Fodor, A. J. (1983). *The Modularity of Mind*. MIT Press-A Bradford Book.

Fry, H. (2018). *Hello World: How to be Human in the Age of the Machine*. Penguin Random House.

Gadsby, C., & Evans, J. (2019). *Dynamically Different: Create Spaces That Spark Learning*. Independent Thinking Press @greatgadsby Blog. https://www.clairegadsby.com/

Gladwell, M. (2005). *Blink: The Power of Thinking Without Thinking*. Penguin.

Godfrey-Smith, P. (2016). *Other Minds: The Octopus and the Evolution of Intelligent Life*. William Collins.

Griffin Occupational Therapy News. (2018, March 24). What Is the Proprioceptive System? The Senses- King George V Lodge HP6 5FB; Company Number: 07140862; HCPC Registration: OT45274. https://www.griffinot.com/what-is-proprioception/. Accessed 04 Dec 2018.

Hall, E. (2018). *Aristotle's Way: How Ancient Wisdom Can Change Your Life*. Bodley Head.

Hawkins, J. A. (2017). *Feelings and Emotion-Based Learning; A New Theory*. Palgrave Macmillan.

Heathcote, D., & Bolton, G. (1995). *Drama for Learning: The Mantle of the Expert Approach to Education*. Heinemann Educational.

Heyes, C. (2017). The Chandaria Lectures—Lecture 1 Cognitive Gadgets, Fri December 8; Institute of Philosophy, School of Advanced Study: University of London. https://www.youtube.com/watch?v=xRbyMehnUfA. Accessed 15 June 2020.

Hohwy, J. (2013). *The Predictive Mind*. Oxford University Press.

Jarvis, P. (2019). *Birth to Three Really Does Matter: In Search of Security*. Friday April 5, 2019. http://histpsych.blogspot.com/2019/04/birth-to-three-really-does-matter-in.html?m=1. Accessed 24 Oct 2019.

Jarvis, P. (2019). *Not Just 'Once' Upon a Time*. Institute of Childhood and Education, Leeds Trinity University, UK; Published in: Genealogy 2019, 3, 44; doi:10.3390/genealogy30300441—Licensee MDPI, Basel, Switzerland. This article is an open access article distributed under the terms and conditions of the Creative Commons Attribution (CC BY) license. http://creativecommons.org/licenses/by/4.0/. Link: www.mdpi.com/journal/genealogy. Accessed 01 Jan 2020.

Johnson, M. H., & de Haan, M. (2015). *Developmental Cognitive Neuroscience: An Introduction*. Wiley & Sons Ltd.

Jackson, J. C., et al. (2019, December 20). Emotion Semantics Show Both Cultural Variation and Universal Structure. *Science, 366*(6472), 1517–1522. https://doi.org/10.1126/science.aaw8160. Science (print ISSN 0036-8075; online ISSN 1095-9203) is published by the American Association for the

Advancement of Science, 1200 New York Avenue NW, Washington, DC 20005. https://science.sciencemag.org/content/366/6472/1517 on December 30, 2019.

Karmiloff-Smith, A. (1996). *Beyond Modularity: A Developmental Perspective on Cognitive Science*. MIT Press.

Karmiloff-Smith, A. (2014, November 7). *Inside the Psychologist's Studio*. The Association for Psychological Science. https://youtu.be/qXmEe_m2Z9M. Accessed 21 May 2020.

Kelly, L. (2017). *The Memory Code: The Secrets of Stonehenge, Easter Island and Other Ancient Monuments*. Pegasus Books Ltd.

Koster-Hale, J., & Saxe, R. (2013, September 4). Theory of Mind: A Neural Prediction Problem. *Perspective Journal, 79*(5), 836–848, Open Archive: Elsevier Inc. https://doi.org/10.1016/j.neuron.2013.08.020. Accessed 18 June 2020.

Langer, S. K. (1967). *Mind: An Essay in Human Feeling* (Vol. 1). The Johns Hopkins University Press.

Leader, D. (2016). *Hands: What We Do with Them—And Why*. Penguin Random House UK.

Lingard, et al. (2008, August 23). Grounded Theory, Mixed Methods, and Action Research. *BMJ, 337*. University of Toronto.

Lisa Feldman Barrett-Website. https://lisafeldmanbarrett.com/multimedia/?t=Uzaan1HyJZ

Lukianoff, G., & Haidt, J. (2018). *The Coddling of the American Mind: How Good Intentions and Bad Ideas Are Setting Up a Generation for Failure*. Allen Lane Penguin Random House UK.

Manson, M. (2015). *The Four Stages of Life: As We Transition Through Life, We Experience Traumas, Reshuffle Priorities, and Lose Relationships. But This All Allows Us to Reach Our True Potential*. https://getpocket.com/explore/item/the-four-stages-of-life?utm_source=pocket-newtab-global-en-GB. Accessed 19 July 2020.

Mareschal, D., Johnson, M., Sirois, S., Spratling, M., Thomas, M., & Westermann, G. (2007). *Neuroconstructivism* (Vols. I & II, Developmental Cognitive Neuroscience). Oxford University Press.

Menary, R. (Ed.), & Clark, A. et al. (2012). *The Extended Mind: Life and Mind: Philosophical Issues in Biology and Psychology*. Massachusetts Institute of Technology (MIT).

Morasso, P. (2005). *Consciousness as the Emergent Property of the Interaction Between Brain Body & Environment: The Crucial Role of Haptic Perception*. Università di Genova Istituto Italiano di Tecnologia | IIT Department of Robotics, Brain and Cognitive Sciences. www.consciousness.it/iwac2005/Material/Morasso.pdf. Accessed 30 July 2020.

Murphy, K. (2020). *You're Not Listening: What You're Missing and Why It Matters.* Penguin Random House.
Murphy Paul, A. (2021). *The Extended Mind: The Power of Thinking Outside the Brain.* Houghton Mufflin Publishing Company.
O'Mara, S. (2019). *In Praise of Walking: The New Science of Why We Walk and Why It Is Good for Us.* Penguin Random House.
Perry, B. (2020, August 25). *Stress, Trauma, and the Brain: Insights for Educators.*
1. The Neurosequential Model. https://www.youtube.com/watch?v=_3is_3XHKKs
2. How Stress Impacts the Brain. https://www.youtube.com/watch?v=COMwI2akgqM&feature=emb_rel_end
3. The Power of Connection. https://www.youtube.com/watch?v=oElS6AGwuxU. Accessed 10 Sep 2020.
Perry, B. D., & Szalavitz, M. (2006). *The Boy Who Was Raised as a Dog: And Other Stories from a Child Psychiatrist's Notebook.* Hatchette Book Group.
Pinker, S. (2002). *The Blank Slate: The Modern Denial of Human Nature.* Allen.
Pinker, S. (2015). *The Village Effect: Why Face-to-Face Contact Matters.* Atlantic Books.
Polanyi, M. (2009). *The Tacit Dimension.* University of Chicago Press.
Robertson, J. (2014). *Dirty Teaching: A Beginner's Guide to Learning Outdoors.* Independent Thinking Press Crown House. https://linktr.ee/creativestar
Robertson, J. (2017). *Messy Maths: A Playful, Outdoor Approach for Early Years.* Independent Thinking Press Crown House @creativestar.
Rogers, C. R. (1951). *Client-Centred Therapy.* Constable.
Rose, H., & Rose, S. (2012). *Genes, Cells and Brains.* Verso.
Rostomyan, A. M. (2020). *Background Knowledge.* Corporate Communications Specialist, Yerevan State University.
Schaffer, H. R., & Emerson, P. F. (1964). The Development of Social Attachments in Infancy. *Monographs of the Society for Research in Child Development, 29* (Serial No. 94).
Shotter, J. (1985). *Social Accountability and Selfhood.* Blackwell.
Shotter, J. (1993). *Conversational Realities: Constructing Life Through Language* (Inquiries in Social Construction Series). London: Sage Publications Ltd.
Skinner, B. F., Ferster, C. B., Cheney, C. D., & Morse, W. H. (1957). *Schedules of Reinforcement.* The Skinner Foundation.
Suzanne Zeedyk, M. (2006, June 7). From Intersubjectivity to Subjectivity: The Transformative Roles of Emotional Intimacy and Imitation. *Infant and Child Development Research Article Wiley Online.* https://onlinelibrary.wiley.com/doi/epdf/10.1002/icd.457. Accessed 10 June 2019.
Szalavitz, M., & Perry, B. D. (2010). *Why Empathy is Essential and Endangered: Born for Love.* Harper Collins.

Taylor, T. (2016) *A Beginner's Guide to Mantle of the Expert: A Transformative Approach to Education*. Singular Publishing @inspireimagination Blog. https://www.mantleoftheexpert.com/

The Story of Human Emotions—Human Emotions Are Complex, Powerful, and Sometimes Unexpected. 7 VIDEOS UK: BBC Radio 3. https://www.bbc.com/ideas/playlists/the-story-of-human-emotions. Accessed 08 Oct 2019.

Thompson, M. C. (1998). Is Everyone Gifted in Their Own Way? by Presentation, Indiana Association for the Gifted 1998 Annual Conference: Royal Fireworks Press Website; publishers for gifted and talented children. https://www.rfwp.com/pages/is-everyone-gifted-in-their-own-way/. Accessed 08 Dec 2019.

University of Rochester. (2020, November 20). Memories Create 'Fingerprints' That Reveal How the Brain Is Organized. *Neuroscience News*. https://neurosciencenews.com/memory-brain-organization-17297/. Accessed 23 Nov 2020.

Velasc, C., & Obrist, M. (2020). *Multisensory Experiences: Where the Senses Meet Technology*. Oxford University Press.

Walker, M. (2017). *Why We Sleep: The New Science of Sleep and Dreams*. Allen Lane.

Walker, M. (2019, July 12). *Why Sleep Matters*. Talks at Google. https://youtu.be/c1yGw_hfEfk. Accessed 10 Apr 2020.

Watson, L. (1999). *Jacobson's Organ and the Remarkable Nature of Smell*. Allen Lane Penguin.

Wenger, E. (1998). *Communities of Practice*. Cambridge University Press.

Wenger, E., McDermott, R., & Snyder, W. (2002). *Cultivating Communities of Practice: A Guide to Managing Knowledge*. Harvard Business School; McGraw-Hill.

Whitehead, A. N. (1929). *Process and Reality: An Essay in Cosmology*. Macmillan Company.

CHAPTER 3

The Plastic Brain and Its Educational Development

3.1 Introduction

Education with its many complex and varied settings, where learning is the 'raison d'etre,' provides many diverse opportunities to research human intelligence processes. This chapter looks at ideas, theories and discoveries about psychology and learning contributed by 'education' psychologists and teacher theorists to the current cannon of belief. Many of them have had ideas we should note and research into in order to revise our assumptions about intelligence to improve our education systems. In 2018 Stephen Hurley interviewed me on Canadian-based VoicEd education radio. See Interview Endnote.[1] As a former teacher and popular education presenter/expert, he asks perceptive questions. He asked me about the 'uncoupling' of feeling and emotion. He asked, "We often conflate them and they become synonymous in our language and the way we use them. What is the difference between those two words? What can we learn from separating those words? What might that allow us in terms of understanding not only ourselves, but ourselves in the world?"

Great questions I had not previously addressed and for me a demonstration of the value of listening to other people! My theory that people are unique and their feelings and emotions likely to make sense to them is

Special thanks to Chris K. Pearson for 'Too Old to Suffer' (2014), Video and Poem Transcript, a poem about emotions for his daughter.

J. A. Hawkins, *Brain Plasticity and Learning*,
https://doi.org/10.1007/978-3-030-83530-9_3

evidenced by Stephen's insightful analyses as he identifies a possible route of inquiry which might help a research investigation. These are key questions. Following on from the previous chapters in which we looked at neuroplastic physicality, after reflecting on Stephen's questions, an important underlying question might be "Why *do* we conflate physical feelings, mental feelings and emotions?" The answer may have some physical basis. For example, an important underlying reason for our 'confusion' may be the speed and variability of reactions throughout the brain and the body…between 2 and 200 miles per hour. I will consider some further possible answers in the next section.

3.2 What Are Feelings and Emotions, and Can They Enable Teaching?

The English language uses the words 'feel,' 'feeling' and 'emotion' with variable meaning. They are words that represent fluid shapeshifting sensations to describe essential, powerful actors in thinking. These words can be doing words (verbs), names for kinds of thoughts (abstract nouns) and descriptors (adjectives and adverbs). Dictionaries define them with different emphasis. Feelings are *inputs*—sensations through sensory organs, for example, nose, tongue, skin, ears and eyes. They are also *outputs*—mental responses (realisations, thoughts and intuitions). Both kinds of feelings perform rapid feedback loops that inform our thoughts and behaviour.

In the human body physical sensations *constantly* make connections between the brain and body through the nervous system. They are essential to cognition—a random, unavoidable and necessary human neuroplastic process constantly impacting and registering on our awareness at various levels as axons and neural nets connect. It might be helpful to separate them into *input* and *output* feeling*s* and think of an emotion as a strong surge powerfully combining both. We register *emotions* strongly through the body and label them with common names, but the mixture of contributory feelings and outcomes affecting our behaviour can often only be explained in retrospect. This process helps us adjust our general and specific conceptual models of the world.

An emotion arrives in a flood of recognition—a conglomeration of feelings both physical and mental that produces a strong glow of energy. It can have a powerful effect as hidden meaningful connections are played out in *neural net performances* such as are seen 'lit up' in an electromagnetic scan. According to neuroscientist Lisa Feldman Barrett, *each* emotion is *'made uniquely'* even if we habitually recognise and associate it with our emotionally affected concepts in particular situations (Feldman Barrett, 2017). Individuals generally ascribe emotional events with different and common meanings, but emotional intelligence is also a dynamic predictive and transformational skill we learn as we adapt our conceptual frameworks (Goleman, 1996).

Although commonly designated with names, emotions have different meanings for different people. They are 'learned' and held in memory and yet variable—changing in a moment from vague to clear to powerfully felt. They may not be identical or perhaps when explained completely understood, but they progress thinking, help us predict and can often be shared. A poem called 'Too Old to Suffer' by Chris K. Pearson, a maths teacher, illustrates this point. He wrote the poem about emotions for his daughter, but when he presented it to his class, their close attention and response is palpable as they relate to his performance. See Special Data Contributions.[2, 3]

In his poems Chris explains to his daughter his own view about how to handle emotions and in the video we see and feel his class relating to his experience in the moment. It provides us with a clear example of the 'magic' of emotional communication. This is just as 'intelligent' as knowing a fixed body of scientific facts and although hard to describe, just as productive for human learning development. We are seeing here how a class relates to their teacher as a human being and is learning from him as a potential role model. He shows them he is not ashamed to express his feelings and how (as he advises his daughter) he has found a good way or model of how a person might be able to deal with some of their emotions.

The same words and their use to convey meaning are used in different combinations to describe sensations that mean variable things to people in different situations. We see this in the arts and in scientific texts where accuracy of description requires the use of affect related feelings to convey meaning. In understanding many phrases and adjectives (e.g. words like cold, hard and bitter) we draw on our sentient knowledge and its emotional associations. We cannot fully understand or appreciate a given text without this facility for extending and connecting our thoughts to our own bodily experience. Feelings and emotions in various combinations ('mixed feelings') communicate *vital information* and shortcut thinking time as we reconfigure our thoughts. We learn different *appropriate* language frameworks, words, phrases and ways of being in different subject disciplines and cultural situations.

Gardener talked about these in his Multiple Intelligences Theory and sometimes we become *emotionally* invested in these different disciplinary thinking and working modes and methods (Gardener, 1993). Feldman Barrett tells us the world would be a sensory blur of undifferentiated sensations if the brain could not process, construct and appreciate these kinds of frames of reference. The brain's job is to predict possible outcomes, fill in missing details and find regularities and irregularities and manage responses. We experience a world of objects, people, noises, and events and not mere "blooming, buzzing confusion." The brain has a great deal to sort out and process. Feldman Barrett cites William James (1884, 1894, 1918) the famous educational theorist who said that emotional brain processes essentially informed by the senses are variously combined.

She talks about granulation (breaking down to expose constituent parts), conceptual combinations and the fact that these are reconfigured and differently created every time we experience and perceive the world. She supplies us with some new science-based language and concepts to use when discussing the relatively new field of emotion studies. She talks about the fact that we may have a limited or finite body budget for dealing with situations we face. Presumably this variable coping capacity explains different responses to difficulty at given points in our lives. Emotions are not just reactions but are *active* constructions and predictions in brain and body. Even when hidden, they are constructed from interpretations of sensory input, established and remembered *concepts*. Scientists have defined concepts as mental representations of a category. *Embedded emotions and emotional feelings help us make logical connections, adjustments and corrections to our general emotion model of the world.*

Our bodies operate constantly in this way even when we know they are probably incorrect and our hypotheses prove wrong, for example, reacting to an unexpected sound. Feldman Barrett talks of the Theory of Constructed Emotion where meanings and predictions contribute to emotional thinking and behaviour. She describes how the brain faces a sensory 'bombardment' on a walk in a city park as we see people, flowers, trees, bushes, vehicles, houses, dogs, birds, insects and the skies above. We are assailed by information from sounds, tastes, smells and touches (clothes, wind and sun) as we guess about and perceive events happening–children playing, people sitting, walking and jogging, holding hands, talking and shouting. Traditionally, categories exist outside in the world, while concepts exist in the mind, but then there are concepts about categories. For example, you have concepts of colours (even if you are partially colour-blind) (Dennett, 1991, 2017).

We make casual assessments of behaviours and relationships between people. Some of these assessments are right and some wrong, but they allow us to navigate life safely. A great deal of information is processed in the moment. The continuous sensory signals affecting our brains are variable and ambiguous. The brain can be tricked—for example as in optical illusion art pictures where there are two ways of 'reading' the image. Perceived and learned concepts help us create explanations and simulations of reality. We experience a world *of our own construction*. The brain brings to awareness more than the information that is available externally. In milliseconds our brains are constantly and rapidly categorising and predicting outcomes from each sensory encounter. We would be unable to think so efficiently without the feelings and emotions we feel. We lose much information if we *don't consider* our own and other people's feelings and emotions.

In summary according to neuroscientist Lisa Feldman Barrett each emotion is unique (2017). It is logical to assume that each feeling about a thought is liable to a similar variability—this is a concept to conjure with! It means we really do each experience events differently and apportion different meanings to 'reality.' Thinking is a *metaphysical* activity—the way humans interpret and rationalise their world. Why do we find it so hard to accept others think differently? It is time to understand, embrace and develop our true capabilities. Education systems have conditioned our brains to believe paradoxically that we should expect the world to be straightforward even though we know that really it is not. This has often been our traditional educational way of thinking as we planned lessons. If we are to understand and make the most of our plastic capabilities perhaps we should change our modus operandi and our attitudes towards 'knowledge.' Here is a light-hearted interview with Hannah Critchlow, a neuroscientist about being human. See Interview Article Endnote.[4]

3.3 Emotional Development in the Early Years: Baby, Toddler and Child

Early years' child psychologist Pam Jarvis (2016) tells us healthy newborn babies are born with ancient primate reflexes. These include crying with hunger, seeking the nipple, the 'startle or Moro reflex,' floating and paddling in warm water, grasping an adult's finger, recognising the mother's voice and searching to identify the mother's face. After birth a baby's brain and body embarks upon an extensive neuronal programme storing experiences and developing mental and body awareness. The brain uses the central nervous system forming ever more intricate pathways embedding rapid neural net capacity throughout the body. Potential bodily capacity responds to external stimulation and repetition and is tuned in to close human interaction. At this early stage, a full complement of billions of neurons is already present, ready and waiting to be activated. See Academic Articles Endnote.[5]

Synapses between neurons are appearing and disappearing at a greater speed before the age of three than at any other stage of life. Jarvis says this process is commonly called ‘blooming and pruning.’ This ‘wastage’ may be occurring to free ‘capacity’ to develop other abilities prioritising more pressing needs and dealing with them. The brain grows rapidly making increasingly complex connections that rapidly increases its capacity. As a child develops the ability to focus attention without being distracted and with self-regulation (inhibitory behaviour) improves. This behaviour triggers cognitive growth and abilities, skills and intelligence increases exponentially in favourable conditions. Individuals are variable and have their own key experiences and growth moments. There are general patterns of skill acquisition, but norms can be wide in kind and quantity. The beginnings of language development are activated at any time up to three, four and even five years of age and beyond.

A recent tweet on Twitter by a teacher asked for advice. She was worried her child might be ‘backward in talking’ at two years old. There were over a hundred reassuring comments about speech delay from other teacher parents. Many had raised a late language developing child of their own and most turned out to be highly successful in later life. They assured her a child’s talking ability varies enormously starting between one year to five and even seven in one case. They advised that if a child responds to words with understanding, makes noises or babbles consistently and is checked for hearing loss or other disabilities, most children will eventually ‘catch up.’ For example, we know that Einstein was a late language developer and would certainly not have fitted into a ‘normative’ range of ability. See Article Endnote.[6]

Babies are born with the ability to recognise the full range of sounds known to the human voice. Survival involves responding and strengthening links to sounds heard most and meaning most. Sound rhythms accompany socially accepted behaviours and the reading of facial expressions (physiognomic perception). Haptic sensation and predictive awareness all play a part. Children respond to love and empathy in close primary relationships. They learn from the intersubjective meanings conveyed between the people who surround them. They develop psychosocial ability—'reading' other people helps them to survive. In the case of some skills like language acquisition, such embedding and wastage may create a problem for older adult scholars trying to learn a foreign language or lose an accent later in life. If the environment is complicated, enjoyable, rewarding and socially engaging more neural nets and more capacities and abilities are created. For example, children of bilingual parents may learn both languages early and speak both with ease.

There is evidence that this exposure and even the addition of a third language increases general intelligence in later life (Dean Hobbs, 2016). Those neural connections and skills not used or needed on a daily basis become 'forgotten.' For example, even on the simple level of sharing or not sharing toys. Some children have serious difficulties even with one language, but they may still have ability potential if given the right kind of opportunity. Everyone is on a continuum through life in this regard with widely variable physical and genetic dispositions and unknown potential. *If we assume they are incapable, it is more likely they will be.* Jarvis insists that the natural way to learn about the world, develop and grow in early years is through spontaneous play-based interaction with peers and adults.

Jarvis evidences her argument by citing the pioneering practices of Rousseau-inspired early education pioneers—for example, Pestalozzi, Froebel, Montessori and Malaguzzi mentioned in her book *Early Years in Context* (Jarvis et al., 2016). Many of these educational giants started their work more than a century ago. They were educationalists and child psychologists who researched children's learning development. Their methods were based on supporting children with love and understanding. Although they lived in traditional times, their work clearly showed that providing a lively holistic environment and supporting children to follow their interest while inculcating an essential knowledge framework encourages optimum intellectual development. These schools and foundations continue to be popular around the world and still follow and develop their methods. Educationally minded parents who can afford it often seek them out.

An article by Amy Brown, a professor of child public health responds to simplistic notions about training children in ways that ignore human feelings and cause long-term emotional and developmental damage. She cites a television programme which recommended that new parents use dog training methods to 'manage' their children and mentions that some self-help books for new parents recommend behaviourist conditioning methods. She points out that such an upbringing can have serious consequences for both parents and children for family life and some individual's long-term mental health. She says babies and toddlers need loving support from their primary caregivers. When they cause disruption to adults, they are often just trying to understand their own emotions and behaviour as they grow and learn about their world. Their 'challenging' behaviour is part of developmental process as they communicate, process and respond to what is going on within and around them. See Article Endnote.[7]

Such behaviourist methods include using food rewards, unvarying timetables, withdrawing attention and isolating children in response to 'unacceptable' behaviour and commands. Brown points out that when parents are responsive to their babies' needs, their children go on to have fewer behavioural problems and better social interactions with others. Encouragement and on the spot adaptation to adult feedback of the situation in the moment is vital for babies. Explaining possible and likely consequences is the best form of discipline. It motivates the developing infant to achieve neural control over bodily functions, for example, making mental responses such as recognising people, manipulating objects and learning to sit and walk. Food rewards particularly can lead to an unhealthy attitude to food leading to eating disorders and obesity. See Academic Paper Endnote.[8]

'Dog training' does not help children to understand why they are required to conform or to act in a particular way since it is dependent on parental whim. Such behaviour does not encourage rational thinking and confidence in their own ability to understand and cope with social situations. This can lead to insecurity, worry, self-rejection, high levels of anxiety and other emotional problems. Baby books that suggest only feeding at set times, leaving your baby to cry for long periods without comfort and withdrawing eye contact discourage maternal behaviours that keep babies safe and healthy. New mothers who trust these books experience high levels of stress when the guidance doesn't work. They can become anxious and frustrated and feel like a failure. They have higher levels of stress and depression and are less likely to cuddle the baby.

The old advice tells them to delay responding to their baby's cries, although this increases infant stress hormones and encouraged them to put the baby to sleep in another room, which is against modern safe sleep guidance. These old techniques were not only harmful, they only worked in getting infants to sleep 15–20% of the time. Children grow up to feel more secure and self-confident when their needs are met with *suitable levels of challenge, empathy and encouragement.* This real time '*interactive biological dynamic'* is the same process that a parent is stimulating in supporting a baby to develop and grow. Such a childhood helps them develop imagination, long-term and short-term memory skills, emotional intelligence and higher IQ scores. *Education at any age should be essentially a continuation of the same kind of support given in a suitable and appropriate manner.* See Academic Endnotes[9] and [10].

3.4 Teenagers and Young Adults Growing Up and Developing Beyond

John Abbott's book *Over Schooled but Under Educated: How the Crisis in Education Is Jeopardising Our Adolescents* (2009) gives a useful historical summary of human cognitive development. It tells how our brains and bodies may have evolved, but still retain survival reflexes, abilities, instincts and tendencies from the past which affect our thinking and behaviour now. It summarises how human brains have developed from the dawn of human history (2010-ch. 3). The potential for activation of these neural nets is retained in our inherited DNA and physical cognitive make-up and they continue to affect us throughout our lives. As in the case of emotions, some traditional scientists thought of such instincts as an encumbrance, an inferior relic of the past, but they may also be a resource. Human nature is perhaps retaining them because we need them in order to cope with the unexpected and continually changing environmental conditions we face.

Some instincts such as the need for physical human contact, social support and some element of 'spiritual' connectivity to the natural world are fundamental and restorative. Many of us have recently become more acutely aware of this because of the physical and mental isolation caused due to 'lockdown' in the current world Covid-19 pandemic. Abbott's chapter 3 is titled 'Human Nature: A Brain for All Times' and helps to put the adolescent brain and our own adult brains into historical context as we have adapted to different environments. For example, our tendency to subjugate our own interest to that of a 'tribal' group, our need for social acceptance and status, and 'automatic' fight or flight responses. We might relate such information to a range of teenage and young adult activities from the 'gang' cultures of violent crime and 'county lines' drug distribution to finding your 'tribe of friends' in interest groups through social media today. This is a time when adult identities are being sought and explored.

'Adult' interest groups with widely differing demographics are also influenced in a similar way, for example, political parties, religious factions, sports fans, conspiracy theorists and terrorists. These phenomena are in turn shaped by our social environments often largely created by other people. Those things we learn to value through life depend on upbringing, education, experience, interests and work we have to undertake. Even when we have adult agency, we are acting, developing and moving forward through a landscape of social-environmental change. Our reactions and ideas going forward tend to be based in the experiences, social and intellectual habits established in our youth. Sarah-Jayne Blakemore's book *Inventing Ourselves: The Secret Life of the Teenage Brain* (2018) tells us that psychologists have identified three reasons why adolescence can be seen as a distinctly separate phase of human development.

This is because of the strong amount of influence these three behaviours exert at this time. They are 'risk-taking,' 'self-consciousness' and 'peer influence.' Blakemore points out that these factors are prevalent in young mammals and in human historical anecdotes from Socrates and Aristotle in ancient Greece through Shakespeare to the present day! Blakemore quotes a study of data collected from 11 countries around the world from 20 researchers led by Laurence Steinberg published in *Developmental Science Journal* in January 2017. They found that adolescence is a time of immature self-regulation and heightened sensation seeking. The countries involved were China, the United States, Columbia, Thailand, Cyprus, Sweden, India, the Philippines, Italy, Kenya and Jordan. The meta-analyses found that 'sensation-seeking' became stronger from the age of 10 onwards, peaking at 19 and fell away during the 20s. In this article written in 2019 Jarvis agrees with these conclusions.

In an earlier paper Blakemore and Mills (2014) describe this type of teenage behaviour (with some additions and editing) as an increased concern with peer judgement—'acting out' to court admiration, self-denigration, embarrassment, low confidence, foolish risk-taking, social ineptitude and a lack of sensitivity to other people's feelings. This causes social vulnerability (and considerable irritation to older people!). They point out it is a period of *sociocultural processing readjustment* necessary to equip young adults to cope with the social complexity of their different communities. However, this does not mean that as adults we should treat teenagers with excessive tolerance but rather that we should consider how we can challenge and support them to *self-actualise and self-develop*—both for their own happiness and society's long-term benefit (Rogers, 1961). They point out the *"audience effect impact"* on teenage behaviour that reduces the ability to reason in the moment. See Article Endnote.[11]

The 'audience effect' can be present at any age, but tumultuous and dysregulated behaviour is generally more pronounced in youth than in older age groups. This may be why good advice to teachers often includes the suggestion that they de-escalate confrontational situations by disciplining 'difficult' pupils away from the 'audience' situation in the classroom. Addictive, tumultuous and dysregulated 'emotional' behaviour is a frequent result when (from their own point of view) young people face challenging events and threatening environments. Teaching strategies such as shaming and threatening pupils with sanctions in front of their peer group is counterproductive to creating a conducive learning environment for all. This may seem counter-intuitive to some who label themselves as 'traditionalists,' but these approaches are in fact modelling confrontational behaviour as acceptable, further embedding *aggressive* neural responses.

Such behaviour will almost always exacerbate pupil (and teacher) problems long-term. Having said this when a pupil uses physical violence in the moment, some appropriate predetermined form of temporary physical restraint and removal will be necessary to protect others. Verbal threats should be avoided if at all possible and replaced with a calm explanation of available options. Isolation is not a solution for young humans since their natural adjustment to adult life involves being able to be socialised and learn appropriate behaviour within their community. Trainee teachers should be given instruction and understanding in the kinds of environmental problems that impact on young people and different kinds of emotional ways in which they respond (Hawkins, 2017, chapter 4-school refuser stories).

As explained below, it should be understood that young adults including trainee teachers of this age are themselves in the process of completing a critical formative stage in their own cognitive development. Scientific research has found that adolescence is a period when a process of huge biological change takes place. It is a period of development when the cerebral cortex is learning how to perform its full function as the centre of conscious thought. The cerebral cortex is the outer layer of the cerebrum, which accesses memory, controls and alters behaviour in the light of experience. As previously referenced, this essential neuroplastic organ only tends to achieve fully operational functionality by the age of approximately 25. If intelligent young adults know this they can be enabled to seek support and help support each other. They can also be supported by Person-Centred Counselling (Rogers, 1961). See Video Presentation Endnote.[12]

3.5 Human Brains Are Neurodiverse and Therefore Variable

In his book *Soft-Wired: How the New Science of Brain Plasticity Can Change Your Life* (2013), Merzenich expresses his disappointment in the slow response of practitioners and lay people to his groundbreaking discoveries. Our new knowledge about the existence of neuroplasticity makes it inappropriate to concentrate solely upon fixed traditional and limiting diagnostic procedures. We need to innovate and adapt programmes to suit groups and individuals. We should be more aware of the plasticity factor and ongoing real-time biological dynamic. This natural force is constantly informed and stimulated by feeling, emotion and mental motivation. In any programme of work or syllabus, it is worth considering how they might be stimulated and/or recognised in more appropriate ways. More research is needed in these areas.

As mentioned in Chap. 1, Michael Merzenich (1942–) is one of those key neuroscientists who proved brain plasticity. He was able to discover and demonstrate this by recording the effects of the growth and deterioration of neural nets in the laboratory. He performed surgical experiments to micro-map monkeys' brains showing the animal brain firing and directing physical movement in real time. In another surgical experiment through scanning and photography he proved beyond doubt that neuroplasticity continues through life beyond childhood growth to adulthood. He has treated a variety of unusual and challenging disabilities and revealed neurotypical and neurodiverse performance evidence using scientific mixed-method research analysis. He has also demonstrated neural diversity combining objective scientific study with practical demonstrations and therapy.

Merzenich's research approach has enabled him to discover how neuroplasticity impacts individuals in similar and different ways. Such groundbreaking research and development is advancing medicine and medical technology in previously unimaginable ways. New interactive non-invasive physiotherapy methods and medical interventions using digital technology are now being developed. Merzenich has continued to study neuroplasticity through his practice as he endeavours to help disabled patients rebuild their lives. In order to achieve this he empathises and supports patients in practical ways as he researches. He engages their collaboration and encourages patient feedback. Neuroplastic potential may vary but all humans have the potential for physical improvement and learning through social and contextual interaction.

Environment has a *'real time interactively dynamic biological effect'* on the brain at any age. If humans are stuck, it is quite possibly because habit or attitude has made them unresponsive. The 'right' stimuli, dynamic repetition and connections have not been encountered and made. We may not all achieve great things but almost all of us have the potential to improve under the 'right conditions.' In this human wisdom and 'common sense' may succeed where 'science' fails. For example, Perry's case study about the mother 'Virginia' who could not help her child (Laura) to thrive and grow until she was shown how to give mother love by an 'expert' mother (Perry & Szalavitz, 2006, see chapter 2). The ability to empathise and interact so as to support neural activity to enable learning in others is a skill that can be learned. This can now be evidenced by the concrete and authentic scientific evidence and meta-analysis of *practical* neuroscientists through the data produced by professional practitioners and others. It can be properly acknowledged and used in a variety of medical and teaching situations.

Existing and specific traditional therapies and 'alternative' approaches from medicine and education can be scientifically action researched. They can be validated or rejected and used appropriately to suit patients, clients and learners. Regulatory bodies can and should be researching and updating their information to reflect new types of neuroscience discovery. One example of the effect of mental health upon a physical condition is that of anxiety upon hypermobility. According to Jessica Eccles, a psychiatrist in the United Kingdom, this is a condition of the joints which is badly affected by anxiety and is apparently generally overlooked in schools. This is probably due to lack of training given to teachers and education resource management about the possibility that a child who does not appear disabled may still require specialist help.

See Video Presentation Endnote[13] and Article Endnote.[14]

Pupils who have other types of illnesses may have the same difficulty. For example, I tutored a girl, a former champion swimmer for her county, whose school labelled her as disaffected because her condition was not recognised by her doctor. She had chronic fatigue syndrome (CFS) also known as myalgic encephalomyelitis (M.E.). When she was finally diagnosed by a consultant physician her problems started to be addressed. She and her parents had suffered many months of unnecessary additional anxiety as they tried to get help. Children whose problems are ignored in this way are not only not helped, but their conditions are prone to further deterioration. This is a serious problem. There may be no awareness by teachers and a lack of specialist therapists available to inform and research collaboratively with teachers, parents and children themselves (Hawkins, 2017). See Article Endnote.[15]

In education there are also implications for 'normal' children, because 'normal' children may not always be able to conform to 'normal' adult expectations. The resulting adult environments created for them may be counterproductive to remedying their problems. A collaborative practitioner research approach to the job within a conducive professional environment is more likely to solve problems as they occur. Expertise can thus be developed and useful knowledge disseminated to professional managers and others—a 'bottom-up' approach. Our understanding of the complexities of neurodiversity owes a great deal to research undertaken by neurodivergent humans themselves. As mentioned in Chap. 1, Barbara Arrowsmith-Young (2013) is a Canadian educationalist currently who runs a school for children with specific learning difficulties. She was born with multiple learning disabilities and discovered how to overcome them herself. She devised a repetitive exercise with flash cards as a means of stimulating her own brain to develop new neural connections. See Video Website & Demonstration Endnotes [16] and [17].

She found that after a period of intense practice in this form of spatial recognition her brain responded and started to function in a number of additional ways as she explains in her book (2013). She is a trailblazer. She started her school in 1980 and it has taken a long time for her work to achieve recognition. She has gone on to educate other learners with special educational needs. Her programme is now used by other schools such as the school run by Howard Eaton (2010, 2018) and also helps adult stroke and accident victims. As well as writing about her recovery from severe learning disabilities, she has written case studies of children she helped. She believes interventions have to be structured and adapted to fit each student's unique situation, current learning ability and potential.

This may seem difficult, but her research methods have provided her with information about how this may be done. Arrowsmith-Young says there is no 'one size fits all' method for helping a person rebuild their brain. She has developed 19 possible proven methods to date for doing exactly that. They are tried and adapted to create a personal sequence of specific interventions. Peter Kinderman wrote a blog article pointing out that the diagnosis of mental health difficulties has always been problematic. He advises that we step back from diagnostic systems in psychiatry due to their unfortunate tendency towards human error, poor reliability, lack of usefulness, validity and reliable epistemology. See Blog Article Endnote.[18]

As far back as 1854 an article in the *Times* cautioned about the problems of defining the difference between sanity and insanity. Human nature is hard to analyse. Normal behaviour can include prejudices, passions, vanities, vices and a whole range of 'eccentric,' obsessive behaviours thought abnormal by fellow humans. Neuroplasticity could help us understand and accept diversity. It explains why human beings are in many ways similar, but at the same time different. In other words, '*We are all part of the human family, but brain plasticity makes us all unique.*' Children might benefit from being more generally thought of as not 'disabled' but more correctly thought of as individual, atypical and neurodiverse. More than that they all need unbiased love, positive regard and our respect as inexperienced future adults.

We need a more nuanced approach to 'disability' where professionals learn how to support and build neural capacity in a positive way, finding out what individuals can already do, might want and choose to do. This approach would involve starting from where they 'are at' and giving preplanned but adaptable encouragement. A culture of collaborative practitioner research within a community of practice that includes the stakeholder, pupils, student, patient and family is necessary (Wenger, 1998; Wenger et al., 2002). It is time for perspectives to change. There is an opening up of research possibilities across the scientific, medical, therapeutic and educational communities. Indeed, Merzenich, the famous neuroscientist, has himself entered into the field of education to develop neuroplastic intervention programmes that can be tailored to individuals (2013). See Endnote.[19]

3.6 Emotional Memory Models Are Important for Information and Motivation

Memory is a very important aspect of education, not an automatic, but a complex one. Until recently very little was known about the brain's physical memory systems and processes. In 1953 a man called Henry Molaison, who suffered from a serious form of epilepsy, had both parts of his hippocampus removed by a surgeon named William Beecher Scoville. This was an invasive and 'never to be repeated' medical operation. The hippocampus is so named because of its sea horse-like appearance. Henry's resulting severe memory difficulties contributed to groundbreaking research in neuroscience and clinical psychology. Before Henry suffered this personal tragedy, the whole brain was deemed to be some kind of largely unexplored storage organ, even by experts. Henry's operation proved the hippocampus to be a part of the brain fundamental to processing recent memory.

Hilde and Ylva Ostby explain this in their book *Adventures in Memory: The Science and Secrets of Remembering and Forgetting* (2018). These sisters from Norway explore some of the latest implications of memory research. As collaborating authors, they contribute their separate professional perspectives and expertise to the subject. They have careers as respectively—an acclaimed writer and expert in literature and a neuroscientist. Henry became an important volunteer research subject, before and after death when he generously donated his brain to science. When he woke up from his operation the last two or three years of his memory were missing—from then on he lived only in the moment. He could no longer store new memories even though some of his long-term memory remained. He had to be reminded his father had died. However, he still remembered him and was distressed to hear every time he was told he was dead.

He estimated his age by looking in a mirror and guessed the yearly seasons by looking out of the window. He was able to find ways to cope with his problem. For example, he was still able to register immediate memories, as when talking, but he could not remember the conversation afterwards. Neuropsychologists studied many aspects of memory using Henry's experience to discover different 'kinds' of memory. However, it is important to use one's own judgement using these labels. Some of them (devised by different people) are talking about the same systems. Experts currently believe the hippocampus may be a 'hub' receiving and editing for long-term memory storage. Although Henry no longer had a hippocampus, he continued to perform tasks using other systems.

Henry's long-term memory retained some procedural memory. Canadian neuropsychologist Brenda Milner discovered this. She asked him to draw and complete a mirror image. They were both surprised when he did this task symmetrically by drawing in the opposite direction. He retained his implicit memory for everyday 'automatic' tasks like brushing hair and tying shoelaces. Such tasks involve muscle memory and perhaps body memory through repetition and proprioception. It would be interesting to know if Henry Molaison remembered qualia. These *evocative* feeling sensations are types of particular tastes, smells and pains, colours—bodily reactions that recall a place or time to mind. Daniel Dennett (1991), American philosopher, notes that although the term is seldom used, we experience them all the time. Qualia are 'thoughts,' 'ideas' and propositional memories *about* sensations that may be true or false, that is, in a person's opinion form their own experience.

Qualia experiences accumulate in memory and can be recalled. They are propositional attitudes. *In other words they are remembered beliefs about experience rather than what it feels like in actually experiencing.* Understanding and communicating words (semantics) are affected by our assumptions based on these 'subconscious' memory systems. Talented writers and poets explore these boundaries and phenomena, which mix and overlap as they describe experiences. Artists, musicians, choreographers, dramatists, film-makers, virtual technologists and other creative people are using and constantly playing with these types of memory on many levels. These memories and thoughts are controversial and are often disagreed upon, because humans have common and different responses to memorable qualia. Groups of us like particular physical sensations that others do not. Experts are employed to recognise and recreate them, for example, cheeses, scents and wines.

There is a practical aspect to recognising dangerous tastes and smells, for example, poisons and deadly fumes. A strong smell is mixed with household gas in the United Kingdom to alert users to danger. In psychology there has been a debate about the importance of qualia, with different definitions of the term. Some philosophers emphasise or deny different types. The nature and existence of qualia remains contentious amongst some traditional scientists, because they are not uniformly verifiable and objectively measurable. Jeffrey M. Zacks, psychologist and neuroscientist, tells us about his studies into how we respond to film. Our senses, feelings and emotions can create and recall qualia in emotional, moving, exciting subconscious and important ways (2015). They exist and are verifiable and we can record them. We can compare them and agree or disagree on their nature with other people. By contrast, temporal or short-term memories are fleeting thoughts, about 20 seconds long which tend to come and go.

These can be recorded in real time with high-resolution brain scans of neurons firing in the prefrontal cortex when remembering a car registration, date, password, passcode or phone number. From a teacher's point of view it is well worth considering how short-term memories are converted and stored in 'long-term memory.' The embedding of neural nets is not easily predictable and may even be circumstantial and environmental. *This is a judgement skill we develop in teaching and apply working with different subject matter and students.* The Ostby sisters relate the story (and re-enacted the research) of an experiment in 1975 by British researchers Baddeley and Godden with divers learning underwater and on land. After various combinations of trials both experiments revealed that a list of words is best learned and remembered in the same location. Words learned underwater were best remembered underwater and similarly on land, even though underwater seems a distracting and alien environment.

Henry's case showed there are important kinds of subconscious and conscious long-term memory types. For example, implicit memory is the basis for automatic and semi-automatic behaviours and skills like putting the kettle on, driving a car, swimming or brushing your teeth. These instinctive actions take place semi-consciously in the body-brain. They are embedded. It turns out that people with amnesia can still do some of these things, even when they can't remember doing them before. Recently researchers have made an important discovery about the biomechanisms behind learning and memory. Depending on the number of synapses, and their proximity, information is processed and stored differently. A team led by Roberto Araya at the University of Montreal has studied synaptic plasticity looking at dendritic spines (tiny protrusions on neuron branches). They found morphological transformation and think this may be the underlying mechanism for learning and memory. See Endnote Neuroscience Article.[20]

Emotional 'blocks' to memory can behave in similar ways to implicit memory. This was evident in some of my own research, when the memory of events had affected school refusers and 'automatically' influenced ongoing behaviour after the causal events were 'forgotten.' These were likely retained in long-term memory continuing to restrict cognitive development and affecting life chances. Some of these events were traumatic incidents and others were habitual environmental conditioning, but both types of causes made strong enough impressions to affect a person's personality and behaviour long-term. Emotional models are important to understanding how memories work. This appears to be the missing ingredient in John Sweller's discussion of Cognitive Load Theory. Because of their impact upon how we absorb information and how we recall it, the teacher's professional empathy and ability to read student responsiveness and maintain attention and interest is important to efficiency. See Academic Paper Endnote.[21]

Cognitive load theory was originally designed to give a framework to help teachers present information and encourage learners in activities that improved intellectual results. His initial ideas were researched through the design of a variety of instructional rules based on the idea of reducing the working memory load and encouraging the construction of concept frameworks, schemas and models. It assumed working memory was limited with only partial auditory, verbal, two- and three-dimensional visual capacities. It also assumed long-term memory to be 'effectively unlimited' with schemas that vary in their degree of 'automation.' This is quite a fixed, mechanistic concept of learning. However, the principle of creating frameworks to support the learner appears to be a sound one, provided one accepts some variation in 'take up.' In 2019 Sweller reconsidered his theory with the help of a group of colleagues. This paper reviews the theory and the instructional designs it generated. See Academic Paper Endnote.[22]

The search for physical traces of memory in the brain started with the work of Lomo and Bliss (1973) into synaptic plasticity mentioned in Chap. 1 (Sect. 1.3). They reported that when synaptic cells were given an extra high-frequency electrical stimulus, they developed a long-term enhancement response. Per Anderson, another researcher, had discovered that neurons sent electrical signals. Lomo caused tiny electrical signals to be sent from one part of the brain to a rabbit's hippocampus. The researchers discovered that the receiving neurons had a 'memory' when they still responded as the signal was reduced. This was called 'long-lasting potentiation,' which became known as long-term potentiation (LTP). As the Norwegian Ostby sisters explain it (2018), Terje Lomo (1935–), a Norwegian physiologist and neuroscientist, and his colleague British neuroscientist Timothy Bliss (1940–) had discovered the physical secret of memory on a micro level.

Although the discovery was ahead of its time, their demonstration of long-term potentiation is now accepted as the dominant synaptic model of how the mammalian brain stores memories. Complicated memories and their associations do not stay entirely in the hippocampus. They are spread out across the cortex. Lomo concluded that the day's memories are 'sorted' and established in the cortex while we sleep. This is supported by more recent research (Walker, 2017). This process is established and reinforced by sleep. Stress and lack of sleep prevent the memory functioning at its optimal level. Interestingly Lomo also reported that his experiments with later rabbits did not always work to establish long-term potentiation when they were not 'happy.' The neocortex (outer layer of the brain), also called the neopallium, isocortex or the six-layered cortex, is a set of layers of the mammalian cerebral cortex involved in higher-order brain functions such as sensory perception, cognition, generation of motor commands, spatial reasoning and language.

The neocortex is believed to be an area for long-term storage. The amygdala is another 'hub' or 'processor' which is responsible for the perception of emotions assigning and infusing memories with emotional significance. The amygdala releases hormones in response to anger, fear and sadness, causing physical defence mechanisms to come into play such as the instigation or controlling of aggressive behaviour. Emotionally charged memories provide us with our own special and individual form of filing, attribution and retrieval system. We learn information from each other, transform it and make it our own. The information we store may or may not be accurate and some of it requires continuous or periodic checking so that it can be amended. Emotional blocks to learning can happen when we are unable to cope and/or feel overwhelmed by '*tagged*' memories of past experiences.

As a former classroom teacher I know that setting up a happy and positive environment is an important element for promoting confidence in learning. However, this does not negate the necessity to strive for excellence. There are positive learning behaviours (Claxton, 2002) and factors such as motivation, mindsets (Dweck, 2006, 2017), goals, aims and visions. These ideas are not negated but enhanced by research into the emotional conceptual side of brain plasticity. For example, the knowledge that neuroscience research *proves* a variable amount of repetition is required to embed learning is useful. It is sometimes worthwhile to *persist* in the repetition of presenting facts and repeating tasks provided we keep positively engaging the attention and motivation of learners—*perhaps by explaining differently*. A great deal more meta-analysis of existing practices through systematic collaborative research is needed in many areas. For this reason, provided they can explain and evidence their work, teachers should be trained and allowed to develop their own ability to make professional judgements.

Doidge talks about a type of meaningful 'practising by repetition' approach to neural reconfigurement (and presumably embedding memorable learning) in his book (2015, Page 198–9). He tells us about a research project by Alvaro Pascal-Leone, a neurologist and expert in non-invasive brain stimulation using transcranial magnetic stimulation (TMS) to map the motor cortex. Pascal-Leone monitored a group of blind learners who practised to learn Braille every day. During the first six months, their brain maps always appeared strongest on a Friday, less evident during the week and returned to the baseline on Monday. After six months the Monday brain maps started to grow and after ten months were established although never as strong as the Friday maps. *Following a two-month break, the brain maps were still evident and established and appeared as they had been before the break.* Doidge calls this the 'tortoise-and-hare' effect.

Soft rewiring for different brain deficits is a process that has tremendous potential for positive remedial education for improving memory and might inform our work with mainstream students and clients. We can make a start on understanding what may be the causes of problems, and share our practice so that we can try out and adopt possible solutions. A professional approach requires adaptation, reinforcement and confirmation by observing and recording the motivation, involvement and feedback of learners while we teach. This should be followed by retrospective reflection and analysis. This approach is supported by the work of physio and occupational therapists done with senior citizens and patients recovering from illness—apart from any actual medical problems, these therapies help to stave off depression and get the circulation, body and brain working properly.

Another method of establishing 'strong' memory might be through 'strong emotional association' and also bodily stimulus via the senses. Novel situations can create strong memories and teachers can evidence these by recording learner voices both in and outside the classroom. Teaching already provides a great deal of evidence to suggest that situations, contexts and tasks creating effects of strong sensory stimulus, being in novel and exciting situations can create an attentive buzz of sensitivity that aids memory, curiosity and motivation, for example, theatrical drama, creating a treasure hunt of clues, storytelling, music, humour and outdoor pursuits. It is currently thought by neuroscientists that the prefrontal cortex plays an important part in processing thoughts and behaviours that use implicit memory. This is the front part of the brain responsible for higher-level thinking. It is thought to be an important control centre.

The brain may choose to uses all kinds of memory to enable an 'appropriate' behaviour response to external events. *The* prefrontal cortex *is generally, doing its best (given the information available to it) to make sense of the world.* Teachers should be aware that emotional trauma, stress and negative emotions like fear, embarrassment and worry prevent its efficient working. See Chap. 6: Sect. 6.5. Long-term remembrances are stored in explicit memory. They have been divided and subcategorised as 'episodic' and 'semantic' memories, but are not *necessarily* usefully separated in practice. Episodic memory describes factual memories of life events. For example, remembering events and highlights of a holiday. Semantic memory is conceptual knowledge about people, things and events—'objective' knowledge about the world: for example, knowing that New York is a big city on the East Coast of the United States.

Henry Molaison's case research tells us the hippocampus is an important information processing centre that receives information and somehow coordinates and physically enables neuronal connections between storage facilities in the brain. Electromagnetic imaging shows the hippocampus, neocortex and amygdala 'lighting up' as memory processing takes place. There is an interplay between them which involves memory making and retrieval during the efficient sorting of new information. These emotionally charged and tagged memories are unique in detail for each individual. In the waking brain a great deal of short-term knowledge is accumulated constantly. The hippocampus has an important job, considering and rejecting large amounts of information and keeping some of it in reserve for recall for various periods of time. Emotional decisions are involved.

As far as we know the hippocampus also has a role in interfacing, absorbing and filing useful long-term information, communicating it to and retrieving it from other parts of the brain. Perhaps the connections it makes with the amygdala might account for some of the emotional reasoning behind human thinking. The information we absorb both consciously and subconsciously is assigned a particular feeling, 'qualia' or 'type' of emotional tag, which may help retrieval at a later date. It seems the neocortex also makes these kinds of decisions, but is generally engaged in more complicated reasoning. As language has developed, humans have developed an intermittent stream of inner dialogue which gives voice to our unworded thoughts.

If in education we regard language as the only form of thought we are liable to discount, reduce and 'dumb down' our own and other people's intelligence abilities. It seems likely that quite a lot of human knowledge is flagged by its human creators with signs and indicators of human emotion. Such knowledge can be read back on different levels and in different ways and given additional meaning. Even the kinds of facts which require us to take an 'objective' view of material reality can be used and manipulated by emotional reasoning. Human experiences, from which emotionally 'tagged' memories arise, have been similar enough for different kinds of human knowledge, subjects and interests to be created and established. For example in the arts where particular performances are appreciated in different ways and for different purposes and critiqued as best practice.

Emotion and feelings are an important intellectual processing system, not just an important communicator of what we value and a reminder of why we value it. As previously mentioned, Lynne Kelly has researched the methods of indigenous peoples for remembering important historical and cultural knowledge. Many of these practices are still maintained today. The indigenous people (Aborigines) in Australia in particular have continued their cultural practices in the face of colonial oppression. She writes about how she has learned to use ancestral memory techniques and proved their effectiveness. She talks about how she has learned to use the Memory Code technique. As she walks around her neighbourhood, she uses the different houses to remind herself of a list of names of countries in the world. She has learned the names of wild birds by designating an object that reminds her of the name.

She uses a stick with shells and beads stuck on in order to remind herself of the main points in a story. Kelly has researched ancient memory technologies all over the world that are associated with physical places. Amongst a variety of other sophisticated technologies of their time, Indigenous peoples helped themselves to memorise and sequence information by walking through the landscape with various hand-held devices. Eleanor Maguire also tells us that memories are more than personal autobiography, because we also share communal memories in many ways. She believes they are in effect 'the glue that holds society together and the basis of culture.' This aligns with Carl Jung's ideas about the collective unconscious (1991). She is a neuroscientist who connects the 'semantic' facts she observes through her photographic imaging research with the 'episodic' knowledge of human experience. See Podcast and Video Endnotes [23] and [24].

3.7 Measuring Intelligence Is About Assessing Actions, Not Just the Retrieval of Facts

Traditionally the brain was known to consist of billions of interconnecting neurons but there was no detailed evidence available about how memory systems operate. The realisation that numbers of neurons can exist *regardless* of 'operational intelligence' still seems to be a difficult concept for a lay person to grasp, perhaps because it is an underlying human assumption that more is better. It is tempting to conflate this idea with the *building up of and accumulation of knowledge*. However, *it is the operational use of that knowledge which is really important*. We know that memory is important, but we also know that skills in adapting and applying knowledge are important for deep learning, especially now that we have computers as an external and additional memory bank.

Traditionally we have often been overly impressed by quantitative statistical analysis of factual knowledge, but it is actually of little use without an insightful qualitative assessment about how to use such information. The ability to do this is ultimately the best measure of intelligence when a task is seen and proved to be well done. Similarly the ability to perform less 'expert' tasks within a field of knowledge may be a measure of *potential* intelligence in that area. Some exam formats are designed to show the students' insight, extended and 'joined up thinking' by asking unexpected and challenging questions about both supplied and remembered texts and materials. These require professional, experienced human examiners to use their judgement in marking. Other exams and tests require 'stock' answers to show an understanding of subject language, structure and concepts. Both of these types of examination are necessary, but the second type has gained much greater popularity and over emphasis in government education and other management systems in the last 20 years. This is probably because they can be automated and are cheaper to devise and administer. Meanwhile qualitative judgement skills may have been neglected with consequent negative results.

Unsurprisingly during the past 20 years, an overemphasis on memorising over meaningful processing has caused exam systems to fail both students and their future employers. In this section I discuss two academic articles. They show some of the underlying problems, assumptions, emotional confusions and reasons for the damage done (even inadvertently) to education systems worldwide. Viatcheslav Wlassoff's article provides a useful example of how psychologists have tended to think about intelligence as he honestly expresses some of the puzzlement and confusion around this subject. Psychologist's attitudes towards exams have sometimes caused 'cognitive dissonance' within learners and teachers. There is a 'disconnect' between what we know in everyday life by custom and practice and what we have been trained to think we ought to be able to prove through 'scientific' examination.

Intelligence as a concept is definitely real, but it is only measurable and able to be properly evidenced in relation to people, perceived facts, contexts and events. Measuring intelligence is not just about the retrieval of factual knowledge from memory but evidencing the ability to actively combine and *deploy* knowledge in predictable, productive and innovative ways. Exams and tests help provide a basis for this by narrowing the field down to a subject area. People taking exams and tests are proving that they 'are able' or 'are not able' to demonstrate a specific range of ability. They are remembering and demonstrating specific kinds of knowledge requested by that test on a particular date at a particular time. However, exams may not indicate 'deep knowledge' and/or deployment abilities long term. Wlassoff cites a study published by John Protzko in 2015 showing that an IQ or intelligence quotient intervention that was initially successful can quickly *fade* in children's memory. See Article and Academic Paper Endnotes [25] and [26].

This was called the Fade-Out Effect. He showed that in this meta-analysis of 39 randomly controlled tests (RCTs) with 7584 participants, *teaching to the test had little long-term effect*. The control groups stayed the same, but those tested after the interventions did not retain the information so well when tested again. Protzko thought his research showed intelligence adapted to environmental demands both upwards and downwards and thought the findings disappointing but not surprising. Two questions come to mind! Why was he disappointed and why was he not surprised? The failure of the research may have been a useful exercise if it raises such questions. For example, are we wasting pupil teaching time on exam revision of facts that will be forgotten in the long term *if they are not made meaningful and relevant to those pupils?* Wlassoff says that IQ tests are frequently criticised by researchers for being too limited to reflect the real intelligence level.

In his article, Wlassoff refers to learning styles as they relate directly to Gardener's 'intelligences' categories (1993). This is not a straightforward matter. The idea that people have only limited learning styles has been disproved. Learning styles and competencies exist within and across subject areas using a range of skills and strategies linked to purposes. He says there is no measure of intelligence that can be applied to different learning styles and intelligence types that would be universally accepted. He seems to assume this is desirable. *One might reasonably ask why such 'measures' of general intelligence across all fields of human endeavour might be considered desirable?*

Wlassoff says that our physical body strength responds to the presence or absence of stimuli or training in exactly the same way and says the reasons for this are a mystery. He was puzzled by Protzko's research results, perhaps because he is not considering learner perceptions and motivations. Neuroplasticity offers a sound scientific explanation for this. Neural nets are now proved to require motivation triggered by feelings and emotion, producing a strong incentive and/or focus and different levels of concerted repetition in order to become active and established.

Wlassoff thinks that genetic factors underlying intelligence are not sufficiently studied and taken into account because there is an element of 'social taboo' regarding this issue. He says this is the case even though we all *know* that levels of intelligence are genetic. He says we *prefer* to assume that most 'normal' people are more affected by social and environmental conditions than inherited ability. He thinks it's time we accepted the role of genetics in determining intelligence and think more about what can be done to help the less able. This is essentially the traditional view that has underpinned and shaped traditional attitudes and practices both in psychology and in education. See Academic Endnote.[27]

It is true that genetics are important, but should perhaps not be used to support this old fashioned rationalisation. It is an interesting fact that DNA has been discovered to be an 'adaptable' process currently being researched in the field of epigenetics. This is the study of heritable phenotype changes affected by environment that affect humans but apparently do not involve alterations in the DNA sequence. Whatever a person's genetic inheritance, neuroplasticity is now proved to be real and very much influenced by social and environmental conditions. There is hope for improvement at all levels of intelligence in diverse ways and for a better understanding of how to 'treat' degeneration. There is a growing body of evidence that our current approach to intelligence and education is mistaken for two reasons.

First, current methods of assessment have been evidenced by outcomes to be quite often *counterproductive* in the opinion of many employers, psychologists, therapists, parents and teachers. Secondly, new scientific research evidences supports this. It is proved by new groundbreaking discoveries and understandings of neuroscientists and psychologists. For example, Sarah-Jayne Blakemore and Tamsin Ford point out that in the last 20 years the idea the brain stops developing after childhood has been disproved. *Research now informs us that the human brain undergoes protracted and substantial change in adolescence (from 10 to 24 years).* This helps to explain why mental health difficulties often start before the age of 18. Neuroscience research shows the prefrontal cortex undergoes 'striking and prolonged' changes during adolescence, which are involved in a lot of high-level cognitive functions, such as decision-making, planning, social interaction, emotional regulation, self-awareness and the inhibition of inappropriate or risky behaviour.

The changes in grey and white matter observed taking place are thought to show important neurodevelopmental processes, which are the basis of neuroplasticity. The ability of the brain to change according to environmental input in adolescence is also happening in parallel to observed changes in cognition, social behaviour and mental health. For this reason this article questions if 'high intensity, high stakes, national exam' courses like GCSEs (General and Advanced Certificates of Secondary Education) undertaken from 14 to 18 years in the United Kingdom are entirely appropriate. Consequently, maybe they should be reimagined and restructured, particularly as they are taking place during a scientifically observable sensitive period of brain plasticity development that stabilises in the mid-20s. See Academic Article Endnote.[28]

3.8 Conclusion

Neuroscience research supports the conclusions of a wide range of experts in all the areas above. However, there is a great deal more research work needed involving meta-analysis of existing human knowledge in many fields. We know there is localisation of brain functions in the cerebral cortex, which has many distinct regions. It contains sensory areas, motor areas and associated areas that are involved in language processing, imagination, intelligence and personality. The term limbic brain was invented to explain some of these processes, but it still remains open to interpretation. The concept of the limbic brain is now considered by some to be more of an interactive processing system. The idea evolved from the analysis of physical functioning parts to a complex interactive model where subcortical parts of the brain connect. This activation involving the production and dissemination of feelings and emotions although to some extent observable (by MRI scan and behaviour) and a 'useful' concept, seems unlikely, given human individuality, ever to be fully understood.

Teenagers and dysregulated adults often behave in ways that are hard for everybody impacted by their behaviour to comprehend, particularly those who are not themselves experiencing the same rollercoaster of emotional responses and events. The synapses are being individually tuned particularly during adolescence, becoming dormant or 'pruned back' as each individual deliberately and inadvertently develops her or his fully functioning adult human brain. Each brain will have similar qualities to other brains, but will be made up in a manner customised by the genetic potential, environment and unique mix of events experienced by each person. Adolescence is a developmental period of formative biological and social transition. Sadly years later many adults may still not have achieved their full capability and potential. All of the above involves developing connections to create a fully functioning 'limbic system' which has the basic dormant potential to cope through life.

The cerebral cortex and limbic brain develop exponentially by embedding neural connections. They do this according to external stimulus as internal capacity is stimulated by an inevitably unique sequence of events. These events range from large-scale emotionally charged 'one-offs' to small-scale repetition events over time. Human beings have evolved and developed *a capacity for adapting to survive* in a semi-predictable but also frequently uncertain world. This explains why humans are paradoxically both predictable in some respects and unpredictable in others. This is the nub of humanity's constant dilemma and one which we must acknowledge if we are to survive. In speculating about the potential implications of neuroscience research for society, it is important to research and consider social environments and rewards in analysing human ability, cognition and behaviour.

Notes

1. *Interview Podcast* Dr Jennifer Hawkins in Conversation with Stephen Hurley Canada: VoiceEd Podcast: https://www.spreaker.com/user/voicedradio/in-conversation-with-stephen-hurley-dr-j_2. Accessed 6 February 2020.
2. Special Data Contribution

   Chris K. Pearson (2014 'Too Old to Suffer') Data Contribution & Poem Transcript, a poem about emotions for his daughter, 7 October 2014 @ MaticulousChris https://www.youtube.com/watch?v=zgLIuCVK0sc Accessed 28/08/2019 Also Poem 2 Don't Panic Bright Eyes https://youtu.be/HjW_L1Qsq9k
3. Special Data Contribution

   Chris K Pearson, Maths teacher, 7 October 2014 Performance 1. https://www.youtube.com/watch?v=zgLIuCVK0sc. Accessed 28/08/2019.

   Performance 2. https://www.youtube.com/watch?v=wm8VHbvWZCs. Accessed 13/04/2021.

3.2 Feelings and emotions communicate information
Data Contribution
Too Old to Suffer by Chris K. Pearson

My daughter told me she's
unhappy
I said
You're too young to be happy
We have to suffer first
And
I'm too old to suffer
So wipe your tears please
And smile

I'll explain
I'm not talking about physical pain
Which affects all ages the same
I'm talking about emotions
and how feeling hurt
Doesn't hurt
It's a metaphor
And this is what the suffer-
ing's for
Because it is just an emotion
If you've endured it before
Then
If you feel the same thing again
What are you suffering for?
If
You felt it before.

If
You felt it before
Then you know what's in store
The hurt is measurable
And no matter how
displeasurable
We don't have to withdraw
When we feel

When we feel
Angst, inadequacy, anger,
Disappointment, loss, regret
Jealousy, fear, anxiety
This debilitating variety
Of emotional hurt
We don't have to withdraw
Not even when we feel
Combinations of them all

And I have
Felt them all
And I have
Tried to fight
Each one of these emotions
I've suffered
And finally I've surrendered
Without giving up

Which brings me to my point
I'm almost 40
and life has taught me
Internal resistance
To my feelings
Is suffering
So I've stopped resisting
And started feeling
I've surrendered
But I haven't given up

I haven't given up
I continue
To endure
Life's demoralizing smorgasbord
Subconscious demons
Winged and clawed
Who torment my thoughts
And then applaud
As I'm floored
I am flawed

And it's okay
Because I'm too old to suffer
I'm too old not to know
That fighting my own
emotions
Is not the way to go
So I don't
I feel them

While immature souls
Can spend entire lifetimes
Repeating the same battles
With their selves
And suffering
I pity the fools

Truly I do
But you're no fool
And you're not immature
You're young
And you'll be okay too
I promise

4. Emine Saner An interview with Neuroscientist Hannah Critchlow: 'Consciousness is a really funny word' The Cambridge fellow on what it means to be human 18 May 2018 Guardian Neuroscience Saturday interview @eminesaner. https://www.theguardian.com/science/2018/may/18/neuroscientist-hannah-critchlow-cambridge-consciousness-funny-word?CMP=share_btn_fb. Accessed 14/06/2018.
5. Dr Pam Jarvis A range of articles about early years (from 2018) UK: Times Educational Supplement. https://www.tes.com/author/pam-jarvis. Last accessed 12/08/2020.
6. Rishi Sriram 'Why Ages 2–7 Matter So Much for Brain Development: Rich experiences—from play to the arts and relationships—fundamentally shape a young child's development.' Brain Based Learning, Edutopia Online: George Lucas Educational Foundation; 24 June 2020. https://www.edutopia.org/article/why-ages-2-7-matter-so-much-brain-development. Accessed 23/11/2020.
7. Professor Amy Brown (2019) 'Dog training for babies? Only if you want to raise kids who are anxious and unsympathetic' 21 August 2019, Child Public Health, Swansea University. https://theconversation.com/dog-training-for-babies-only-if-you-want-to-raise-kids-who-are-anxious-and-unsympathetic-122150. Accessed 22/08/2019.
8. Netalie Shloim, Lisa R. Edelson, Nathalie Martin and Marion M. Hetherington1 (2015) 'Parenting Styles, Feeding Styles, Feeding Practices, and Weight Status in 4–12 Year-Old Children: A Systematic Review of the Literature': 1Biopsychology Group, Human Nutrition, School of Psychology, Institute of Psychological Sciences, University of Leeds, UK; 2Behavior and Perception, Nestlé Research Center, Lausanne, Switzerland. https://www.frontiersin.org/articles/10.3389/fpsyg.2015.01849/full. Accessed 22/08/2019.
9. Meyer, A., Proudfit, G.H., Bufferd, S.J. et al. (2015) Self-Reported and Observed Punitive Parenting Prospectively Predicts Increased Error-Related Brain Activity in Six-Year-Old Children. J Abnorm Child Psychol 43, 821–829 (2015). https://doi.org/10.1007/s10802-014-9918-1. Accessed 22/08/2019.
10. Neir Eshel, a Bernadette Daelmans, b Meena Cabral de Mello, b & Jose Martines b (2006) 'Responsive parenting: interventions and outcomes' Bulletin of the World Health Organization | December 2006, 84 (12)

https://www.scielosp.org/article/bwho/2006.v84n12/991-998/en/. Accessed 22/08/2019.

11. Pam Jarvis 'Teenage behaviour: 5 tips for teachers: Teenagers' brains go through changes that explain their behaviour – we must try and bear this in mind.' 14-10-2019 *Times Educational Supplement*. https://www.tes.com/news/teenage-behaviour-5-tips-teachers. Accessed 14/10/2019.
12. Bruce Perry (2020) Stress, Trauma, and the Brain: Insights for Educators-No. 4 'Regulating Yourself and Your Classroom' 25 Aug 2020 ThinkTVPBS. https://www.youtube.com/watch?v=nqW2Xv16bWw. Accessed 20/09/2020.
13. Dr Jessica Eccles (2017) 'What is the link between joint hypermobility and anxiety?' https://www.youtube.com/watch?v=Mjo7rdAv5p s. Accessed 26/09/2020.
14. Jane Green (2020) 'Why you need to know about EDS, HSD and hypermobility: Schools need to be more aware of Ehlers-Danlos syndromes (EDS) and hypermobility spectrum disorders (HSD) to better support pupils diagnosed with the conditions' 23 July 2020 The Times Educational Supplement. https://www.tes.com/news/why-you-need-know-about-eds-hsd-and-hypermobility. Accessed 24/09/2020.
15. The Neurodiversity Rainbow Video 'Attention Regulation: The Difference Between ADHD and Autism (Explaining The Neurodiversity Rainbow).' https://www.youtube.com/watch?v=xQc5Feyl2GE. Accessed 26/09/2020.
16. Barbara Arrowsmith-Young (2013): "The Woman Who Changed Her Brain" | Talks at Google. https://www.youtube.com/watch?v=Uk1ZpxuYic8. Accessed 23/02/2020.
17. Howard Eaton Website http://howardeaton.com/videos/ Accessed 23/02/2020; Demonstration by students (2013). https://www.youtube.com/watch?v=epHBDNBPnHg. Accessed 23/02/2020.
18. Peter Kinderman's blog Saturday, 29 June 2013. http://peterkinderman.blogspot.com/2013/06/drop-language-of-disorder-less.html. Accessed 01/05/2019.
19. Dr Michael Merzenich – Website https://www.soft-wired.com/about-dr-merzenich/. Accessed 03/10/2020.
20. Professor Roberto Araya (2020) 'Unlocking the Mysteries of the Brain: How does our brain store information?' 26 August 2020 University of Montreal – Neuroscience News-Open Articles. https://neurosciencenews.com/brain-mysteries-16911/. Accessed 02/09/2020.
21. Sweller, J. (2010) "Element Interactivity and Intrinsic, Extraneous, and Germane Cognitive Load", Educational Psychology Review June 2010, Volume 22, Issue 2, pp. 123–138 Springer.
22. Sweller, J., van Merriënboer, J.J.G. & Paas, F. (2019) 'Cognitive Architecture and Instructional Design: 20 Years Later' Educ Psychol Rev.

31, 261–292 (2019) Springer. https://doi.org/10.1007/s10648-019-09465-5. Accessed 06/08/2019.

23. Professor Lynne Kelly (2016) Podcast about 'The Memory Code.' https://abcmedia.akamaized.net/rn/podcast/2016/07/aim_20160703.mp3. Accessed 12/04/2021 Accessed 21/11/2020.
24. 2021 Professor Eleanor Maguire, University College London (2014) 'The Neuroscience of Memory'

    Wellcome Trust Centre for Neuroimaging: Talk at The Royal Institution, UK. https://www.youtube.com/watch?v=gdzmNwTLakg. Accessed 21/11/2020.
25. Viatcheslav Wlassoff, PhD (2016) 'Children's IQ and Fade-Out Effect', 1 March 2016 – Psychology & Psychiatry e Global Neuroscience Initiative Foundation (GNIF). http://www.brainblogger.com/2016/03/01/childrens-iq-and-fade-out-effect/. Accessed 15/04/2019.
26. Protzko, J. (2015). The environment in raising early intelligence: A meta-analysis of the fadeout effect *Intelligence, 53*, 202–210. https://doi.org/10.1016/j.intell.2015.10.006. Accessed 15/04/2019.
27. Dupont C, Armant DR, Brenner CA (2009). 'Epigenetics: definition, mechanisms and clinical perspective' September 2009 Seminars in Reproductive Medicine. 27 (5): 351–7. https://doi.org/10.1055/s-0029-1237423. PMC 2791696. PMID 19711245 Epigenetics: definition, mechanisms and clinical perspective. Accessed 15/04/2019.
28. Professor Sarah-Jayne Blakemore, Department of Psychology, University of Cambridge

    Professor Tamsin Ford, Department of Psychiatry, University of Cambridge 'Neuroscience, adolescence and GCSEs – a misalignment? GCSEs are a relic from a time before our understanding of adolescent development and mental health' 28/11/2020. https://bigeducation.org/rethinking-blogs/neuroscience-adolescence-and-gcses-a-misalignment/. Accessed 01/12/2020.

References

Abbott, J. (2009). *Over Schooled but Under Educated: How the crisis in education is jeopardizing our adolescents.* Continuum.

Arrowsmith-Young, B. (2013). *The Woman who changed her brain: How We Can Shape Our Minds and Other Tales of Cognitive Transformation.* Random House.

Baddeley, A. (2012). *Working Memory: Theories, Models, and Controversies.* Department of Psychology, University of York; *Annual Review of Psychology, 63*, 1–29. Downloaded from www.annualreviews.org. Access provided by 31.127.140.192 on. Downloaded from www.annualreviews.org. Accessed 01 July 20.

Bignell, S. (2018). *Autism, Asperger's & ADHD: What You Need to Know. A Guide for Parents, Students and other Professionals*. My Child Services Ltd.

Blakemore, S.-J. (2018). *Inventing Ourselves: The Secret Life of the Teenage Brain*. Penguin Random House.

Blakemore, S.-J., & Mills, K. L. (2014). Is Adolescence a Sensitive Period for Sociocultural Processing? *Annual Review of Psychology, 65*, 187–207 (Volume publication date January 2014) https://www.annualreviews.org/doi/full/10.1146/annurev-psych-010213-115202 Accessed 22/08/2020.

Bliss, T. V., & Lomo, T. (1973). Long-lasting potentiation of synaptic transmission in the dentate area of the anaesthetized rabbit following stimulation of the perforant path. *The Journal of Physiology, 232*(2), 331–356. https://doi.org/10.1113/jphysiol.1973.sp010273. PMC 1350458. PMID 4727084 Accessed 03/04/2021

Brooks, D. (2019, January 17). Students Learn From People They Love: Putting relationship quality at the center of education. *The New York Times*. https://www.nytimes.com/2019/01/17/opinion/learning-emotion-education.html. Accessed 23 Jan 2019.

Byrne, J. H. (2017). *Learning and Memory: A Comprehensive Reference* (2nd ed.). Academic Press, Elsevier.

Clark, A. (2001). *Mindware: An introduction to the philosophy of cognitive science*. Oxford University Press.

Claxton, G. (2002). *Building Learning Power: Helping Young People Become Better Learners*. TLO.

Cobb, M. (2020). *The Idea of the Brain: A History*. Profile Books Ltd.

Cromby, J. (2007). Feelings and Embodied Subjectivity. In V. Van-Deventer, M. Terre-Blanche, E. Fourie, & P. Segalo (Eds.), *Citizen City: Between Constructing Agent and Constructed Agency* (pp. 232–240). Captus University Press.

Cromby, J. (2015). *Feeling Bodies: Embodying Psychology*. Palgrave.

Dean Hobbs, R. (2016). *The Global Child: How Experts Would Change Education. Research-Based Acquisition of Languages*. Lap Lambert Academic Publishing.

Dennett, D. C. (1991). *Consciousness Explained*. Penguin Random House.

Dennett, D. C. (2017). *From Bacteria to Bach and back: The Evolution of Minds*. Penguin Random House.

Department of Education (CESE). (2018). *Cognitive Load Theory in Practice Examples for the Classroom November 2018 Centre for Education Statistics and Evaluation Sydney NSW 2001*. https://cese.nsw.gov.au. Accessed 08 June 2019.

Dharani, K. (2015). *The Biology of Thought: A Neuronal Mechanism in the Generation of Thought – A New Molecular Model*. Academic Press, Elsevier.

Doidge, N. (2007). *The Brain that changes Itself: Stories of Personal Triumph from the Frontiers of Brain Science*. Penguin.

Doidge, N. (2015). *The Brain's Way of Healing: Stories of Remarkable Recoveries and Discoveries.* Penguin.
Dweck, C. (2006). *Mindset: Changing The Way You think To Fulfil Your Potential.* Random House.
Dweck, C. (2017). *Mindset – Updated Edition: Changing The Way You think To Fulfil Your Potential.* Robinson.
Eanes, R. (2016). *Positive Parenting: An Essential Guide New York: Penguin.* Random House.
Eaton, H. (2010). *Brain School: Stories of Children with Learning Disabilities and Attention Disorders Who Changed Their Lives by Improving Their Cognitive Functioning.* Glia Press.
Eaton, H. (2018). *The Brain Pioneer: The True Story of How Barbara Arrowsmith-Young Used Brain Science to Help Children with Learning Disabilities.* Barnes and Noble.
Elder, Z. (2012). In I. Gilbert (Ed.), *Full on Learning: Involve Me and I'll Understand.* Crown House Publishing.
Euler, Matthew J. (2018) Intelligence and Uncertainty: Implications of Hierarchical Predictive Processing for the Neuroscience of Cognitive Ability. *Neuroscience and Biobehavioral Reviews Journal.* Homepage: www.elsevier.com/locate/neubiorev. Accessed 08 May 2019.
Family Advice Website. (2019, January 04). *Tracy's Blog 'Toddlers, Meltdowns and Brain Development: Why Parents Need to Ditch Traditional Discipline.'* https://raisedgood.com/toddlers-meltdowns-brain-development-ditch-traditional-discipline/. Accessed 05 May 2019.
Feldman Barrett, L. (2017). *How Emotions Are Made; The Secret Life of the Brain.* Pan Macmillan.
Florance, C. (2004). *A Boy Beyond Reach.* Simon & Schuster UK Ltd.
Peter Ford. (2018, September 7). On Episodic and Semantic Memory: A Caution Against a Prosaic Curriculum. *edsacredprofane.* https://edsacredprofane.wordpress.com/2018/09/07/on-episodic-and-semantic-memory-a-caution-against-a-prosaic-curriculum/. Accessed 04 Mar 2019.
Fry, H. (2018). *Hello World: How to be Human in the Age of the Machine.* Penguin Random house.
Gardner, H. (1993). *Frames of Mind: the Theory of Multiple Intelligences.* Fontana.
Goleman, D. (1996). *Emotional Intelligence: Why it can matter more than IQ.* Bloomsbury.
Greenfield, S. (2004). *Tomorrow's People: how 21s-century technology is changing the way we think and feel.* Penguin.
Greenfield, S. (2014). *Mind Change.* Penguin Rider.
Hale, I. (2018). *Asperger's, Autism and You: Is the epidemic real? Are vaccines the cause?* Institute for Education, Research and Scholarships.

Hattie, J., & Yates, G. (2014). *Visible Learning and the Science of How We Learn*. Routledge.

Hawkins, J. A. (2017). *Feelings and Emotion-based Learning; A New Theory*. Palgrave Macmillan.

James, W. (1884). What is an Emotion? *Mind, 34*, 188–205.

James, W. (1894). The Physical Basis of Emotion. *Psychological Review, 1*, 516–529.

James, W. (1918). *The Principles of Psychology – Vols. 1–2 (Illustrated Works of W. James 1842–1910)*. Pantianos Classics.

Jarvis, P., Swiniarski, L., & Holland, W. (2016). *Early Years Pioneers in Context: Their Lives, Lasting Influence and Impact on Practice Today*. Routledge.

Jung, C. G. (1991). *The Archetypes and the Collective Unconscious (Collected Works of C. G. Jung1865–1961)*. Routledge.

Katz, B. (2019, March 28). A Scottish Woman Doesn't Feel Pain or Stress. Now Researchers Think They Know Why: Genetic Analysis Revealed That Jo Cameron Has a Previously Unidentified Mutation That May Affect Pain Relief, Memory and Mood. *The Smithsonian Magazine*. https://www.smithsonianmag.com/smart-news/scottish-woman-doesnt-feel-pain-now-researchers-think-they-know-why-180971822/. Accessed 29 Mar 2019.

Kelly, L. (2017). *The Memory Code: The Secrets of Stonehenge, Easter Island and Other Ancient Monuments*. Pegasus Books Ltd.

Kincheloe, J. L., Steinberg, S. R., & Villaverde, L. E. (1999). *Rethinking Intelligence: confronting psychological assumptions about teaching and learning*. Routledge.

Kohn, A. (2006). *Unconditional Parenting: Moving from Rewards and Punishments to Love and Reason*. Atria, Simon and Schuster Inc.

Merzenich, M. (2013). *Soft-Wired: How the New Science of Brain Plasticity Can Change Your Life*. Parnassus Publishing.

Novick, J. M. (2020). In M. F. Bunting, M. R. Dougherty, & R. W. Engle (Eds.), *Cognitive and Working Memory Training: Perspectives from Psychology, Neuroscience, and Human Development*. Oxford University Press.

Alex Orlando. (2020, June 14). *What Happens in Your Brain When You Make Memories? Scientists Are Using New Tech to Decode How Our Brains Remember*. https://www.discovermagazine.com/mind/what-happens-in-your-brain-when-you-make-memories. Accessed 04 Oct 2020.

Ostby, H., & Ostby, Y. (2018). *Adventures in Memory: The Science and Secrets of Remembering and Forgetting*. Greystone Books.

Pascual-Leone, A., Hamilton, R., Tormos, J. M., Keenan, J. P., & Catala, M. D. (1999). Neuroplasticity in the adjustment to blindness. In J. Grafman & Y. Christen (Eds.), *Neuronal Plasticity: Building a bridge from the laboratory to the clinic* (pp. 94–108). Springer-Verlag.

Paul, A. M. (2012, February 22). *What Actors Can Teach Us About Memory and Learning the Most Effective Memorization Techniques Draw on Physical and*

Emotional Engagement as Much as They Do Pure Brain Power. @anniemurphy-paul. http://ideas.time.com/2012/02/22/what-actors-can-teach-us-about-memory-and-learning/. Accessed 11 Nov 2018.

Perry, B. D., & Szalavitz M. (2006). *The Boy Who Was Raised as a Dog: And Other Stories from a Psychiatrist's Notebook*. Hachette Book Group.

Plass, J. L., & Kalyuga, S. (2019, March 7). Four Ways of Considering Emotion in Cognitive Load Theory. Springer Science+Business Media, LLC, part of Springer Nature 2019 *Educational Psychology Review*. https://doi.org/10.1007/s10648-019-09473-5. Accessed 06 June 2019.

Professor Amanda Kirby (2019, April 2). Neurodiversity expert 'Could you be misinterpreting a student's behaviour?' In General, Neurodiversity, Teaching strategies: *Cambridge Outlook Magazine*. https://blog.cambridgeinternational.org/could-you-be-misinterpreting-a-students-behaviour/. Accessed 08 April 2019.

Rogers, C. R. (1961). *On Becoming a Person: A Therapist's View of Psychotherapy*. Constable.

Rowlands, M. (2016, November 15). *Memory and the Self: Rilkean Memory*. http://philosophyofbrains.com/2016/11/15/memory-and-the-self-rilkean-memory.aspx. Accessed 07 Aug 2018.

Sealey, C. (2018). What *Your Handbag Can Teach You About Memory. Clare Sealy Talks About Memory in the Classroom in This Sneak Peek of Our NQT Test Guide: 'Your first year in teaching.'* https://www.tes.com/news/memory-techniques-for-classroom. Accessed 12 Sept 2018.

Shalaby, C. (2017). *Troublemakers: Lessons in Freedom from Young Children at School*. New Press.

Shore, S. (2001). *Beyond the Wall: Personal Experiences with Autism and Asperger Syndrome*. Autism Asperger Publishing Co.

Steiner, R. (1920). *Lectures in Stuttgart. Teaching from Humane Knowledge* (Pusch, R., Querido, R., & Amrine, F., Trans.). Amazon Fulfillment.

Youki Terada (2017, September 20). 'Why Students Forget—And What You Can Do About It' Our Brains Are Wired to Forget, But There Are Research-backed Strategies You Can Use to Make Your Teaching Stick. https://www.edutopia.org/article/why-students-forget-and-what-you-can-do-about-it. Accessed 12 Feb 2019.

Tsabary, S. (2013). *Out of Control: Why Disciplining Your Child Doesn't Work and What Will*. Namaste Publishing.

Tyng, C. M., Amin, H. U., Saad, M. N. M., & Malik, A. S. (2017). The Influences of Emotion on Learning and Memory. *Frontiers in Psychology, 8*, 1454. Published Online 2017 Aug 24. https://doi.org/10.3389/fpsyg.2017.01454. PMCID: PMC5573739 PMID:28883804 https://www.ncbi.nlm.nih.gov/pmc/articles/PMC5573739/ Accessed 25 Sept 2018.

Walker, M. (2017). *Why We Sleep: The New Science of Sleep and Dreams*. Allen Lane.

Ward, J. (2019). *On the Fringes: Preventing exclusion in schools through inclusive, child-centred, needs-based practice.* Crown House.

Wenger, E. (1998). *Communities of practice.* Cambridge University Press.

Wenger, E., McDermott, R., & Snyder, W. (2002). *Cultivating communities of practice: a guide to managing knowledge.* Harvard Business School; McGraw-Hill.

Wexler, N. (2019). *Writing and Cognitive Load Theory.* https://researched.org.uk/writing-and-cognitive-load-theory/. Accessed 29 June 2019.

Whitebread, D. (2017). Quality in Early Childhood Education: The Contribution of Developmental Psychology. In *International Handbook of Early Childhood Education* (pp. 319–334). Springer.

Willingham, D. T. (2009). *Why Don't students Like School? A cognitive scientist answers questions about how the mind works and what it means for the classroom.* Hoboken NJ.

World Health Organization International Report. (2017–2018). Health Behaviour in School-aged Children (HBSC) Survey in Europe and Canada: Key findings from 2017–2018. WHO. Website: www.euro.who.intSpotlightonadolescent. Accessed 23 Jan 2019.

Zacks, J. M. (2015). *Flicker.* Oxford University Press.

Zadina, J. N. (2015). *The emerging role of educational neuroscience in education reform.* Published by Elsevier España, S.L.U. on behalf of Colegio Oficial de Psicólogos de Madrid. This is an open access article under the CC BY-NC-ND license http://creativecommons.org/licenses/by-nc-nd/4.0/

CHAPTER 4

System Planning: Teaching Problems and Solutions

4.1 Introduction

The next three chapters discuss narrative and theoretical evidence from teachers commenting on current educational debates. These teachers consider the feasibility and benefits of teaching approaches from different perspectives. They include insights gained by teachers doing their own psychological research into their work and problems faced in managing and improving teaching. I collected many additional reference links to these, which back up, evidence and inform the text and are referenced in each of the chapter reference lists. I want to express my thanks to all those whose contributions I chose to directly reference in the text and apologise to those I could not include from the many I read.

Human brains learn and adjust to complex circumstances every day and so it is extraordinary traditional scientific researchers have ignored the evidence of emotional and sensory aspects of thinking. It appears they did so because they could not categorise these processes using methods of comparison with norms, averages and standardisations. Some politicians and education managers still have not caught up. Yet we know (and always knew) motivations are enabled by feelings and emotions. Neuroplasticity research has proved that these processes are present! They are unique to us all individually, sometimes puzzlingly involuntary, but certainly not *necessarily* automatic and even apparently random.

J. A. Hawkins, *Brain Plasticity and Learning*,
https://doi.org/10.1007/978-3-030-83530-9_4

Neuroplasticity appears to have evolved to help us to understand and cope in our complex world. Now that we understand it exists, it is time for teachers to engage in the process of explaining back to science how this works for learning. Many teachers, coaches and tutors have developed their own and their learners' neuroplasticity anyway, but understandably were not trained to realise their own success *in those terms*. In this chapter I look at these issues from their perspectives. I am looking at teachers' and parents' own opinions in order to open up a debate about how we might utilise the fact of neuroplasticity to advance understanding in education.

4.2 Educational Communities as They Relate to National Visions

Any school or educational institution is a community of human beings within a national ecosystem. Such establishments are funded by government education departments and endowed by philanthropic trusts and business partnerships. Schools and colleges are affected by politics at every level, for example, lecturers, teachers, department heads and system managers and those who provide, gift and manage the financial resources needed to fund them. They all have different roles, frames of reference, experience, purposes, stated aims, beliefs and needs. All these people have their own models, concepts, inferences and beliefs about their work. It is a complex world.

In any education system there are liable to be attempts to rearrange, circumnavigate, update or subvert rules, regulations, principles aims and objectives by those with power *for a wide range of personal and civic reasons.* Even well-established structures and traditions are affected by *existential challenges* in the changing cultural climate of the day. In this section I am going to consider two national education systems. They are Finland, ranked as an excellent educational system in the world and the United Kingdom, with a good reputation historically, but where currently teachers frequently disagree with government policy and their retention is generally low. In both countries education institutions have *evolved over time* with civically responsible democratic checks and balances in the form of senior leader teams of rectors, managers, governors and trust committees and laid down procedures. To some extent this is an inevitable evolution where different experiences, perspectives, curiosity, innovative thinking and knowledge meet.

In the United Kingdom, for example, the problems and challenges in setting and following a consistent vision that are faced by system organisers over time, both political and functional, are discussed in the current new book called *Our Passion for Schooling* by Tim Brighouse and Mick Waters (2021). They interviewed secretaries and ministers of state, chief inspectors, civil servants, school leaders and head teachers about the changes in British education since 1976, which have led to the current centralised system. This system has relied upon the honourable behaviour of all participants, but although 'centralised,' there appears to have been no clear central cohesive national goal around the purpose of education. In this country not all citizens themselves are taught about how and why they all have a human right to education under the law. If they were this might provide a better check upon consistency of delivery. See 'The Nolan principles of governance' Endnote.[1]

In Britain some state schools are funded and managed by local state education authorities and others by 'self-managing' academies. The latter are given taxpayer funds to run a sector of state schools with an independent business model and fewer restrictions. The two systems operate side by side and are both actually semi-autonomous and are both subject to external control by government inspection. All schools are required to have their own vision and mission statements to show their own individual values and beliefs to their community. Schools in both sectors tend to vary in education quality. Democratic processes are not consistent across the system and neither are the educational principles, aims, values and outcomes. Olli-Pekka Heinonen is the Director General of the Finnish Education Agency and former Minister of Education. He tells us about the principles behind their different educational approach in preparing their citizens for the jobs and challenges of the future (2020). See Podcast Presentation Endnote.[2]

Finland became an independent country 170 years ago. The stated aims of the government at that time still influence the present day. Their principle of equality for the whole population was set out to raise the educational level of all of them by developing education through all of their languages and cultures. Education is seen as a means to achieve and improve Finland's national status in the world. This goal appears to have been achieved judging by the excellent quality of its education outcomes. For example, the country's system has performed well in global school system results awarded by the Organisation for Economic Co-operation and Development (OECD). The organisation monitors early education through to vocational upper secondary and adult education and training. See Report Endnote.[3]

In Finland the schools work towards 'every learner achieving their personal best.' Three stated goals feed into each other and are interconnected. These are as follows: (1) an equal society with full gender equality (fairness, justice); (2) safety and stability (well-being, security); (3) education and innovation (child-centred and adaptable). Equality means that all pupils of all abilities are achieving the best they can in their circumstances *for themselves.* The system *aims* to nurture its citizens so all prosper to the best of their ability, though this may not always be achieved. *Everyone, including children, parents and older students know this.* It helps to make all stakeholders accountable. In Finnish society it is generally believed by political parties that treating citizens 'equally' feeds into the social capital of the country. This agreement has prevented polarised policy changes and encouraged long-term planning for improvement.

Different governments as they change political affiliations have consistently supported the education system towards this aim. This has created a confident education culture and trust in the system. In around 1990, teachers groups and unions asked for teachers to be trained up to Master's certificate level in education. As a result a better qualified workforce led to the reforming of education legislation 1991–1997. This established the current system through legislation where *all* teachers are required to reach this level and are expected to take responsibility for pedagogy. There are, of course, always problems arising, new challenges and developments. However, the fact that 'Trust' has been declared a central pillar has aided collaboration and solidarity. The aim of Finnish education has been to create *'a network of autonomous actors.'* 'Autonomous' in this context means self-management and freedom for teachers to research, provided they deliver central government goals.

Heinonen tells us the individuals participating have to feel a lot of trust in each other to make the system function. For example, an acceptance that national levels and aims for learning and teaching outcomes will be achieved in different ways. Schools are run by the self-governing principalities. They are provided with teachers who are trained in similarly autonomous universities. When teachers have asked for more freedom and autonomy to teach, the government have changed the rules and made more room for individuality and manoeuvre at the local level. Personal learner agency extends to the pupil and student levels. A feedback loop between the teacher and pupil creates a process of continuous evaluation considered vital in the Finnish system. The teachers trust learners to work hard and in return the child '*reveals what he knows and what he doesn't know.*'

'*When there is trust the child tells the teacher when he doesn't understand.*' This creates a learning loop and the opportunity for both quantitative and qualitative evaluation. There are signs that insightful teachers in the UK education system have wanted to develop similar pedagogical approaches to those in Finland. For example, according to Nikki Booth, there has been over 40 years of research evidence about formative assessment (otherwise known as Assessment for Learning)—a very similar approach to the Finnish one described by Heinonen. Much of this has been based on the work of Paul Black and Dylan Wiliam (1998). Black and Wiliam have defined formative assessment as *all* activities undertaken by teachers and their students that modify the teaching and learning activities in which they are engaged. Teachers and pupils are involved in a dialogue (oral or written) that moves the learning process forward. See Article Endnote.[4]

Wiliam says that the term formative assessment was misinterpreted by some teachers. He wishes, with hindsight, they had called it responsive teaching (Twitter, 2017). A responsive assessment approach is used to gather evidence and give feedback and has been found to improve learner and teacher performance and learner outcomes. *It also seems to align very well with scientific discoveries about neuroplasticity.* The education system in the United Kingdom lacks the vision and clear goals for learning present in the Finnish system. Government goals do not include creating an education system aiming to improve the mental well-being and self-betterment of the whole population currently operating in Finland. Increasingly called for with parliamentary petitions by the public, it tends only to happen in an ad hoc, incidental and piecemeal fashion. There is no explicitly stated intention or system in place to create equality of opportunity throughout the system or the population.

The Office for Standards in Education (Ofsted) inspects education for learners of all ages. They do it by determining if external examination results of yearly cohorts have achieved the annual level of statistical formulae required and if not why not. Meanwhile the detail of each student's personal learning journey other than by grades in core subjects, their challenges and successes are not recorded. Inspectors are expert former heads and senior leaders, but they are not permitted to offer advice on the professional development of schools. Inspectors are asked to analyse Standard Attainment Test scores in each subject area. They are marked externally in order to 'award' a whole school description based on the numerical data. The judgements are general labels such as 'outstanding,' 'good enough,' 'requires improvement' and 'failing' in both sectors of state-funded education. League Tables are then published for parental scrutiny. See Endnote.[5]

Teachers and their unions have complained about the imposition of the narrow aims and targets set. Teachers have to adjust to sudden system changes, finding themselves judged against fashionable criteria of the day as different governments adopt new policies, without consulting them. Politicians responsible have sometimes paid educationalist experts who agree with their policies to write reports. Teachers have found by complying with some of these requirements they are inadvertently 'teaching to the test' and ignoring pupil well-being. There has been an assumption of duty of care to every child, but the system is not sufficiently flexible. A current book by Lisa Cherry explains how this might be achieved (2021). Christopher Bagley, an educational psychologist working with schools explains how drawing on the views of students might enable the education system to be less coercive and more nurturing. See Blog Article Endnote.[6]

Jeremy Hannay, then a headmaster for 12 years at a recently Ofsted inspected 'outstanding' primary (elementary) school, speaks eloquently about this in a blog article called "The Apocalypse of English Education" (2020). He tells us the recent pandemic causing thousands of children to isolate at home has exposed the 'flawed underpinnings' of the current model of education. Hannay claims convincingly the UK system is based on a competitive market business model under a mistaken belief that such an approach will drive improvement—an external and remote assessment system that is generally based on unsupported parental choice and 'comparable measures' like standardised tests. Hannay says in his experience the Ofsted system and frameworks are disadvantaging those who lack money, emotional support and educationally privileged backgrounds. See Blog Article Endnote.[7]

The Finnish pillar goal of equity is a practical and emotionally intelligent objective on the part of their democratic government. This encourages learners by directing and enabling them to achieve their personal best. This commitment is universally understood across agencies within education and by stakeholders, including teachers, parents and learners. There are advantages to such a system:

- Everyone in the system is challenged to achieve the equality objective.
- Accountability is implicit between the actors at all levels, for example, senior management, middle management leaders, teaching staff and students.
- The participants are repeatedly made aware they are respected.

- Such a goal tends towards social cohesion and progress in resolving issues arising.

This system aims at freedom, autonomy and flexibility through motivation around its central goal rather than fixed authoritarian state control. Challenges are addressed by bringing relevant stakeholders together to define a solution and laws are passed after consultation has taken place. The UK government does have a system of parliamentary committees which might perform this function if it was extended and used consistently. Such a system might make adaptability and implementation easier, because everyone knows why laws and policies have been introduced and what challenges are to be addressed. The Finnish government believe the transfer of knowledge is only part of the job of the education system. Its fundamental role is to prepare children for a future that is complex and unpredictable. Finland and Canada currently take an 'individual human rights' approach to education (Kidd, 2014). These two countries' test scores in maths, reading and science are nevertheless generally amongst the highest in the world. In the next section I take a look at some of the issues around existing subject curriculum structures and suggestions proposed by teacher leaders. Article Endnote.[8]

4.3 Learning Purposes and Political Purposes

A multitude of purposes and visions drive educational institutions and schools of thought. The purposes of education are many, but learning works best when knowledge is discovered, renewed and repurposed for everybody's benefit. Inevitable tensions arise where so many ideas meet and intersect. This necessarily involves the educational community in research and development. Educational philosophies affect the practical work of teachers in the classroom especially when they are judged by them. However, they are still forced daily to make pragmatic decisions. If everyone takes part in the dialogue as part of the system, it might better benefit that system. Noam Chomsky (1928–), American linguist and philosopher, explains ways educational systems change and vary their understanding of purposes for learning. Currently 92, he tells us about educational influences, themes and tropes affecting his life in education in the West (Chomsky & Polychroniou, 2017).

He says the 'traditional' interpretations arising from the Enlightenment ended in an educational movement believing in inquiry and innovation. Scholars followed their own agenda, researching how and what to learn aiming to find the exciting and new. This involved searching the 'riches of the past to internalise ideas and carry them on a quest for understanding.' Learning was a journey of discovery and reinterpretation. Chomsky tells us that after 1960s demonstrations, authorities were concerned youngsters were becoming too independent. They considered this a crisis of democracy because institutions were failing to sufficiently 'indoctrinate' the young. They declared young people needed discipline, a framework to work within, to conform, follow orders and to be trained not to challenge authority. Arguments were made by the powerful for more controlled vocational training and Chomsky believes they imposed debts on students 'to trap them into a life of conformity.' He does not mention capitalist corporation motives for creating a 'useful' workforce. See Video Endnote.[9]

There were tensions around whether to train students to pass tests or to engage in creative inquiry. Two schools of thought about purpose emerged and were labelled as two different ways of looking at the world. These 'sharp differences' identified by Chomsky existed together following different perceptions and personal interpretations on a multidimensional level within institutions. The two arguments and approaches were labelled traditional education and progressive education. Over the years they 'shape shifted' to provide different convenient 'cover stories' and justifications for different purposes fulfilling different government and management aims, opinions and agendas. Still argued about today by teachers on social media, in reality there has been a blurring of these binary positions through custom and practice. They turn out not to be extreme opposites in practice, but have coexisted in a range of practical variations for different purposes. See Journal Article Endnote.[10]

The varied methods of learning used under these umbrella terms are not mutually exclusive. They depend upon choice, judgement, balance and suitability for purpose. Chomsky points out that scientific research would not have progressed if scientists did not challenge, imagine and question doctrine and authority in order to innovate. People argue a free market approach through private education disadvantages the poor and gives easy privilege to the rich. On the other hand to appreciate educational opportunity (having paid for it by working and holiday jobs, or winning support by gaining a scholarship) can aid motivation and contribute to effort in gaining skills leading to better employment. Both state and private systems require structure and regulation. The Finnish model does this by adopting a central principle capable of inspiring society. It can also be reinterpreted in different ways by teachers.

Chomsky asks us to consider whether the aim of education is to increase Gross Domestic Product (GDP) or to get new ideas towards scientific and cultural progress. He points out information technology in the 1950s was funded by public money in universities resulting in the billionaire capitalist fortunes of today. He says that without educational systems that encourage invention there would be no new technology for economic gain. We might make a list by adding health care and environmental remedies to this. Perhaps we need an age of 'New Enlightenment' where old models and approaches are examined and good ideas reimagined, adapted and combined. Currently the world and its man-made environmental crisis need creative ideas to imagine and find solutions. Ideas from the past might help us.

Der Bauhaus in Germany was an educational institution for young adults that insisted that every student registered for a full legal apprenticeship, which included manual instruction in a historic craft. Its far-sighted instigator, director, architect and educational philosopher Walter Gropius (1883–1969) envisioned its workshops as 'laboratories' for creating well designed, technically viable prototypes. Natural and old materials were explored and repurposed as well as modern ones including steel, concrete, slate, wood, clay, glass and brick used to design in new ways (1955). The principle of producing 'form that followed human function' was used to design buildings and mass-produced manufactured articles. The philosophical elements of the vocational and intellectual Bauhaus vision might still be incorporated into future subject departments in a variety of education institutions.

Taking a much more hands-on approach by researching ancient crafts and knowledge might be rewarding—not just for dealing with plastics and pollution. An appreciation of creative thinking starts with arts in all subjects in school and can contribute to satisfying and constructive pastimes. An understanding of arts and sciences in world cultures could encourage a more respectful coexistence through sharing, encouraging social integration amongst and across populations. Gropius' vision was that 'spontaneous art should permeate' the life of citizens in all areas of society. It would tend towards social, holistic and personal well-being. Unfortunately people unjustly associate his name with weaker, inferior quality commercial forms of 1960s buildings and design. See Blog Article Endnotes.[11,12]

Rudolf Steiner (1861–1925) was an educational thinker whose ideas about humane schooling became the inspiration for modern-day Steiner Waldorf schools. He also applied his ideas to agriculture, architecture, medicine and social reform. Steiner believed that teaching is an art and that teachers should not 'impose' fixed programmes of external knowledge upon learners. He believed they need to have 'a real sympathy with a child's nature' and a wide understanding about the stages of the growing child. Steiner said that teachers should learn 'on the job' through the actual variable process of teaching different students every year. The teacher's job was to develop children according to their personality and individual capability. Steiner developed the idea of encouraging each child to express themselves with confidence through creative activity and independent thinking. See Endnote.[13]

Ken Robinson (1950–2020) had a vision of creative schools and the idea that everyone is born with natural learning potential. This concept can now be evidenced by the scientific fact of human brain plasticity. Robinson's research around the world over many years led him to the conclusion that schools 'actively discourage' pupils' natural creative abilities. His philosophy challenges assumptions about teaching. He points out underlying problems such as the industrialisation of schools. In spite of his popularity with teachers, many Western governments today still favour uncreative subject school systems (Robinson & Aronica, 2015). This may be partly because creativity is seen as unproductive and its developmental value to intelligence is not understood by politicians. Some government ministers have only valued children themselves as potential employees for the capitalist interests that lobby and fund them.

Similarly officials have persisted in seeing children as 'blank slates' (Pinker, 2002) and teachers as child trainers and technicians. Some governments believe teachers need directing in methods of teaching. They want to achieve utilitarian outcomes like the production of types of workers. Schools are viewed according to a business model—an investment towards raising levels of monetary gain. In a neoliberal economy the accumulation of money is seen as the principal measure of success and value in all areas of life. A wasteful acquisitive culture of materialism has been spread by 'technically developed' countries around the world. Children are seen as a commodity and children's tests and scores in exams as a product. These practices have arisen from a pervasive and controlling neoliberalist culture developed for profit by capitalists and politicians over the past 30 years. Unfortunately, these kinds of educational environments create negative as well as positive effects for everyone including the rich.

Under such systems teachers are restrained from developing their own personal professional learning. While some people will happily discover that their exam results and talents fulfil government agendas, others will inevitably be prevented from discovering ideas and abilities. People become unfulfilled, discouraged and deprived with attendant social consequences, for example, unethical scientific research agendas, crime and mental health challenges. A good education mitigates against such problems. Many teachers in the United Kingdom have become disaffected with these politics. Human learning processes are naturally driven, evolving to cope with an ever-changing world. Educational institutions change and evolve, but core human needs and values are a constant. Cultural histories are diverse, and educational systems can also be diverse and multicultural.

Education systems should enrich individuals' lives and through them society at large for everyone's benefit. They can empower and enable human beings to adjust and thrive in different settings (Bhopal, 2018). They have often developed naturally to perform this fundamental role in society. Democratic governments should listen to their teachers in order to keep fundamental democratic values alive and to maintain relevance. To do this democratic education systems have to be based on cultural wisdoms and include the aspirations and voices of the people. For example, Finland has a system which is organised so a professional class of teachers can communicate their concerns up through the government system.

The system works on the principle that every pupil is valued. For example, everyone in a class is ready to move up to the next class even when differentiated achievements reveal five-year differences. This allows for different ability and maturation levels to coexist as they do in society. Everybody is valued regardless of ability. There are ways to teach with social connection to promote confident personal identity. This government and education system recognises, acknowledges and supports this challenge as a central goal. In the United Kingdom there is an absence of trust in the system because the emotional well-being of all stakeholders is not considered of practical importance. It is not factored into the system. However, one type of equal opportunity intervention has been developed by expert teachers.

It is called differentiation achieved by personalising learning—setting work differently to match student needs. Teachers are now researching to support personalised learning online. This work accelerated by the pandemic has highlighted how much we need personal human contact! There are experienced head teachers who understand how to achieve human educational goals. Many become government inspectors, but as already mentioned are currently prevented from giving advice. They are skilled and knowledgeable about how to support other teachers. Tom Sherrington and Chris Chivers are two of these. They are insightful and understand how to lead schools humanely to achieve good outcomes. They can support the long-term process of developing schools in their local community cultures to contribute to a stable and successful society.

Excellent education does not happen when restricted by the short-term political aims of standardisation in order to produce statistical outcomes on numerical graphs. Even though these are outcomes politicians may understand much more easily than even one child, they cannot contribute in isolation to a well-rounded education. Such systems trigger mental health difficulties and are not conducive to co-operation, democratic progress and long-term innovative solutions to world problems. Schools should be supported to interpret the curriculum to standards their pupils can reach even above set baseline standards. Good teachers given the opportunity can produce outstanding results, explain, justify and evidence their work. See Blog Article Endnotes.[14,15,16]

The snobbery of treating other people as less worthy and intelligent has not changed, but passed from the aristocracy to dictators, religious leaders and a powerful billionaire elite. This culture tends to value people by their lifestyle, possessions and consumer potential rather than for what they know and contribute to society. It is worth noting that currently there are problems for modern children of all 'classes' who may be living in a 'cultural desert of neglect.' This happens where parents influenced by advertising have materialistic lifestyles and tend not to engage with or discuss the world of ideas and knowledge. The old normative national reference points set out in programmes aimed at improving 'social mobility' no longer apply. Not only because of multi-ethnic human connectivity, but because education is a human right.

Education should create a holistic, richer, mentally stimulating humane environment. Different cultures are knowledge rich with expertise around ways we think, philosophise and feel about the world. A good education gives a 'spiritual' aspect to human life that contributes in many different ways to human well-being and happy prosperous societies. Michael Merrick says it is a tragedy when young people are denied access to their cultural inheritance. He is talking about this as a problem for underprivileged children like the traditional 'working class,' who were encouraged to leave their family and culture in order to 'improve' themselves. Merrick points out there is an 'implied denigration' of 'working-class' culture when we assume students must leave homes behind when they become educated 'to join a privileged elite.'

The same can be said of immigrant children, who should be taught to appreciate their heritage as a means of integrating and educating their peers. It is time to stop talking in terms of class. Most people no longer regard themselves as 'working class' in the old Western way. They regard themselves as entrepreneurs and small business people working in the 'gig' economy. Currently it is a socially redundant term. Populations are much more diverse and with a shared multi-culture via the internet, potentially much richer and more complex. Wealthy people today have excessive power but long term they will lose out in today's materialist world. All members of society need a good life, a good education, to value their culture, learn about and respect others for the sake of universal well-being opportunity. See Article Endnote.[17]

4.4 Teaching Values and Principles for Curricula Planning

Rich knowledge is important for all learners in the modern world, especially now it is accessible through rising multiculturalism and the internet. The problem of providing a 'rich knowledge' connected curriculum is recognised by Finnish educators. They are addressing it through their developing system of collaborative teacher research. Olli-Pekka Heinonen, director general of the Finnish Education Agency, explains teachers are proud that their schools combine good learning outcomes in terms of knowledge with pupils' good feelings about their well-being. The government policy and the teachers' professional aim is to take care of both at the same time. He understands, "*If the wellbeing is not there the learning will not be there—if they are both there you give the possibility of the individual being able to take care of his (or her) own wellbeing in the future.*"

Heinonen appreciates that the whole system depends on trust. This seems to be greatly facilitated by the 'equality' goal that is expected to be interpreted and acted upon by all professionals involved. It provides an evaluation check that is built into the system at every level. When a problem in one school is referred to the education ministry, an attempt is made to strengthen the school's own capacity and resilience to solve their problems. Teacher leadership is considered important. Finnish political parties are in strong agreement about schooling aims. Changes in government do not affect long-term planning in education. It is not politicised. Students with more diverse needs are funded and provided for at the start of their education. It is likely their neuroplasticity growth will be enabled to develop at an individual pace. The vocational system is competence based. This makes a personal learning path onwards through life possible as achievements are recorded.

The goal is to educate the whole person, not just give qualifications to enter the labour market, but to see education as a holistic challenge. Qualified teachers continue training, doing collaborative research with school and university colleagues. They experiment with their national core curriculum and how to teach skills across subjects (transversal competencies) and they make recommendations. The law requires educational outcomes to be judged on evidence of 'knowledge and skill combined with competencies'—'not only what you know but *what you can do* with what you know.' It is a core goal of the curriculum. In higher education young people and adults are assessed to find out what they can do. This involves a collaboration between teacher and student to find out and address missing skills and knowledge. Students learn on the job; they can be supported in an organisation with teacher involvement. They are moving from an education system approach to a 'learning system approach.' See Academic Paper Endnote.[18]

This combines formal education with non-formal and informal education. Developing the country's human capital for each individual brings the labour market and education system into a close dialogue with each other. Heinenon says, "There is always the possibility for a young person to make other choices—no dead ends—whatever you choose to do you can still do a PhD!" The aim is to help each child to *'grow into humanity'* and become responsible members of society and to have the skills and knowledge needed in life. In view of their vision the current stated challenges are:

1. The diversity of children with their diversified needs today (schools are struggling to face this new situation).
2. How can they maintain a cohesive educational policy and still aim towards self-governance and institutional 'autonomy?'

How can they strengthen the learning culture of schools and universities so that they are able to tackle the challenges and solve the problems appropriately themselves? See previous Academic Paper Endnote[19] and Podcast Presentation Endnote.[20]

Human history shows us how human environments provide the 'food for thought' that stimulates the different types of intelligence that contribute to human knowledge. We have individual gene pools that are variously stimulated by our environmental experience. According to Abbott (2009), the learning debate should not be around 'nature versus nurture' but about 'nature *via* nurture'—meaning that neural configurations are activated, developed and embedded through positive endorsements in our environment. This idea fits in with Finnish ideas about helping schools to produce an interactive, supportive and reactive environment to solve learners' learning problems. Nurturing encompasses providing an appropriately stimulating environment with teachers listening, responding sympathetically, but also challenging, intriguing and assessing humanely.

According to the Nature via Nurture Theory, the brain already has nascent predispositions that respond to different environments. These environments may provide conditions for the development of dormant talents, skills and general intelligence (Plomin, 2018). Although generally more active in youth, brain plasticity means that our intelligence has the potential to be activated (or deactivated) at any point in life. There is no one overall measurable general intelligence that is rigidly pre-set unless we make it so by embedding such a belief. We *may* have important potential predispositions built into our bodies and brains at birth, but sadly without environmental stimulation, we may remain unaware of them and they may never develop. On the plus side teachers can use their professional understanding to collaborate and action research on the job to engage students and inspire their interest by creating good environments. Social community and teaching programmes that contribute to diverse cultural interactions are a useful way to inculcate and extend knowledge.

Educational improvements can be devised to create an exciting nurturing environment. Some children and older learners are being trained in methods exposing them to stimulating new ideas. However, as we saw with the 'progressive' and 'traditional' approaches they are often used to fuel conflicting arguments and under researched. In many countries this has happened as different solutions have been favoured by politicians and education officials. However well researched by academics under specific conditions with particular pupils, they always require adaptation in practice with different pupils and settings. Educational programmes sometimes cause conflict.

Here are some possible problems that deserve our consideration:

- They have encouraged a belief that all educational research involves sorting children into normative categories and judging them by normative outcomes.
- Although contributing to a 'tool bag of methods' some have tended to be too prescriptive and unadaptable in practice.
- Methods and philosophies are unconnected. We have failed to create systems that encourage co-ordination and consistency through collaborative discussion, research and development—no universal goal.
- They have been made with restrictive assumptions and ideas about outcomes.
- Creativity and innovation have not been valued sufficiently as learning processes.
- They do not acknowledge the value and meaning of the process of self-actualisation, equality of opportunity and personal development.
- They value reductionist outcomes that evidence normative philosophies of 'utilitarianism' and 'instrumentalism' based on uninformed assumptions.
- They interpret well-being and happiness as adapting and complying with what is imagined may be required by a future employer. See Blog Article Endnote.[21]

Even when educational ideas are soundly based and well presented, they are sometimes misinterpreted by educational managers, politicians and commercial companies. They are seen as commercial quick-fix solutions even sometimes against the advice of originators. Carol Dweck has complained publicly about misinterpretations of her Growth Mindset Theory research based on her explanations of attitudes to learning. It was not at all her intention it should be interpreted as a quick fix formula. She has emphasised that simply telling people to be more confident and try harder is not enough. She says unless it involves teachers (and parents) teaching for conceptual understanding through feedback it does not work. Dweck believes that learning is about embracing challenges and working to overcome failure confidently. Teaching works best when researched and managed by teachers on the ground, where professional research, development and dialogue are part of the system of organisation, for example, in learning communities. See Video Endnote.[22]

Education is a complex professional field and management solutions imposed without training, research and feedback fail. There are master crafter teacher training developers with specialisms in specific teaching approaches that advance teaching skills, research and knowledge, but their use has been inconsistent. These experts are insightful teacher trainers, but unlike Finland many countries have no official provision or rationale for teacher research systems in which these lead research teachers can train others. The attainment standards laid out in the National Curriculum of the United Kingdom are a useful teaching guide. However, one such teacher, Martin Robinson, advises that three tenets of ancient Greek education might help teachers to plan and organise their curricula more successfully to inspire students in different age groups and settings (2013). He tells us these three areas of expertise, called The Trivium, can help us plan, practise and access knowledge by using different brain capabilities.

He explains how these three arts, 'frames of reference' or 'schools' of expertise can be deployed to expedite learning and how the skills they use are necessary for an integrated holistic rich knowledge education. They are:

- Grammar, which involves 'learning about the way things were or are'
- Dialectic, involving 'questioning the way things were or are'
- Rhetoric, showing 'how things are or could be'

They involve different sets of skills and thinking positions that point intellectually in different directions. Robinson says they create tensions, though often conflicting, that give us three different ways of 'thinking' about the world. We could picture them as the three points on the corners of a triangle. If we animated the model it would not always be equilateral, but 'on activation' would be constantly straining and bulging out of symmetry! In many areas of the British school system this has become a long thin narrow triangle with the long point heading almost exclusively towards Grammar (knowledge and testing).

Robinson argues that a balanced approach is needed in order to develop, exercise and expand what I would term 'brain plasticity'—the ability to appreciate and take different points of view. Robinson's book is an example in itself of the art of written rhetoric—persuasive, knowledgeable, benevolent and helpful. His second book is written by teachers he trained in planning their school curriculum. The Trivium-based curricula data evidences how this was achieved, each chapter written by a head or school leader. However, words do sometimes need redefining and 'grammar' and 'progressive' are two such. For example, Sam Gorse (Robinson, 2016, p. 42) demonstrates the word 'progressive' used as a term that can encompass both innovation and knowledge (grammar). This creates a balanced approach across all subjects and Robinson explains how this gives a better framework for teaching basic and rich knowledge in a fresh and innovative way. Teacher Anoara Mughal explains the dialogic aspect in her article here. See Blog Endnote.[23]

4.5 Culture and Language Acquisition: Its Influence on Learning

The underlying argument of evolutionary psychology is that brains and intelligence evolved learning how to solve problems in ancestral environments up to the present day. This meta-theory incorporates all aspects of psychology, since it involves studying evolutionary adaptations. David Geary (1957–) is a cognitive developmental and evolutionary psychologist who argues that social interaction is the main driver in developing brain structures. This process explains behaviours occurring in all cultures For example, the creation of words and communication behaviours in verbal and written languages, social group cultures, the ability to infer others' emotions, identify family members and co-operate with others (2005). In his iconic book *The Language Instinct*, Steven Pinker says that language is so intimately woven into the fabric of human life that it is hardly possible to imagine life without it. Language plays a major role in human thinking, but we can still think without it (1994).

Everybody experiences the difference between what they wrote and what they meant to say. Feelings and emotion models help us make meaning, stimulate and connect thoughts. We frequently express our thoughts intermittently through an internally voiced commentary. Although we tend to discount them, feelings, emotional models and behaviour are generally driven by unworded thoughts 'rational' to the person who experiences them (Hawkins, 2017). Lev Vygotsky, the much revered and criticised learning theorist, said that human thought is worded *and* silent. He explained that thought and language connect and relate, but are not necessarily the same thing. They are separate processes which have an internal mobile relationship to each other (Vygotsky in Rieber & Carton, 1987). For example, we talk about the gist of an idea as we look for the best way to express it in words and we invent new words for new concepts, ideas, tasks, discoveries and events. We are capable of 'reification,' giving names and expressions to abstract thoughts as we use words to create meaning.

The languages we use in internal and external dialogues facilitate learning as we work ideas out. Finding and expressing shades of meaning appropriately is paramount for thoughts to be created and communicated to an audience. For example, dejavue (France), zeitgeist (Germany), hygge (Denmark), sobremesa (Spain) and schadenfreude (Germany) are adopted by English speakers because they express meanings that have no single word in English. Many languages are 'agglutinative' where one word can contain smaller words (morphemes) added together to create a meaning that would be a phrase or sentence in other languages, for example, German. Mandarin Chinese has many characters (or symbols) that may be combined to create shades of meaning in different combinations. These create subtle meanings in nuanced poetry and prose. Human learning potential, knowledge, languages and achievements are incredibly complex.

As teachers we should remember there are other cultures than our own deserving of respect. Co-operative world social and cultural systems are rich and have been developed and lost over millennia, unintentionally, deliberately, haphazardly and incidentally. We are super social animals and changeable as we develop social and personal skills and behaviours. One person cannot embrace all knowledge. Different people are good at different things. Some of us learn to be experts in particular areas and our DNA adapts a little in the process. We use language to switch in and out of frames of reference in different situations. We have a variable mix of abilities and characteristics derived from natural propensities and experience. Language has been a major factor in enabling humans to develop knowledge and flexibility of thought. Languages and their founding cultures leverage cognitive abilities such as innovative ideas in science, poetry and literature.

There are racial languages, specific social group languages and specific subject languages where words are used in different contexts. The biological phenomena and enabler of all of these learning behaviours is neuroplasticity. This process, by enabling development has given rise to an enormous multilingual body of knowledge. Robert Dean Hobbs, American linguist and modern foreign language expert, has taught around the world. His book *The Global Child* (2016) tells us the ways in which immigrant children often suffer because their native language is disrespected and unsupported at school. A child's native language has emotional resonance for that child. Dean Hobbs' research also shows how an education that respects multiculturalism can benefit everyone even should they remain monolingual. His goal has been to determine factors towards constructing a multilingual model of education.

His research included a meta-analysis of psycholinguistic, neurolinguistic and sociolinguistic studies in various countries around the world. His book analyses language education experiences from Asia, Africa, Europe, the United States and South America. He used qualitative and quantitative research, engaging in online interviews to record the perceptions of multilingual teachers. His research was based on their personal perspectives, observations, knowledge and experience. He found children with two or more languages had higher academic outcomes in all subjects whatever their background.

Multilingual teachers and their students face many challenges in their work because of dominant language protocols and prejudice. He found that directors of successful businesses realised the value of paying attention to diversity by hiring multi-ethnic, multilingual employees and selling products and services aimed at a multicultural market (Shea, 2008). For example, targeting the mother tongues of consumers can help to increase sales. Research indicated that people often needed to be taught their own mother tongues for work because they had not learned them at school. Some Spanish speakers in the United States were losing the ability to speak Spanish because of a lack of support in schools and yet longitudinal studies of 42,000 children over 20 years demonstrated the relationship between academic success and first language support.

Acknowledgement of a child's first language is important because social and emotional interactions are important for learning development. Enjoyment and motivation, curiosity, encouragement, respect and empathy are important *drivers* and their opposites in the form of boredom, disinterest, disengagement and rejection are similarly *demotivating* for learning. World government policymakers need to know about the benefits of providing more foreign language learning opportunities for developing children's intelligence. Research shows some governments tend to overly mandate their own language in preference to encouraging children to be multilingual. They are prejudiced and don't appear to understand how the process of learning languages helps learning development. He advises education leaders and curriculum planners to note the benefits of a multilingual and a multicultural education.

Dean Hobbs believes all students as individuals, broader society, trade, business and diplomacy in general would benefit. Howard Gardner (1943–), a developmental psychologist and educational researcher at

Harvard, produced the 'Theory of Multiple Intelligences' in 1983. He says humans have developed a range of different intelligences with their own specific languages and terminology useful for learning about areas of knowledge. We use and develop them to research, analyse, plan and justify our work to others and we invent them, for example, computer terminology. They are interactive, social, emotional and physically active languages that encourage learning in different ways in different subjects. Gardener's multiple intelligences is an important theory, but there is overlap and 'creative intelligence' seems to be missing. These abilities facilitate our social life and work opportunities in complicated ways and contribute to future career flexibility, health and well-being. However, personally I would prefer the word 'competencies.' See Website and Article Endnotes.[24,25]

I mentioned them in my last book as follows: "Gardner identified eight distinct intelligences. *Naturalistic intelligence, which involves an affinity with, understanding and knowledge about nature. Linguistic intelligence involving language such as poems, political speeches and newspaper articles. Logical-mathematical intelligence, using the ability to calculate and solve abstract problems. Spatial intelligence, used to do mapping, navigating, engineering, architecture, design and sculpture. Musical intelligence, where people create and interpret different types of sound. Bodily- kinaesthetic intelligence, involving using one's own body to create products, dance, do representational art, understand health care and all kinds of sports. Interpersonal intelligence, in understanding other people's moods, desires, skills, motivations and intentions. Finally, intrapersonal intelligence involves a person's ability to recognise, assess and develop his own character and abilities*" (Hawkins, 2017).

Unofficial subject languages and communities are currently very important. In the United Kingdom science teacher Dan Roberts explains how he has missed subject counselling and support. Years ago he had continuing professional development (CPD) from other local school science leaders managed by his Local Education Authority (LEA). At the time his science department was the worst performing in the county. He learned from and he greatly valued the advice and support of experienced science teachers. All the local science teachers met including those from the new 'independent' academy schools (state funded). All schools had to send their teachers to training days for professional subject development by experts. They had a chance to share good practices and gain support. As schools with low exam scores became academies the old professional development system broke down and LEA services became fragmented. See Blog Article Endnote.[26]

However, teachers in Dan's area still feel the need for support. In 2018, twenty schools around the city of Plymouth created high school subject hubs to improve the quality of teaching and student outcomes. The schools showed commitment, determination and trust towards working together and putting 'the children in our city first.' Unfortunately the fragmentation of the education system and schools' individual 'priorities and challenges' led to the hubs being disbanded the following year. At the time the schools' priorities and challenges became increasingly driven under government pressure to improve league tables (school exam result lists). They were judged by parents (especially less-educated parents) following government advice to be the best measure of school excellence. Schools' viability and future funding are dependent on these exam scores and pupil numbers. The government believes competition between schools increases performance.

It makes sense, of course, to judge schools to some extent by the success of their exam results. However, when there is no provision for the emotional and developmental support of teachers' own learning, there are consequences. The results tend to deteriorate and may remain unachieved at their potential level. Teachers become demotivated and resign from the profession. Dan believes it 'incredibly important' for teachers and subject leaders to have the opportunity to develop their subject-specific pedagogy. He says in order to do this effectively it must be with other subject professionals. When he wrote this article in 2019 they had re-formed a hub with six local schools and the help of a 'teaching school' to create a Secondary InspirED Subject Community (SISC) for each subject area going forward. Each subject community will plan their own professional development sharing conferences focusing on improving the quality of teaching resulting from government-funded exam board training.

Senior subject leaders come together democratically to regularly review, monitor and evaluate each subject community's effectiveness. He hopes despite the challenges they face, subject communities will set strategic priorities and drive forward an improvement agenda. Dan is passionate in his belief that a teacher's subject knowledge is vital to teaching effectiveness and learning outcomes for young people. Professionals share good practice across many subject communities through national and international subject associations and grassroots organisations—for example, UK Twitter subject communities like @researchED1, @TeamEnglishFE, @BrewEd2017, @LitdriveUK, @Team_English1, @EnserMark (Geography), @imagineinquiry (Drama/Creative thinking), @RogersHistory and @teachmeeticons.

Alex Gordon (@MrAWGordon) recently gave a webinar presentation explaining the operation of 'Learning and Teaching Councils' in schools. See Webinar Endnote.[27] This kind of professional group engagement is useful because as Wenger says it is important for learning that people express who they are, in a forum where active participation takes place. This is brain plasticity working at optimum levels—a dynamic trajectory model of learning motivating professionals towards system improvement as collaborative learners. Wenger says opening up a dialogue between inside and outside perspectives (teachers from different schools) produces a more developed community with networks that help 'leverage dormant capabilities' (Wenger, McDermott, & Snyder, 2002). Wenger (1998) believes professionals gain agency and expertise by discussion, creating a balance between theory and practice. There are many scenarios and teacher solutions. It is a challenging job. Those teachers who are successful find much satisfaction in the profession. Teaching does not benefit from quick fixes; teachers need support and respect from government and society for their professionalism. See Article Endnote.[28]

4.6 Assessment Planning and Ideas About Intelligence

Intelligence (embodied in neuroplasticity) is not predictable and cannot be accurately measured. It would be of limited use if it was (Kincheloe et al., 1999). Neuroscience tells us intelligence is dynamic, natural but not always predictable because it is 'operational.' It can be accumulative, but may also be in abeyance, dormant and underdeveloped. It can fail to develop and also degenerate—through illness, disability or underuse. Some aspects of development are *similar* in individuals, but microbiological factors make each human brain unique. Everyone is at different stages in their development and has unique potential even if it is not achieved. Good teachers know passing tests and exams, though they tend towards it, do not *necessarily* lead to cognitive competence in adult life. For at least around the last seventy years there has been a simple understanding of intelligence based on behaviourist psychology.

On being given restricted information humans can be manipulated. Learners may only need to absorb limited sets of information to succeed in some areas of life, but this makes them less resilient to change. Remembering and giving back facts exactly as presented is not *necessarily* a sign of intelligence. It is often forgotten exams have always given and requested fresh information to demonstrate skill set abilities. Unexpected exam questions are asked around a taught set of skills and information to show if the student can improvise to produce an answer. In this case answers cannot be marked exactly the same by different teachers. Examiners require a high level of professional subject knowledge and wisdom in their various subjects at all levels. A high standard can be set through regular teacher subject moderation and research across cohorts, checking and comparing results across a range of work. Groups of teacher examiners can consult and result objections be considered.

More oral/aural/practical skills exams could be developed to promote complex higher-order practical intelligences, for example, music, poetry, creative writing, culinary arts, drama, sport, scientific invention and engineering. People with various academic inabilities, 'ordinary' children and also gifted children would benefit. It would cost more short term than marking multiple-choice questions by algorithms, but it would be valuable long term. Such a system would tend towards higher quality cognitive activity in students and greater professional integrity in teachers. The extreme opposite, the exclusively normative framework, tends towards 'dumbing down' because potential is hard to predict and high achievement harder to encourage, perceive and credit. At its best teaching for learning is a complex transactional, transformative human process—one we need in order to survive and move forward in our dynamic world.

Neuroscientists are in the early stages of explaining how this works, but good teachers have been proving it on the ground for years. Matthew Cobb's descriptions in his recent book on the history of human perceptions about the brain support this view (2020). We know that human intelligence is greatly influenced by individual genetics, previous experiences, needs, opportunities, contexts, cultures and the language of different influential adults, peers, cohorts and groups of pupils. It seems Finland has hit upon one method of philosophising and acknowledging the problem of 'intelligence' and learning in a suitably humane way. They appear to have worked out how to enable the process of brain plasticity throughout their education system. That they are continuing to work on it is evidence—an implicit and practical acknowledgement of the process itself!

The Finnish Director General of Education admits they don't always achieve their goals and are constantly facing new challenges and changes in circumstances. However, they have devised a system that proclaims an intention to be adaptable and benevolent in supporting citizens. Meanwhile in the United Kingdom, in 2017 the Ofsted published a research report that found secondary schools were shortening key stage 3, the level at which a competence in literacy is generally attained. It appears pupils developing slowly for a variety of genetic and environmental reasons may be disadvantaged. The report does not offer any possible connections to the reality of lifelong brain plasticity. There is no apparent acknowledgement of the educational sector of Further Education either. This existed for years beneath degree level university study through apprenticeships, day release and adult evening classes. See Article Endnote.[29]

Over the years, Further Education has been greatly reduced and although now starting to be recognised again is only slowly recovering. The report found some schools deterring low attaining pupils from doing vocational courses, because certain exams such as EBacc (English Baccalaureate) subjects at GCSE (General Certificate of Secondary Education) score more highly in league tables. This view is supported by a current national website advising secondary pupils and their parents about requirements for university entry. There is nothing wrong with the information given there, but it gives the impression that this is the only 'successful' prospect. Researchers also found that some primary schools were putting too much emphasis on practising for SATs (Standardised Attainment Tests) by 'teaching to the test.' When I taught teenage special needs students in the 1980s and 1990s they were discouraged from entering written exams. Lower results, although an achievement for them, might reflect badly on the school. The evidence shows this is still happening. See Website Endnote.[30]

In 2018 Helen Ward, writing in the *Times Educational Supplement* reported that following this report it was the intention of the Ofsted to 'downgrade' exam results *as a measure of school quality*. Report details were leaked saying that exam results and 'outcomes' would be replaced by 'quality of education' measures. It said that low attaining pupils were not receiving enough consideration. It seemed likely that 'exam factory' schools that narrow the curriculum would be marked down. This binary thinking and currently popular 'outrage' response assumed by someone (unreferenced) is currently a problem in our society. It is wrong because it is not an either-or problem, but a problem that requires insight and continuous inferential thinking. See Endnote.[31] What is actually needed is that these tests should be evaluated by teachers in a mixed-method manner within a system that feeds their findings upwards so that the system becomes self-correcting. A great deal of practical research is needed on the ground by teachers and teacher academics working together in order to find ways this may be done. See Article Endnote.[32]

Teachers have developed a multiplicity of ways to measure intelligence. Tests and exams include a wide range of subject languages and structural information, knowledge frameworks and skills. These subject disciplines have been developed in different ways by groups, fields, faculties, departments and schools of thought. Neuroscience can now prove to us that they certainly involve visual, auditory and kinaesthetic modes and ways of learning. This is the interactive, multisensory learning involved in proprioception. See Chap. 3. Around the time of the 2017 Ofsted research report mentioned above, the new chief inspector, Amanda Spielman, chose to review the curriculum as her first priority. It is likely she looked at the report just published by the inspectorate. She said, “School leaders need to recognise how easy it is to focus on the performance of the school and lose sight of the pupil.” Spielman also admitted the United Kingdom’s inspectorate was responsible, when she added, “I acknowledge that inspection may well have helped to tip this balance in the past.” See Article Endnote.[33]

This explains how and why the inspection system in the United Kingdom has failed to achieve improvements because the political arm of government has ‘effectively’ used it to disable and disenable teachers in the education system. This has caused a prolonged period of tension and disconnection between parents and teachers in the field, education experts and management bodies. This problem appears to be based on an inability to respect the professionalism of teachers, to listen to them and to conceptualise a system with a central goal acceptable to everyone. The government has not trusted the teachers and has not developed and appreciated them as a resource. They have also failed to hold teachers to account effectively by asking them to fully research, self-regulate and explain their work. Even more importantly they have not been prepared to listen and support improvement strategies advised from below.

There is no point in requesting teachers to seriously listen to parents and young people if they are not supported with the resources to do so. Teachers need to be part of a formal consultation process where concerns are reported directly upwards about learners, parents and carers concerns. Pertinent points need to be evidenced in the decision-making at the top. At the time of writing there is still a great deal of complaint and criticism of the United Kingdom’s assessment and examination systems from head teachers, teachers and psychologists who care about and understand the situation in the classroom. Psychologist and Early Years expert of 30 years’ experience Pam Jarvis has written extensively both academically and in the educational press about the detrimental effects of the proposed new baseline assessment of children in the Early Years. Jarvis is worried about the highly top-down prescriptive effect of the new 'Early Adopter' Early Years Foundation Stage (EYFS) curriculum for England (2020a, 2020b). See Article Endnotes.[34,35]

Children in the United Kingdom enter the formal school system as young as four years. Instead of training teachers to assess, the government favours inappropriate formal external testing over teacher assessment. Government managers promote this approach without seeming to understand the particularly intimate, individual nature of young children or that work needs to be researched, developed and adapted by teachers to different learners on the ground. The business spreadsheet approach to efficiency has undermined the whole system. For example, in the 'pandemic' summer of 2020 the minister for education, Gavin Williamson, stated that national exam results could be estimated by teachers and would then be automatically lowered by algorithm across the nation. He did not understand this would have a detrimental effect on learners and teachers morale. This is an example of the effect of how ignorance at the top affects individuals on the ground.

Even before the Ofsted inspectorate's formation in 1992, political parties were not able to agree upon the education system's central focus. In the United Kingdom there is no long-term vision that transcends political party views. Here are a few of the possible dangers and reasons why a 'marketisation' approach to education may appear useful, but is actually not. At each point, it is liable to reduce innovation, self-reliance and productivity. It also causes resentment and confusion amongst the participants and the population.

- It is a convenient way to control people and condition them to conform to government agendas and targets.
- It may make marking and assessment less time consuming, easier and cheaper, particularly when done automatically by algorithm.
- It may not require and therefore reduce the range of quality resources available to train and examine teachers.
- It could simplify the demands of the professional's and teacher's roles.
- Dumbing down education might make groups of people easier to ignore and/or manage.

All of the above are long-term self-destructive behaviours for happy, productive, sustainable human societies. Here are two professional examples below from education and psychology. As referenced earlier, headmaster Jeremy Hannay, a primary (elementary) school head, has passionate feelings about "The Apocalypse of English Education." He explains in detail how he feels the government has failed to understand the nature of the educational process. He still hopes it may be rejuvenated and repurposed. He had spent 12 years refusing to follow government directives in his school. He was not even impressed when his school recently passed its inspection with an 'outstanding' designation! See Article Endnote.[36]

Turning to high school education, a recent article considers "Neuroscience, adolescence and GCSEs—a misalignment? GCSEs are a relic from a time before our understanding of adolescent development and mental health." Professor Sarah-Jayne Blakemore, a psychologist, and Professor Tamsin Ford, a psychiatrist, have explained why they have come to this conclusion. They tell us that according to their research into adolescent anxiety, depression, eating and substance use disorders, these first appear before the age of 18 years and yet few young people can access and receive specialist mental health support at this critical time. There is evidence that mental health problems are increasing amongst young people in the United Kingdom.

Young people are telling researchers that exam stress and fear of academic failure are their biggest worries. This doubled between 2015/2016 and 2018/2019 according to a survey of calls to a children's phone call helpline. Exam stress and fear of failure has become the eighth most common reason for calls made, but it can obviously also exacerbate other problems and reduce resiliency to them. Two-thirds of teachers responding to a National Union of Teachers survey agreed that pupils became overly stressed or anxious before these public exams. See Article Endnote.[37]

4.7 Conclusion

We have plenty of historical narrative and physical environmental evidence of our ability to create nurturing communities and cultures that work. Unfortunately, human societies also create adverse social conditions that produce unpleasant, cruel, unhappy, deprived and unproductive social environments and national systems. Especially those governed and created by those who have little empathy for their fellow citizens. It is time to work collaboratively around the world to create local and national environments and curricula systems that work for everybody's long-term benefit. Understanding our own brain plasticity in many of the complex aspects of our lives might help us to do this. Experts and teachers in classrooms are calling for a review of the system.

While I was writing this chapter I contacted Neil Hawkes, a former head teacher from Oxford in the United Kingdom. He is the founder of the International Values Education Trust (IVET). He has a following of likeminded educationalists around the world. He told me he would like me *"to emphasise that for humanity to flourish, e.g. achieve the United Nations' sustainable goals, it needs to raise its level of consciousness. To do this education must focus on the development of ethical intelligence, which gives folk the capacity to ethically self-regulate their behaviour. This is naturally achieved in values-based schools."* See Website Articles Endnote.[38,39]

There are other reasons for the ways in which some education systems are failing children and older students, but the ways in which exam systems have sometimes operated have definitely been a part of the problem. If this is not addressed, it seems likely that ultimately a country's intellectual ability will tend to suffer instead of achieving its full potential. Feelings and emotions about learning would be better expressed, observed and allowed for within a 'multilayered dialogic system.' Many problems could be resolved in and by schools and teachers themselves. This is suggested by a grass roots movement of teachers in various countries around the world called 'Flip the System.' This was started by Jelmer Evers and Rene Kneyber in the Netherlands who wrote a book (Evers & Kneyber, 2015); teachers in the United Kingdom got together and followed on by contributing their own book (Rycroft-Smith & Dutaut, 2018).

The spontaneous appearance of this movement evidences the need for change. It shows that extreme exam conformity and fixed approaches are problematic. We need to acknowledge that it is possible to adopt a universally beneficial principle that improves our national education systems with an inclusive culture while appreciating and learning about cultural systems in other countries. Education should exist within a dynamic and flexible system reliant upon autonomous professionals within it reporting upwards to senior leaders through to political management. They should be asked to explain how different educational systems of organisation fit in, deliver and contribute to the fulfilment of a national vision. The problem lies in getting the political establishment to appreciate this practical solution and to listen wisely.

Notes

1. The Nolan Principles of Governance

   It is fair to say that good governance is crucial to all organisations whether they are in the public, private or third sector. It is therefore useful to look at already established and well-used guides to governance. The Nolan Principles were set up in 1994 as a guide for boards of public sectors organisations.

   - Openness: Board members must be transparent in all actions and they must declare and record all relevant personal and business interests and must be able to explain actions.
   - Honesty: Board members must ensure that they always act in the best interests of the organisation and that all activities are transparent and accountable.
   - Objectivity: Board members must consider all matters on their merits and base decisions on the information and advice available and reach their decision independently.
   - Integrity: Board members must actively support and promote their values and must not be influenced by personal interests in exercising their roles and responsibilities.
   - Accountability: Board members must take responsibility for and be able to explain their actions, and demonstrate that their contribution to governance is effective.
   - Leadership: Board members must uphold the principles and commitment to delivering good outcomes for tenants and other service users and lead the organisation by example.

   https://www.governanceprinciples.scot/nolan-principles Accessed 02/06/2020.
2. Olli-Pekka Heinonen, Director General of the Finnish Education Agency, "Lessons from Finland: The Finnish education system is among the best in the world, regularly appearing at or near the top of PISA and other surveys. How do they do it and what can we learn from them?" 10 March 2020 Edinburgh: The Scottish Parliament https://scottishparliament.podbean.com/e/lessons-from-finland-olli-pekka-heinonen-director-general-of-the-finnish-education-agency/ Accessed 16/06/2020.
3. OECD (2019), Education at a Glance 2019: OECD Indicators, OECD Publishing, Paris, https://doi.org/10.1787/f8d7880d-en. Accessed 16/12/2020.
4. Nikki Booth 'What is formative assessment, why hasn't it worked in schools, and how can we make it better in the classroom?' September 2017

https://impact.chartered.college/article/booth-what-formative-assessment-make-better-classroom/ Accessed 06/02/2020.
5. About us – Ofsted – GOV.UK www.gov.uk/government/organisations/ofsted/about Accessed 16/12/2020.
6. Christopher Bagley "Visioning a new education system: reduce exam tyranny and empower teachers and pupils" 4/4/2019 http://www.psychchange.org/blog/visioning-a-new-education-system-reduce-exam-tyranny-and-empower-teachers-and-pupils Accessed 10/08/2019.
7. Jeremy Hannay "The Apocalypse of English Education–The End & The Beginning" 04/06/2020 https://noordinaryclassroom.home.blog/2020/04/06/the-apocalypse-of-english-education-the-end-the-beginning/comment-page-1/ Accessed 19/08/2020.
8. Claudia Civinini 'Pisa education rankings: How did the UK perform? How do 15-year-olds across the UK compare in maths, science and reading skills to their peers across the world?' 3rd December 2019 Programme for International Student Assessment (Pisa): Organisation for Economic Co-operation and Development (OECD) https://www.tes.com/news/pisa-education-rankings-how-did-uk-perform Accessed 18/12/2020.
9. Noam Chomsky 'The Purpose of Education' 19 September 2020 Everyone Academy on Facebook https://fb.watch/2nl7erMy__/ Accessed 22/12/2020.
10. Emily Robinson 'Progressivism' RSA 25 May 2018 https://medium.com/rsa-journal/progressivism-6c1df1281da7 Accessed 03/01/2019.
11. Dr James Shea, Dr Gareth Bates, Dr Steve Connolly (Editors) 'New Wave Progressive Teachers' Posted on 27 October 2020 Peer Reviewed Education Blog: University of Bedfordshire https://peerreviewededucationblog.com/ Accessed 31/10/2020.
12. Tim Oates 'The Progressive/Traditional Debate-'Skills versus Knowledge: A curriculum debate that matters-and one which we need to reject' September 2018 https://impact.chartered.college/article/skills-versus-knowledge-curriculum-debate-matters-one-need-reject/ Accessed 26/09/2020.
13. Steiner Waldorf Schools Fellowship What is Steiner education? https://www.steinerwaldorf.org/steiner-education/what-is-steiner-education/ Accessed 29/12/2020.
14. Chris Chivers 'Differentiation As Informed Dialogue?' 4/12/2015 https://chrischiversthinks.weebly.com/blog-thinking-aloud/differentiation-as-informed-dialogue Accessed 24/01/2019.
15. Tom Sherrington 'Rescuing Differentiation from the Checklist of Bad Practice.' 24 January 2019 https://teacherhead.com/2019/01/24/rescuing-differentiation-from-the-checklist-of-bad-practice/ Accessed 09/11/2019.

16. Simon Smith 'Lip Service (All surface, No depth)' 13 January 2020 https://smithsmm.wordpress.com/2020/01/13/lip-service-all-surface-no-depth/ Accessed 14/01/2020.
17. Michael Merrick 'Social mobility and the threat to knowledge' 16 March 2018 The Times Educational Supplement: London https://www.tes.com/news/school-news/breaking-views/social-mobility-and-threat-knowledge Accessed 5/8/2019.
18. Kirsi Pyhältö, Janne Pietarinen and Tiina Soini (2018) Dynamic and shared sense-making in large-scale curriculum reform in school districts, The Curriculum Journal, 29:2, 181–200, DOI: https://doi.org/10.1080/09585176.2018.1447306 https://www.tandfonline.com/doi/abs/10.1080/09585176.2018.1447306 Accessed 12/06/2019.
19. Janne Pietarinen, Kirsi Pyhältö and Tiina Soini (2019) Shared Sense-making in Curriculum Reform: Orchestrating the Local Curriculum Work, Scandinavian Journal of Educational Research, 63:4, 491–505, DOI: https://doi.org/10.1080/00313831.2017.1402367 https://www.tandfonline.com/doi/abs/10.1080/00313831.2017.1402367?src=recsys&journalCode=csje20& Accessed 11/11/2020.
20. Olli-Pekka Heinonen, Director General of the Finnish Education Agency, "Lessons from Finland: The Finnish education system is among the best in the world, regularly appearing at or near the top of PISA and other surveys. How do they do it and what can we learn from them?" 10 March 2020 Edinburgh: The Scottish Parliament https://scottishparliament.podbean.com/e/lessons-from-finland-olli-pekka-heinonen-director-general-of-the-finnish-education-agency/ Accessed 16/06/2020.
21. Roger Titcombe 'Schools should be genuinely child-centred. Part-2 Why-this-is-always-threatened-by-marketisation' https://rogertitcombelearningmatters.wordpress.com/2019/07/21/schools-should-be-genuinely-child-centred-part-2-why-this-is-always-threatened-by-marketisation/ Accessed 13/09/2019.
22. Dr Carol Dweck 'Teaching a Growth Mindset' 7 July 2020 USA: Ecubed, Stanford University https://youtu.be/Aywci2TNiJ8 Accessed 10/08/2020.
23. Anoara Mughal 'Let's Talk About Dialogic Talking' 2018https://www.inspiremetacognition.com/post/let-s-talk-about-dialogic-talking Accessed 18/01/2021.
24. Howard E. Gardner (1983) The Official Authoritative Site of Multiple Intelligences USA: Harvard Graduate School of Education https://www.multipleintelligencesoasis.org/ Accessed 17/01/2021.
25. Rum Tan 21 October 2015 https://smiletutor.sg/how-to-harness-develop-your-natural-strengths-multiple-intelligences-theory/ Accessed 24/11/2020.

26. Dan Roberts 'The Importance of Subject Communities' 3 September 2019 http://www.danjjroberts.co.uk/2019/09/03/the-importance-of-subject-communities/ Accessed 16/10/2019.
27. Alex Gordon 'Using a Teaching and Learning Council and Pedagogical principles for Teaching Remotely' 9 January 2021 https://www.bigmarker.com/teachmeeticons/TMHistoryIcons-Remote-learning-in-History?bmid=0c49e32c0b3f Accessed 09/01/2021.
28. Cindy Lamothe How 'Intellectual Humility Can Make You a Better Person' 3 February 2017 https://www.thecut.com/2017/02/how-intellectual-humility-can-make-you-a-better-person.html Accessed 27/11/2018.
29. Sian Griffiths, Education Editor "School inspector threatens D-minus for 'exam factories.'" 12 August 2018 The Sunday Times: London https://www.thetimes.co.uk/article/schools-inspector-threatens-d-minus-for-exam-factories-rvp3jrnbd?%2D%2Dxx-eta=denied_for_visit=0&visit_number=0&visit_remaining=0&visit_used=0 Accessed 20/10/2018.
30. The Uni-Guide 'What is the English Baccalaureate?' The Student Room, UK https://www.theuniguide.co.uk/advice/gcse-choices-university/whats-the-english-baccalaureate Accessed 21/01/2021.
31. INFERENCE, inference—definitions, derivations and similar meanings

    Inference—noun

    1 An act of inferring, especially of reaching a conclusion from facts, observation and careful thought
    2. That which is inferred, especially a conclusion

    inferential adjective—of or based on inference
    Words with similar meanings:
    deduction, conclusion, assumption, presumption

    1. Surmise, conjecture, extrapolation, construction, interpretation, reading.
    2. Corollary, consequence.

    infer—verb (inferred, inferring)

    1. To conclude or judge from facts, observation and deduction
    2. (colloquial) to imply or suggest. (from Latin inferre, to bring in)

    inferable or inferrable adjectives. Capable of being inferred or deduced
    Words with similar meanings:

deduce, conclude, derive, extrapolate, assume, presume, surmise, gather, understand

Reference—O'Neill, M. (editor et al.) (2001) Chambers Concise Dictionary and Thesaurus Edinburgh, UK: Chambers Harrap Publishers.

Inference—noun

1. Conclusion: conclusion drawn from evidence of reasoning
2. Logical Reasoning Process: the process of reasoning from a premise to a conclusion
3. Implication: something that is implied

Derivation: (late sixteenth century. From medieval Latin 'inferentia', from Latin 'inferre' [see infer])

inferential adjective, inferentially adverb

infer— verb (inferred, inferring, infers)

1. To conclude something on the basis of evidence or reasoning.
2. To suggest or lead to something as a conclusion
3. To imply or suggest something
4. To make a reasonable guess at something

Derivation: (early sixteenth C. From Latin 'inferre', literally 'to bring in' from ferre 'to carry')

inferable adjective; inferably adverb; inferer noun

Reference—Encarta World English Dictionary (1999) New York: St. Martin's Press, Bloomsbury

32. Jon Severs 'Inference: why comprehension is not just about vocabulary and knowledge. Professor Jane Oakhill explains how to teach comprehension skills, such as inference, in schools' 08 May 2019.

    https://www.tes.com/news/inference-why-comprehension-not-just-about-vocabulary-and-knowledge Accessed 08/08/2019.
33. Helen Ward 'Ofsted 'to mark down exam-factory schools': New inspection framework could introduce new 'quality of education' measure' 12 August 2018 https://www.tes.com/news/ofsted-mark-down-exam-factory-schools Accessed 15/08/2018.
34. Pam Jarvis 'Education for Greater Equality' 27 July 2020 https://www.edgethinking.org/post/the-narrative-of-curriculum Accessed10/01/2021.
35. Pam Jarvis 'Every Child Really Does Matter: in search of developmental recognition' 15 July 2020 https://histpsych.blogspot.com/2020/07/every-child-really-does-matter-in.html Accessed 15/08/2020.

36. Jeremy Hannay "The Apocalypse of English Education–The End & The Beginning" 04/06/2020 https://noordinaryclassroom.home.blog/2020/04/06/the-apocalypse-of-english-education-the-end-the-beginning/comment-page-1/ Accessed 19/08/2020.
37. Professor Sarah-Jayne Blakemore, Department of Psychology, University of Cambridge & Professor Tamsin Ford, Department of Psychiatry, University of Cambridge 'Neuroscience, adolescence and GCSEs – a misalignment? GCSEs are a relic from a time before our understanding of adolescent development and mental health' 28/11/2020 https://bigeducation.org/rethinking-blogs/neuroscience-adolescence-and-gcses-a-misalignment/ Accessed 01/12/2020.
38. United Nations 'The 17 Sustainable Development Goals' https://www.un.org/sustainabledevelopment/ Accessed 21/01/2021.
39. Dr Neil Hawke International Values Education Trust: Positive Values https://ivetfoundation.com/positive-values/ Accessed 21/01/2021.

References

Abbott, J. (2009). *Over Schooled but Under Educated: How the Crisis in Education Is Jeopardizing Our Adolescents.* Continuum.

Bagley, C. (2019, April 4). *Visioning a New Education System: Reduce Exam Tyranny and Empower Teachers and Pupils.* http://www.psychchange.org/blog/visioning-a-new-education-system-reduce-exam-tyranny-and-empower-teachers-and-pupils. Accessed 10 Aug 2019.

Barshay, J. (2019). *Columnist for The Hechinger Report 'Why Content Knowledge is Crucial to Effective Critical Thinking'.* https://www.kqed.org/mindshift/54470/why-content-knowledge-is-crucial-to-effective-critical-thinking. Accessed 21 Mar 202.

Bhopal, K. (2018). *White Privilege: The Myth of a Post-Racial Society.* Policy Press.

Black, P., & Wiliam, D. (1998). *Inside the Black Box: Raising Standards Through Classroom Assessment.* King's College School of Education.

Blakemore, S.-J., Department of Psychology, University of Cambridge; Professor Tamsin Ford, Department of Psychiatry, University of Cambridge. (2020, November 28). *Neuroscience, Adolescence and GCSEs – A Misalignment? GCSEs Are a Relic from a Time Before Our Understanding of Adolescent Development and Mental Health.* https://bigeducation.org/rethinking-blogs/neuroscience-adolescence-and-gcses-a-misalignment/. Accessed 01 Dec 2020.

Bold Beginnings Report. (2017). *Bold Beginnings: The Reception Curriculum in a Sample of Good and Outstanding Primary Schools.* Ofsted. https://assets.publishing.service.gov.uk/government/uploads/system/uploads/attachment_data/file/663560/28933_Ofsted_-_Early_Years_Curriculum_Report_-_Accessible.pdf. Accessed 15 Feb 2021.

Booth, N. (2017, September). *What Is Formative Assessment, Why Hasn't It Worked in Schools, and How Can We Make It Better in the Classroom?* https://impact.chartered.college/article/booth-what-formative-assessment-make-better-classroom/. Accessed 06 Feb 2020.

Brighouse, T., & Waters, M. (2021). *Our Passion for Schooling*. Independent Thinking Press.

Burns, N. (2020, December 8). *5 Common Mistakes When Teaching Pupils How to Learn Dialogic Answers This. Metacognition Is the New Trendy Thing in Schools – But It's Not as Simple as It First Appears*. 3:00 pm. https://www.tes.com/news/metacognition-5-common-mistakes-when-teaching-pupils-how-learn

Cambridge Primary Review Trust & York University. (2020). *Dialogic Teaching: Improving Children's Learning by Improving the Quality of Classroom Talk: The Programme Uses Video Review, Print Materials and In-School Mentoring to Support Teachers' Practice Across English, Maths and Science Lessons* (EEF Summary, Key Stage 2, 4985 Pupils in 76 Schools). Independent Evaluator Sheffield Hallam University Education Endowment Foundation. https://educationendowmentfoundation.org.uk/projects-and-evaluation/projects/dialogic-teaching/. Accessed 21 Mar 2021.

Carey, A. (2021). *Disadvantaged School Ditches the Curriculum and Gives Kids Free Rein*. https://www.theage.com.au/national/victoria/disadvantaged-school-ditches-the-curriculum-and-gives-kids-free-rein-20210213-p5726w.html. Accessed 16 Mar 2021.

CHARTERED COLLEGE BLOG-Metacognition Useful-books, Resources & Teaching-tips by Anoara Mughal. https://chartered.college/metacognition-useful-books-resources-teaching-tips @CharteredColl

Cherry, L. (2021). *Conversations that Make a Difference for Children and Young People: Relationship-Focused Practice from the Frontline*. Routledge.

Chitpin, S., & Karoui, O. (2021). Culturally Responsive Pedagogy: A Canadian Perspective. *johepal, 2*(1), 45–62. http://johepal.com/article-1-92-en.html. Accessed 14 Apr 2021.

Chivers, C. (2015, December 4). *Differentiation as Informed Dialogue?* https://chrischiversthinks.weebly.com/blog-thinking-aloud/differentiation-as-informed-dialogue. Accessed 24 Jan 2019.

Chomsky, N. (2020, September 19). *The Purpose of Education*. Everyone Academy on Facebook. https://fb.watch/2nl7erMy__/. Accessed 22 Dec 2020.

Chomsky, N., & Polychroniou, C. J. (2017). *Optimism over Despair*. Random House Penguin.

CHRIS CHIVERSTHINKS. https://chrischiversthinks.weebly.com/blog @ChrisChivers2

Civinini, C. (2019). Pisa Education Rankings: How Did the UK Perform? How Do 15-Year-Olds Across the UK Compare in Maths, Science and Reading Skills to Their Peers Across the World? 3rd December 2019 Programme for International Student Assessment (Pisa): Organisation for Economic

Co-operation and Development (OECD). https://www.tes.com/news/pisa-education-rankings-how-did-uk-perform. Accessed 18 Dec 2020.

Cobb, M. (2020) *The Idea of the Brain: A History*. Profile Books Ltd.

Counsell, C. (2018a). *In Search of Senior Curriculum Leadership: Introduction – A Dangerous Absence*. https://thedignityofthethingblog.wordpress.com/2018/03/27/in-search-of-senior-curriculum-leadership-introduction-a-dangerous-absence/. Accessed 19 Feb 2021.

Counsell, C. (2018b). *Senior Curriculum Leadership 1: The Indirect Manifestation of Knowledge: (A) Curriculum as Narrative*. https://thedignityofthething-blog.wordpress.com/2018/04/07/senior-curriculum-leadership-1-the-indirect-manifestation-of-knowledge-a-curriculum-as-narrative/. Accessed 19 Feb 2021.

Counsell, C. (2018c). *Senior Curriculum Leadership 1: The Indirect Manifestation of Knowledge: (B) Final Performance as Deceiver and Guide*. https://thedignityofthethingblog.wordpress.com/2018/04/12/senior-curriculum-leadership-1-the-indirect-manifestation-of-knowledge-b-final-performance-as-deceiver-and-guide/. Accessed 19 Feb 2021.

Crehan, L. (2016). *Clever Lands: The Secrets Behind the Success of the World's Education Superpowers*. Unbound.

CURRICULUM IMPACT Mary Myatt. https://www.marymyatt.com/blog @MaryMya

DAN J ROBERTS. http://www.danjjroberts.co.uk/category/international-schools-3/ @Danjjroberts.

Dean Hobbs, R. (2016). *The Global Child: How Experts Would Change Education. Research-Based Acquisition of Languages*. Lap Lambert Academic Publishing.

Deputy Grocott. https://deputygrocott.blogspot.com/

Didau, D. (2015). *What If Everything You Knew About Education Was Wrong?* Crown House Publishing Ltd.

Didau, D. (2017, April 24). *Whatferentiation Is?* @DavidDidau. https://learningspy.co.uk/research/teachers-think-differentiation/#comments. Accessed 13 May 2018.

Douglas, E. (2018, July 18). *What Role Does Executive Function Play in the Classroom? What Is Executive Function and How Does It Affect Learning?* https://www.tes.com/news/what-role-does-executive-function-play-classroom. Accessed 07 Jan 2020.

Duckett, I. (2020). *Build Back Different: A Bigger, Broader and Braver Curriculum: What Kind of School Curriculum Can Pave an Alternative Road Out of This Pandemic*. Socialist Educational Association. https://socialisteducationalassociation.org/2020/08/25/build-back-different-bigger-broader-braver-curriculum-ian-duckett/amp/?__twitter_impression=true. Accessed 25 Aug 2020.

Dweck, C. (2017). *Mindset – Updated Edition: Changing the Way You Think to Fulfil Your Potential*. Robinson.

Dweck, C. (2020, July 7). *Teaching a Growth Mindset*. Ecubed, Stanford University. https://youtu.be/Aywci2TNiJ8. Accessed 10 Aug 2020.

EDUCHATTER- https://educhatter.wordpress.com

Evers, J., & Kneyber, R. (2015). *Flip the System: Changing Education from the Ground Up*. Routledge.

Fact Sheet: Early Years. (2017). www.suttontrust.com. Accessed 09 Oct 2018.

Fellmayer, J. (2019, April 30). *Functional Grammar and Writing Workshop: A Hybrid Approach to Teaching Literacy to English Language Learners*. English Language Development in an International School (ELD) @learn_thru_it New-Blog. https://learnthroughit.org/2021/03/16/right-now/ https://learnthroughit.home.blog/2019/04/30/functional-grammar-and-writing-workshop-a-hybrid-approach-to-teaching-literacy-to-english-language-learners/. Accessed 08 Apr 2020.

Foster, K. (2020, May 29). *Imagining the Future of Higher Education: What Comes Next?* https://youtu.be/5XgJ9Y0ryIc. Accessed 10 Aug 2020.

Gardner, H. (1993). *Frames of Mind: The Theory of Multiple Intelligences*. Fontana.

Gardner, H. E. (1983). *The Official Authoritative Site of Multiple Intelligences*. Harvard Graduate School of Education. https://www.multipleintelligencesoasis.org/. Accessed 17 Jan 2021.

Geary, D. C. (2005). *The Origin of Mind: Evolution of Brain, Cognition, and General Intelligence*. American Psychological Association.

Gilbert, I. (Ed.). (2018). *The Working Class: Poverty, Education and Alternative Voices*. Independent Thinking Press, Crown House Publishing.

Gordon, A. (2021, January 9). *Using a Teaching and Learning Council and Pedagogical Principles for Teaching Remotely*. https://www.bigmarker.com/teachmeeticons/TMHistoryIcons-Remote-learning-in-History?bmid=0c49e32c0b3f. Accessed 09 Jan 2021.

GRAHAM BROWN-MARTIN- https://www.grahambrownmartin.com/blog @grahamBM.

Griffiths, S., Education Editor. (2018, August 12). School Inspector Threatens D-Minus for 'Exam Factories'. *The Sunday Times: London*. https://www.thetimes.co.uk/article/schools-inspector-threatens-d-minus-for-exam-factories-rvp3jrnbd?%2D%2Dxx-eta=denied_for_visit=0&visit_number=0&visit_remaining=0&visit_used=0. Accessed 20 Oct 2018.

Gropius, W. (1955). *The New Architecture and the Bauhaus*. Faber and Faber Limited.

Hall, E. (2018). *Aristotle's Way: How Ancient Wisdom Can Change Your Life*. Bodley Head.

Hannay, J. (2020, April 4). *The Apocalypse of English Education–The End & The Beginning*. https://noordinaryclassroom.home.blog/2020/04/06/the-apocalypse-of-english-education-the-end-the-beginning/comment-page-1/. Accessed 19 Aug 2020.

Hattie, J., & Yates, G. (2014). *Visible Learning and the Science of How We Learn*. Routledge.

Hawkes, N., & Hawkes, J. (2018). *The Inner Curriculum: How to Nourish Wellbeing, Resilience and Self-leadership*. John Catt Educational.

Hawkins, J. A. (2017). *Feelings and Emotion-Based Learning; A New Theory*. Palgrave.

Heinonen, O.-P. (2020, March 10). Director General of the Finnish Education Agency, *Lessons from Finland: The Finnish Education System Is Among the Best in the World, Regularly Appearing At or Near the Top of PISA How Do They Do It and What Can We Learn from Them?* The Scottish Parliament. https://scottishparliament.podbean.com/e/lessons-from-finland-olli-pekka-heinonen-director-general-of-the-finnish-education-agency/. Accessed 16 June 2020.

Olli-Pekka Heinonen, Director General of the Finnish Education Agency. (2020, March 10). *Lessons from Finland: The Finnish Education System Is Among the Best in the World, Regularly Appearing At or Near the Top of PISA and Other Surveys. How Do They Do It and What Can We Learn from Them?* The Scottish Parliament. https://scottishparliament.podbean.com/e/lessons-from-finland-olli-pekka-heinonen-director-general-of-the-finnish-education-agency/. Accessed 16 June 2020.

Helen Ward. (2017). *Sats for Seven-Year-Olds to Be Scrapped from 2023: But a New Reception-Year Baseline Assessment Will Be Introduced from 2020*. https://www.tes.com/news/sats-seven-year-olds-be-scrapped-2023. Accessed 20 Oct 2019.

HYPSYCH Pam Jarvis. https://histpsych.blogspot.com/@Dr_Pam_Jarvis.

Immordino-Yang, M. H., & Knecht, D. R. (2020, May). Building Meaning Builds Teens' Brains Connecting Adolescents' Concrete Work to Big Ideas May Help Shape Their Neural Networks over Time. *Educational Leadership Journal, 77*(8) Science and Technology Learning and the Brain Pages The Neurosequential Model in Active Learning (involving regulation, relating and reasoning) 36–43 Association for Supervision and Curriculum Development (ASCD)- Comprising 113, 573 members – superintendents, principals, teachers, and advocates from more than 129 countries – the ASCD community also includes 71 affiliate organizations. http://www.ascd.org/publications/educational-leadership/may20/vol77/num08/Building-Meaning-Builds-Teens'-Brains.aspx. Accessed 23 June 2020.

Immordino-Yang, M. H., & Knecht, D. R. (2020, May). Building Meaning Builds Teens' Brains Connecting Adolescents' Concrete Work to Big Ideas May Help Shape Their Neural Networks over Time. *Educational Leadership Journal, 77*(8) Science and Technology Learning and the Brain Pages The Neurosequential Model in Active Learning (involving regulation, relating and reasoning) 36–43 Association for Supervision and Curriculum Development (ASCD)- Comprising 113, 573 members – superintendents, principals, teachers, and advocates from more than 129 countries – the ASCD community also includes 71 affiliate organizations. http://www.ascd.org/publications/educational-leadership/may20/vol77/num08/Building-Meaning-Builds-Teens'-Brains.aspx. Accessed 23 June 2020.

Jarvis, P. (2016a-updated 2017). Chartered Psychologist, Historian, Researcher, Teacher and Grandparent 'Engineering Elephants'. https://www.huffingtonpost.co.uk/pam-jarvis/engineering-elephants_b_12135688.htm. Accessed 16 July 2019.

Jarvis, P. (2016b). Chartered Psychologist 'Summer-Borns Find the Living Less Easy'. https://www.huffingtonpost.co.uk/pam-jarvis/summer-winter-seasonal-births_b_11829662.html. Accessed 20 June 2018.

Jarvis, P. (2018). Simplistic Beginnings? The Invisibility of Sustained Shared Thinking in Ofsted Advice Documents. *FORUM, 60*(3), 301312. https://doi.org/10.15730/forum.2018.60.3.301312

Jarvis, P. (2020a, July 27). *Education for Greater Equality.* https://www.edge-thinking.org/post/the-narrative-of-curriculum. Accessed 10 Jan 2021.

Jarvis, P. (2020b, July 15). Every Child Really Does Matter: In Search of Developmental Recognition. https://histpsych.blogspot.com/2020/07/every-child-really-does-matter-in.html. Accessed 15 Aug 2020.

Jarvis, P., et al. (2012). *Research in the Early Years: A Step-by-Step Guide Paperback.* Routledge.

Jarvis, P., et al. (2016). *The Complete Companion for Teaching and Leading Practice in the Early Years.* Routledge.

KEVU Keeping Early Years Unique. https://www.keyu.co.uk/guest-blog/

Kidd, D. (2014). *Teaching: Notes from the Front Line.* Independent Thinking Press.

Kincheloe, J. L., Steinberg, S. R., & Villaverde, L. E. (1999). *Rethinking Intelligence: Confronting Psychological Assumptions About Teaching and Learning.* Routledge.

Koski, J., & Rector and CEO of Laurea University. (2021, April 6). *A Balance of Efficiency and Creativity in University Management.* Rector's Blog, Laurea University of Applied Sciences, Finland. https://journal.laurea.fi/tehokkuuden-ja-luovuuden-tasapaino-korkeakoulun-johtamisessa/#f5c6d4fd. Accessed 29 Apr 2021.

Lamothe, C. (2017, February 3). *How 'Intellectual Humility Can Make You a Better Person'.* https://www.thecut.com/2017/02/how-intellectual-humility-can-make-you-a-better-person.html. Accessed 27 Nov 2018.

Lampert, M. (2001). *Teaching Problems and the Problems of Teaching.* Yale University Press.

Learn Through It Teaching and learning in Brussels, Belgium https://learnthroughit.home.blog/ @Learn_thru_it

LEARNINGMATTERS. https://rogertitcombelearningmatters.wordpress.com/ @rogertitcombe.

LEARNINGSPY-David Didau. https://learningspy.co.uk @DavidDidau.

Lehain, M. (2020, January 23). *There's Nothing Special About Grammars – A Culture of Excellence Can Transform Any School.* https://www.telegraph.co.uk/politics/2020/01/23/nothing-special-grammars-culture-excellence-can-transform/. Accessed 23 Jan 2020.

Lenabellina, High School teacher. (2016). *An Open Letter to Mr John Swinney.* Minister for Education in Scotland. https://lenabellina.wordpress.com/2016/06/12/an-open-letter-to-mr-john-swinney/. Accessed 03 Mar 2021.

Lisa (Leigh) Patel. (2014). Countering Coloniality in Educational Research: From Ownership to Answerability. *Educational Studies, 50*(4), 357–377. https://doi.org/10.1080/00131946.2014.924942. Accessed 01 Mar 2021.

LONG VIEW ON EDUCATION-Benjamin Doxtdater http://www.longviewoneducation.org @doxtdatorb.

LOVE LEARNING BLOG Debra Kidd. https://debrakidd.wpcomstaging.com/ @debrakidd.

Lukianoff, G., & Haidt, J. (2018). *The Coddling of the American Mind: How Good Intentions and Bad Ideas Are Setting Up a Generation for Failure.* Allen Lane Penguin.

Manning, K. (2016, February 1). *The Future of Higher Education.* TEDxBaltimore. https://youtu.be/XfRoM21qHtE. Accessed 12 May 2019.

McGill, R. M. (2017). *Mark. Plan. Teach.* Bloomsbury Education.

Merrick, M. (2018, March 16). Social Mobility and the Threat to Knowledge. *The Times Educational Supplement: London.* https://www.tes.com/news/school-news/breaking-views/social-mobility-and-threat-knowledge. Accessed 5 Aug 2019.

More Than A Score. (2018). *Baseline Testing in Reception.* https://youtu.be/OxwjEPyKW0s. Accessed 01 Mar 2021.

MRGORDON TEACHER https://mrgordonteacher.wordpress.com/

Mughal, A. (2018). *Let's Talk About Dialogic Talking.* https://www.inspiremetacognition.com/post/let-s-talk-about-dialogic-talking. Accessed 18 Jan 2021.

MUSINGS OF A MUGHAL-Anoara Mughal https://musingasofamughal.wordpress.com/ @anoara_a.

Myatt, M. (2018). *The Curriculum: Gallimaufry to Coherence.* John Catt Educational.

MYEDEXPERT- Compelling resources by trusted contributors. https://myedexpert.com/category/podcasts/

Neil Hawke International Values Education Trust: Positive Values. https://ivet-foundation.com/positive-values/. Accessed 21 Jan 2021.

NO ORDINARY CLASSROOM-Jeremy Hannay. https://noordinaryclassroom.home.blog/

Noonoo, S. (2017). What's the Difference Between Project- and Challenge-Based Learning, Anyway? *EdSurge News.* https://www.edsurge.com/news/2017-12-27-what-s-the-difference-between-project-and-challenge-based-learning-anyway. Accessed 21 Mar 2021.

Nutt, J. (2020). *Why Education Needs to Throw Off the Shackles of Data: An Obsession with Data Has Plagued Education – It Marginalises Individual Children and Their Futures.* https://www.tes.com/news/why-education-needs-throw-shackles-data. Accessed 01 Sep 2020.

O'Neill, M., et al. (Ed.). (2001). *Inference – Definitions, Derivations and Similar Meanings Chambers Concise Dictionary and Thesaurus.* Chambers Harrap Publishers.

Oates, T. (2018, September). The Progressive/Traditional Debate-'Skills Versus Knowledge: A Curriculum Debate That Matters-and One Which We Need to Reject'. *Impact, Journal of the Chartered College of Teaching.* https://impact.chartered.college/article/skills-versus-knowledge-curriculum-debate-matters-one-need-reject/. Accessed 26 Nov 2020.

Ofsted GOV.UK. About Us. www.gov.uk/government/organisations/ofsted/about. Accessed 16 Dec 2020.

Alison Peacock, The Lord Baker of Dorking, Sarah-Jayne Blakemore, Rachel Macfarlane, Bill Lucas, Peter Hyman, Olly Newton, Sarah Fletcher, Geoff Barton. (2020). *A Discussion on Assessment by Members of the Rethinking Assessment Group.* WEBINAR https://my.chartered.college/2020/11/webinar-rethinking-assessment/. Accessed 18 Mar 2021.

PEER REVIEWED EDUCATION University Teacher Training Blog. https://peerreviewededucationblog.com/

Pietarinen, J., Pyhältö, K., & Soini, T. (2019). Shared Sense-making in Curriculum Reform: Orchestrating the Local Curriculum Work. *Scandinavian Journal of Educational Research, 63*(4), 491–505. https://doi.org/10.1080/00313831.2017.1402367. Accessed 11 Nov 2020.

Pinker, S. (1994). *The Language Instinct.* Penguin.

Pinker, S. (2002). *The Blank Slate: The Modern Denial of Human Nature.* Allen.

Plomin, R. (2018). *Blueprint: How DNA Makes Us Who We Are.* Penguin Random House.

PSYCH CHANGE BLOG Chris Bagley. http://www.psychchange.org/blog/ @hiddendepths

Pyhältö, K., Pietarinen, J., & Soini, T. (2018). Dynamic and Shared Sense-Making in Large-Scale Curriculum Reform in School Districts. *The Curriculum Journal, 29*(2), 181–200. https://doi.org/10.1080/09585176.2018.1447306. Accessed 12 June 2019.

Random House UK, & Biesta, G. (2013). *The Beautiful Risk of Education.* Routledge, Taylor & Francis.

Ray, J. (2020). Video Report by ITV News Correspondent 'Government to Announce Sweeping Reforms to England's GCSEs and A-Levels', Monday November 30, 2020. https://www.itv.com/news/2020-11-29/government-to-announce-sweeping-reforms-to-englands-gcses-and-a-levels. Accessed 16 Mar 2021.

Reay, D. (2017). *Miseducation: Inequality, Education and the Working Classes.* Policy Press.

Rector's Blog Teaching Research, Laurea University of Applied Sciences, Finland. https://journal.laurea.fi/category/rehtorin-blogi/

Report OECD. (2019). *Education at a Glance 2019: OECD Indicators.* OECD Publishing, Organisation for Economic Co-operation and Development. https://doi.org/10.1787/f8d7880d-en. Accessed 16 Dec 2020.

RETHINKING ASSESSMENT. *A Movement to Value the Strengths of Every Child.* The Chartered College of Teaching. https://rethinkingassessment.com/. Accessed 18 Mar 2021.

Reuell, P. (2019). Harvard Gazette Staff Writer 'Lessons in Learning: Study Shows Students in 'Active Learning' Classrooms Learn More Than They Think'. *Science and Technology-The Harvard Gazette.* https://news.harvard.edu/gazette/story/2019/09/study-shows-that-students-learn-more-when-taking-part-in-classrooms-that-employ-active-learning-strategies/. Accessed 21 Mar 2021.

Richards, C. (2017, December 4). *Ofsted Must Be Careful Not to Lose All Its Credibility in Early Years.* https://www.tes.com/news/ofsted-must-be-careful-not-lose-all-its-credibility-early-year. Accessed 20 Aug 2019.

Roberts, D. (2019, September 3). *The Importance of Subject Communities.* http://www.danjjroberts.co.uk/2019/09/03/the-importance-of-subject-communities/. Accessed 16 Oct 2019.

Robinson, E. (2018, May 25). Progressivism. *RSA.* https://medium.com/rsa-journal/progressivism-6c1df1281da7. Accessed 03 Jan 2019.

Robinson, K., & Aronica, L. (2015). *Creative Schools.* Penguin Random House.

Robinson, M. (2013). *Trivium 21c: Preparing Young People for the Future with Lessons from the Past.* Independent Thinking Press.

Robinson, M. (Ed.). (2016) *Trivium in Practice* (T. Sherrington, S. Gorse, N. Wells, D. Hall, M. Grenier, N. Rose, C. Hendrick, & J. L. Taylor). Independent Thinking Press

Robinson, M. (2019). *Curriculum: Athena versus the Machine.* Independent Thinking Press.

Robinson, M. (2020, July 5). *On Curriculum and Culture.* https://www.youtube.com/watch?time_continue=3&v=EM_1jOhN9ac&feature=emb_logo. Accessed 07 July 2020.

Rycroft-Smith, L., & Dutaut, J.-L. (Eds.). (2018). *Flip the System.* Routledge.

Severs, J. (2019, May 8). *Inference: Why Comprehension Is Not Just About Vocabulary and Knowledge. Professor Jane Oakhill Explains How to Teach Comprehension Skills, Such as Inference, in Schools.* https://www.tes.com/news/inference-why-comprehension-not-just-about-vocabulary-and-knowledge. Accessed 08 Aug 2019.

Shea, J. (2020). Dr Gareth Bates, Dr Steve Connolly (Editors). *New Wave Progressive Teachers.* Posted on October 27, 2020, Peer Reviewed Education Blog: University of Bedfordshire. https://peerreviewededucationblog.com/. Accessed 31 Oct 2020.

Shea, T. (2008). Dismantling Language Barriers. *HR Magazine, 53*(11), 1–4.

Sherrington, T. (2017). *The Learning Rainforest: Great Teaching in Real Classrooms.* John Catt Educational.

Sherrington, T. (2019, January 24). *Rescuing Differentiation from the Checklist of Bad Practice.* https://teacherhead.com/2019/01/24/rescuing-differentiation-from-the-checklist-of-bad-practice/. Accessed 09 Nov 2019.

Sherrington, T. (2020, February 8). *Into the Rainforest of Teaching and School Leadership: 7 Variables Between/Within Schools: Where Are You on Each Scale?* https://teacherhead.com/2020/02/08/7-variables-between-within-schools-where-are-you-on-each-scale/. Accessed 13 Feb 2020.

Shotter, J. (1993). *Conversational Realities: Constructing Life Through Language* (Inquiries in Social Construction Series). London: Sage Publications Ltd.

Smith, M. (2019). *How to Make Your Pupils Curious Enough to Learn. Marc Smith Explains That Curiosity Is a Crucial Motivator That We Do Not Always Use as Much as We Should.* https://www.tes.com/news/how-make-your-pupils-curious-enough-learn. Accessed 16 July 2019.

Smith, S. (2020, January 13). *Lip Service (All Surface, No Depth).* https://smithsmm.wordpress.com/2020/01/13/lip-service-all-surface-no-depth/. Accessed 14 Jan 2020.

Speck, D. (2020). *Exclusive: Teachers 'Need an End to Joyless Curriculum' 'Fundamental Change' Needed to National Curriculum or 'We Do a Disservice to Children', Warns Shadow Education Secretary.* https://www.tes.com/news/exclusive-teachers-need-end-joyless-curriculum. Accessed 19 Feb 2021.

Steiner, R. (1920). *Lectures in Stuttgart. Teaching from Humane Knowledge* (R. Pusch, R. Querido, & F. Amrine, Trans.). Amazon Fulfillment.

Steiner Waldorf Schools Fellowship. *What Is Steiner Education?* https://www.steinerwaldorf.org/steiner-education/what-is-steiner-education/. Accessed 29 Dec 2020.

Stenhouse, L. (1975). *An Introduction to Curriculum Research and Development.* Heinemann Educational.

SUE COWLEY. www.suecowley.wordpress.com @Sue_Cowley.

Tan, R. (2015, October 21). https://smiletutor.sg/how-to-harness-develop-your-natural-strengths-multiple-intelligences-theory/. Accessed 24 Nov 2020.

TEACHER TOOLKIT-Ross Morrison McGill https://www.teachertoolkit.co.uk/blog/ @Teachertoolkit.

TEACHERHEAD-Tom Sherrington. https://teacherhead.com @teacherhead

Tharby, A. (2018). *How to Explain Absolutely Anything to Absolutely Anyone: The Art and Science of Teacher Explanation.* Crown House.

THE EDUCATOR K/12 Higher Education. https://www.theeducatoronline.com/k12/news/new-english-and-maths-syllabus-for-nsw-schools-announced/276072 Australia.

The Nolan Principles of Governance. https://www.governanceprinciples.scot/nolan-principles. Accessed 02 June 2020.

THE PEARL HIGH SCHOOL, Democratic Education https://thepearlhigh-school.org/democratic-education/

The Uni-Guide. *What Is the English Baccalaureate?* The Student Room, UK. https://www.theuniguide.co.uk/advice/gcse-choices-university/whats-the-english-baccalaureate. Accessed 21 Jan 2021.

Titcombe, R. (2015). *Learning Matters.* FGI Publishing.

Townend, J., & Walker, J. (2006). *Structure of Language: Spoken and Written English.* Wiley.

TRIVIUM21c-Martin Robinson https://trivium21c.com/ @Trivium21c.

United Nations. *The 17 Sustainable Development Goals.* https://www.un.org/sustainabledevelopment/. Accessed 21 Jan 2021.

VALUES BASED EDUCATION-Neil Hawkes https://www.valuesbasededucation.com/ @NeilHawkes.

Vygotsky, L. S. (1896–1934). *Collected Works* (Vol. 1, R. W. Rieber & A. S. Carton, Ed.). (1987). New York: Plenum.

Ward, H. (2018, August 12). Ofsted 'To Mark Down Exam-Factory Schools': New Inspection Framework Could Introduce New 'Quality of Education' Measure. https://www.tes.com/news/ofsted-mark-down-exam-factory-schools. Accessed 15 Aug 2018.

Waters, M. (2013). *Thinking Allowed on Schooling.* Independent Thinking Press.

Wenger, E. (1998). *Communities of Practice.* Cambridge University Press.

Wenger, E., McDermott, R., & Snyder, W. (2002). *Cultivating Communities of Practice: A Guide to Managing Knowledge.* Harvard Business School; McGraw-Hill.

Wiliam, D. (2017) @dylanwiliam Wiliam, D. (2013b). *Example of a Really Big Mistake: Calling Formative Assessment Formative Assessment and Not Something Like "Responsive Teaching".* Website www.dylanwiliam.org Twitter blog. Availableat:https://twitter.com/dylanwiliam/status/393045049337847808. Accessed 25 June 2017.

Willingham, D. T. (2019). *How to Teach Critical Thinking* (Occasional Paper Series- Education for a Changing World NSW Department of Education). https://www.education.nsw.gov.au. Accessed 21 Mar 2021.

Wyse, D., & Manyukhina, Y. (2020). *Children's Agency and the Curriculum: Theme 4 – Building Learning Independence.* https://theeducation.exchange/childrens-agency-and-the-curriculum/. Accessed 19 Oct 2020.

CHAPTER 5

Teaching and Learning Processes, Equality and Collaboration

5.1 Introduction

The fact of neuroplasticity gives us a great opportunity to reassess our ability to learn and teach while gaining more understanding of our own human nature. We can explore possibilities for pursuing the principle of universal human empowerment and happiness as stated in the American Declaration of Independence (1776) and the American Declaration of the Rights and Duties of Man (1948). Education is key. Ignorance has contributed to bad, self-serving oppressive behaviour by dictatorships around the world. Fortunately, an opportunity to reconsider and re-educate has emerged because environmental and post-pandemic imperatives are helping to change human mindsets. Everybody's human right to a happy existence is at stake. We need everyone to contribute experience, specific knowledge and skills if we are to improve our world. Luckily we have people with potential and knowledge existing, who can be drawn upon. There are approaches to schooling, which have already proved helpful in producing happy, productive and well-adjusted humans. The endnote video that follows sums up a modern parent's cultural dilemma about doing what is best for his or her child. See Video Endnote.[1]

Special thanks to Kirsty Henderson for 'How it feels to be a new parent of an autistic child' (2020) Twitter, 23 July 2020, 20 tweets, 5 min read.

J. A. Hawkins, *Brain Plasticity and Learning*,
https://doi.org/10.1007/978-3-030-83530-9_5

Elements of proven education approaches can inform us and be adapted to new settings, for example, Pestalozzi, Montessori and Reggio Emelia. In this chapter I look at some of these schooling philosophies, learning methods and psychological approaches. We need teachers who are generally open to the possibilities for themselves of self-development through helping others—teachers who realise that when learners are confident and happy, they tend to be in a good position to learn. Conversely when students are unhappy, they are unlikely to give attention to lessons. This is something we all know is true. Still, there are those who maintain that learners who fail to learn a prescribed programme of work are always at fault. They may appear lazy, but this can no longer be an easy default position unless possible reasons and honest reflections around student voice are considered.

5.2 Student Well-Being and the Teaching Environment

Early learning researchers and practitioners have long worked with and observed the learning development process at the beginning of the human life cycle (Jarvis, 2017). School starts with the Early Years, which has had to learn how to create an innovative, empathic environmental stimulus approach to teaching. They see human brain plasticity in its simplest form—at its most adaptable stage. Early years' settings use approaches conducive to learning development in ways that provide appropriate content and contexts. Their underlying principles are often found beneficial to learning at any age. *Their early years methods can be reimagined and used in other programmes with older age groups*—for example, 'role play,' modelling, storytelling, listening, 'making-up' ('brain-storming'), team-building and physical exercises, learning to be aware of and use information from bodily sensations, games, connecting with nature, project presenting, group discussion, giving opinions and opinion analysis.

Now we are aware of the potential for continuing human lifelong brain plasticity; early years' teachers' work takes on new importance. Many of the early years' principles regarding child empowerment appear educationally sound for older learners, especially in motivating and stimulating dormant capabilities, re-establishing and developing new neural competencies such as retraining for employment, learning a language, taking up a hobby and researching an interest. These kinds of learning experiences may promote mental well-being at any age. However, Sue Cowley has written about the different ways in which infant and early years' education is often regarded as being inferior, instead of deserving of respect for its expertise. See Blog Article.[2]

It is almost as if the simple happiness and uncomplicated communications of young children, who are generally open to learning, is regarded as not legitimately learning, if it is enjoyable. We must overcome this outdated and negative attitude. Nothing could be further from the truth because we all, to some extent, go back to first principles when we learn fresh skills in completely new situations and areas of knowledge! For example, an older person learning to swim, ride a bike or put together a piece of 'quick assembly' furniture has to go through basic stages. People with physical disability perhaps from a stroke, accident or illness are often helped with empathic occupational and medical physiotherapy. Similar approaches are being developed for those who have inherent/genetic and different and/or disadvantaged experiences.

Recognition and further research are needed now that our concepts of inability are challenged. Prejudice has persisted against the early years' sector, often seen as operating at a lower level instead of being an important area of innovative expertise in learning development stages and principles. For example, they have proved although education is a serious business, there is no reason why it can't be achieved through experiences of enjoyment and satisfaction. Some learners enjoy demonstrating their knowledge through tests and exams, but lack other important skills such as perceptiveness, imagination and resilience. This is why potential students to university are usually interviewed—as pointed out by Chomsky in the video referenced in Chap. 4. Their personality and motivation are often considered more important when they are compared with others—in spite of other people's superior exam results.

Traditional psychologists attempted to categorise early child thinking processes and 'mechanisms.' Their evidence was based upon their own experience and their 'analysis' of problematic case studies. Many of their explanations and tropes were embedded in the notion that the human psyche is largely static. We now know that this is not true. Such methods sometimes either provide a false explanation or actually impose, limit and cause the restricted baseline outcomes expected. Johann Heinrich Pestalozzi (1746–1827) was a Swiss educationalist who is credited with laying the foundation for modern education because he recognised that children have unique needs and capabilities. He is also given credit for converting the population of Switzerland to literacy by 1830.

He ran schools in German- and French-speaking regions of Switzerland and wrote about his educational approaches, which involved the idea of learning by 'head, hand and heart.' Pestalozzi believed every aspect of a child's life contributes to the formation of their character, personality and reasoning capacity. Pestalozzi's educational methods were child-centred, being based on learner differences, sense perceptions, preferences and self-motivated activity. He understood the need to give caring support in order to develop character and a good relationship with society. We have tended to forget this in the age of the internet where we constantly expect rapid/instant solutions to long-term, complex and slowly developing problems like environmental pollution.

In the United Kingdom there has been an early years' controversy in education caused by government anxiety about school test results. They are continually pushing for testing and are unconvinced by and distrustful of teacher assessments. Government experts favour numerical assessment and do not measure extra achievements or potential. There is no room or allowance made for crediting additional abilities or explanations for lesser abilities which do not fit in with the test. Potential requires a qualitative assessment by an experienced teacher. Well-trained traditional early years' teachers were ahead of their time. They embraced an empathic, open-minded, adaptable approach to education, which is exemplified in early years' education philosophies.

Maria Montessori (1870–1952) is an example. She was a qualified Italian doctor and an early paediatric expert in learning difficulties. Montessori helped to found schools in the Netherlands, India, Italy, the Maldives, Pakistan, Sri Lanka and London and was also involved in the foundation of the UNESCO Institute for Education in 1951. She believed young children should be given the freedom to explore their environment by using all their senses. She understood the value of encouraging and allowing children to be repetitive in learning—not in a forced, but in an enjoyable manner. They were encouraged to pay attention and focus on the natural world and to compare and judge colours, smells, textures, tastes, weighing, pouring, floating and touching objects. She believed in respecting children's individuality, encouraging self-motivation and autonomy, recognising and encouraging potential in every child.

Human behaviour is developed by studying innate characteristics in psychology and her research ranged from young children to older children. Montessori identified different learning features and 'developmental imperatives' active in each of four development stages: birth to 6, 6–12, 12–18 and 18–24 years. She produced educational approaches up to 12 years although her schools generally teach children from 5 to 7 years. Apparently her legacy was that she influenced 4000 school classrooms across the world. Many of her ideas are still useful today and appear to be compatible with neuroscience. She was interested in the natural ways young children communicate and learn. Her fundamental principles may be applicable with adaptations at later stages now we know about lifelong learning potential.

Pam Jarvis, is an English psychologist and early years' expert who advises a group of educational organisations and 65 universities about early years education. They are opposed to recent UK government policies that persistently favour early testing believing these actually may deter early child development. In 2015 the UK government proposed a 'get ready' for school programme at two years which together with a push to create formal testing at four years has caused concern amongst experts. Jarvis points out that until quite recently children started school at five years (in many countries they start at seven years). She tells us such early conditioning could cause 'deprivation' because this framework goes against the natural human development process.

The human brain is not fully formed at birth and needs a long period of nurturing in an environment with lots of social interaction with close primary caregivers. She explains that humans are linguistic primates whose brains are highly plastic as they develop neural connections in early years. Emotional and social communications and interactions are vital to establishing processing capacities to understand abstract concepts, thoughts and symbolic behaviours. See Chap. 3. Behaviours are complex as people respond to each other in complicated 'performances' of interactions. Jarvis cites psychologist Suzanne Zeedyk who suggests infants' and carers' communications are typically improvised in a 'symbolic dance' or drama—a kind of 'jazz duet' of repetition and variation as babies learn about the world. She tells us that core human skills are rooted in such symbolic communication sequences. See Academic Paper and Article Endnotes [3] and [4].

This is how a child learns how to 'translate' abstract thoughts into complex combinations of symbols that create sequences of meaning in spoken language and written language. These interactions create the ability to make inferential connections that require practice in a free exchange of spontaneous meaning making. Children learn to reason and to be 'intersubjective' through responding, reacting and testing their conclusions through play-based interactions with adults and peers. Jarvis tells us that the process of social/environmental stimulation helps children build neural connections in the amygdala and dorsolateral prefrontal cortex of the brain. These are essential for emotional regulation, concentration and meaningful social understanding.

The Reggio Emilia approach was developed by an Italian teacher named Loris Malaguzzi (1920–1994). It is a philosophy of early years' education aiming to empower young learners. They are seen as curious and able to learn by interacting with their environment, building relationships with others. They are viewed as curious, intelligent individuals, who should be supported to reach their potential. Self-expression, communication, teamwork, thinking and problem-solving are all encouraged. The environment itself acts as a 'third teacher.' This approach encourages the child to build on personal experience and take some control over the direction of the learning curriculum (Cagliari et al., 2016). The learning experience is determined by the type and character of the community and resources in each setting. Teachers observe pupils and talk to parents to understand how each child learns. Teachers plan and adapt the learning process, activities and resources appropriately.

Neural plasticity is encouraged by matching the child's interest, aiming to foster critical thinking and collaboration. Although this is termed a child-centred approach guidance is given as the teacher assesses appropriately. The aim is to create a motivating and encouraging environment structured to enable learning. It is an approach that values the 'non-verbal languages' of children, paying attention to gestures, glances and emotions. Much of the work of early years' educators fits with the idea of 'responsive teaching' or 'formative assessment' previously spoken of and developed by Dylan Wiliam (1998). Children's spontaneous comments are respected, listened to and responded to appropriately. Reggio Emilia activities encourage children to express and demonstrate their knowledge through dramatic play, storytelling, writing, puppetry, painting and sculpting, dancing, constructing and other expressive activities (Taylor, 2016). Article Endnote.[5]

Children revisit the activities including literacy and numeracy and subjects that interest them from different points of view to gain more understanding. Malaguzzi thought that this approach would support children's learning development and understanding of the world. The practical philosophies outlined above have developed and propagated ideas on the ground. Each has its individual history and traditions, but many have been absorbed and spread by practitioners in pockets of good practice around the world. They have some essential qualities in common and have survived because they have been successfully adapted and been generally appreciated by parents, children and teachers. Presumably, this is because on the whole they have been seen to create well-balanced, self-sufficient, knowledgeable, adaptable people able to contribute and gain from society.

Given the right conditions, brain plasticity shows why these methods work. Enjoyable experiences can be 'stage managed' to stimulate through enjoyment, wonder, novelty and surprise. This is the practical basis for the work of creative educational practitioners, many of whom sell their services to schools, for example, the work of Nick Corston. Learners who have been supported emotionally to develop ideas creatively are more likely to question and usefully and accurately adjust their own perceptions of facts (knowledge) to their view of the world. Ergo if students learn to relate new facts encountered into an integrated whole personality, they will likely cope better with adversity in the future and changeable environmental conditions than those who have not. See Video Endnote.[6]

5.3 Every Child Matters and Has Different and Similar Needs

As I researched neuroplasticity I came to realise that *'we are all part of the human family and our brain plasticity makes us all unique.'* Special Educational Needs and Disability (SEND) teachers have direct experience in understanding, supporting and empowering this process. This is why in this section I look at evidence and insights from special needs education. I have the idea these might contribute fresh ideas to mainstream teaching. Good special needs education can evidence the processes involved in plasticity and ways it may fail or can be enabled. If we take seriously that all human beings have unique experiences and preferences in learning, it follows we should explain and acknowledge difference and similarity when teachers are trained. The word 'special' has come to be misapplied. If we accept the reality of brain plasticity in practice it should be universally applied. In other words if every child matters and is unique then every child is special. This principle should underpin all approaches to teaching rather than exist as a label or sentimental cliché. See SEND Acronyms Endnote.[7]

We need fundamental principles and a fair approach to every child. As we teach and parent, it is in our long-term interest to help children to become the best they can be and *want to develop themselves to be.* Just as we adults are challenged in our everyday lives, we should regard learners of all ages as 'challenged' and also 'special' in many aspects of life including those who are 'academic.' If we want good outcomes, we should be encouraging them in a variety of ways. This is important because learners are inclined to reach inaccurate conclusions about themselves based on their treatment and experience. They should be taught that 'failing' and changing one's approach to life in order to learn and move on is normal. Learners need affirmation and positive regard in order to develop their capabilities. Teachers should help all learners develop confidence including those especially talented in certain ways and neglected within the system. The ultimate judge of their own happiness should be the learner, not just by what they say but by how they behave.

Teachers can be catalysts, developing potential by recognising and researching learners' abilities, interests and motivations (as they relate to the core curriculum in school), whatever they may be. Teachers can research children's responses through professional programmes and approaches, adjusting them with insight through experience, feedback and analysis of outcome data. Parents should also be recognised as having a special role as supporters and data collectors. Everyone's opinion is important even when and perhaps especially we don't agree. Education is complicated and of its nature is open to adaptation by diverse kinds of people. Teachers need to be taught to be aware of how they can develop their skills, research and adapt as they teach.

Good teachers already do this even when unacknowledged. Their data and student outcomes are reflected informally to justify their work to themselves. They should be trained to record this research data, to share, listen and pass on their knowledge and justify their work to colleagues. They often need someone to listen to them and they can do this for each other to improve their practice. In Finland the National Board of Education's role is also to listen, receive, check out and coordinate continuous improvement in the school system. The government accepts that a continuous process of system revision and improvement is informed by the work of teachers on the ground. The managers who implement the law believe it is normal for systemic problems to arise and if correctly identified that they can be addressed in the next framework revision in educational legislation. Teachers are required to assess and solve problems themselves, explain, inform, suggest answers, report and discuss with colleagues in the system. See Report Endnote.[8]

In this way they build resilient institutions to provide services capable of adjusting to the ongoing 'assessment of individual needs.' A report (2011) claims Finland's success is explained by its high-quality special education system which monitors the whole system itself. In international comparisons Finland is considered to have prevented the marginalisation of young people with learning difficulties exceptionally well. Early detection of learning difficulties with support for the child with learning disabilities is planned and implemented in collaboration with parents and a multi-professional team. The tailoring of special education services to learners is facilitated by a general trust in Finnish schools giving them freedom to organise educational programmes. A report finds that tailored services have more value for learners and parents than 'mass-produced' services. They have a greater impact because they aim to help people adapt, the better to cope with our rapidly changing and complex world.

Valuing everyone, though it may not be totally achievable, helps *more* people to be included in mainstream society and to be 'useful' to that society. Research has shown preventing marginalisation of one young person saves society a million euros in a person's lifetime. The long-term benefits of tailored education system services should be born in mind when cuts are planned in special education. The report says that effective tailored services requiring more resources than standard services short term, proved to be cost effective and a good investment long term. The system is an integrated and self-integrating, bottom-up system, expected to develop and change over time. Research and development is factored in and ongoing in the classroom around national goals. It is accepted new self-reforming regulatory bodies for accountability may need to be established. Alternative interventions are expected to be found and better solutions developed. Education management supports and coordinates a common understanding of any new requirements.

Everyone is involved in pooling information from research bodies, charitable welfare groups and peer review institutions to working parties of practitioners and researchers. The report says common terms to communicate experience gained should be developed as a continuous ongoing process. One institution should not monopolise initiatives in organising reform and many initiatives should be encouraged. These should arise from specialists and institutions in the system in order to create an effective whole. In the United Kingdom, Elly Chapple has a daughter, Ella who was born deaf-blind. She has written extensively about her experiences as a parent, including her own hearing problems and others in a similar situation. She has delved deeply not only into her own experiences of education and social systems, but into the way we are all different and similar.

Chapple says, "We need to take the time to reflect about our assumptions and give pause for thought and time for space to be and understand what our children are teaching us." She points out that it is often through simple behaviours deep learning happens—it is not just about what we ourselves perceive about how something should be learned. We need to reframe our thinking and understand not everyone learns in the same way. She gives a moving account of how Ella at seven years learned to connect 'ideas,' imagine, learn and communicate. At one developmental point she did it by touching and sorting a box of family shoes twice a day. After nine months she used that knowledge to find her boots, said "boots" and communicated to her mother that she wanted to put them on. See Blog Article Endnote[9] and Video Endnote.[10]

A teacher could not understand why it took so long, but her mother realised she needed to take her time until the learning task was completed. It was her own necessarily unique process to make a breakthrough and establish a skill set. After all, learning to recognise objects, speaking and communicating takes time and requires repetition even when a baby can see and hear—different skills are interwoven and require practice. That set of communication skills led to further development. Elly admires special needs teachers, who have a skilled, adaptable approach. For example, Barbara Miles is a special needs teacher at the Perkins School for the Blind in Massachusetts. Elly Chapple credits Miles with understanding and teaching in a way that 'resonates far beyond the deaf-blind world across the human space.' Let us look at an article by Barbara. See Blog Article Endnote.[11]

There are parallels here for other teachers in other settings with other learners. The fundamental principle of using one's own human insights to form connections and a teaching/learning relationship applies. For example, Miles says that watching a child touch an object or another person tells her how the child is 'perceiving.' When she touches and 'interacts with the child's hands,' she closes her own eyes for a few moments so that she can concentrate on her own sense of touch and 'enter' the child's 'tactile world for a short time.' This practice improves her own skills. If her hands are open and relaxed, the child often responds by exploring or initiating a game. When a child initiates a movement she responds with her own movements, and they make a 'conversation' that develops over time. She says her empathic approach also works with younger children and children who are not curious.

Miles says these kinds of connections are a beginning point and can lead to building a trusting relationship and a child's gradual and growing interest in the world of objects outside her own body—a kind of mental congruence. *The teaching process described above is in essence the process of* mental congruence *with 'knowledge' and ideas.* It can take place between teachers and learners differently, but in some form is probably essential to every 'deep' learning experience. Of course, this kind of teacher behaviour does not apply in all teaching situations, but perhaps the idea of imagining the world from the learner's perspective (perhaps mixing things up with surprise and novelty) and encouraging them to choose how to focus and 'engage' might help, where learners struggle. See PDF Article.[12]

Congruence means being in agreement and harmony about something. It is a term used by Carl Rogers (1902–1987), the humanist psychologist responsible for person-centred counselling. It describes when a person's self-awareness is fairly consistent with reality. Some subject teachers remember their own problems with learning subject concepts and some have forgotten. The first are often more likely to be able to break down and teach the process responsively than the second. The practice of giving positive regard and initiating empathic interaction is a skill that should be taught and practised when teachers are trained, retrained and engaged in research. In the past special educational needs children have been 'diagnosed' and a complicated range of 'disorders,' 'conditions' and 'presentations' observed. There are a range of specialist areas of expertise, but still every child is different and similar.

All children need acceptance within the human 'family,' security, love, an environment that is 'comfortable' to them and experiences of kindly social interaction. Specialist therapy approaches and the 'conditions' they treat are likely to be unknown to mainstream teachers. It would benefit all teachers to spend time training in special needs settings, to be challenged to appreciate how everyone may learn. In these 'live' situations there are approaches and resolutions of care to be learned about and experienced. They cannot be learned in any other way. Trainee teaching students own character development and perspective on their future teaching career choice would be improved. The 'vocational' development aspect of teaching would be better served long term and they would benefit no matter which area of teaching they eventually taught in. Mainstream teaching would be informed and improved. Students would be better prepared for challenges in the future.

Sabbaticals and research projects should be undertaken as routine for the continuing professional development of teachers during their careers, not least for their own mental well-being in a demanding profession. Teachers on the ground and school leaders need opportunities to expand their knowledge with the support of colleagues, university specialists and experts. Initiatives should arise from the needs of learners on the ground.

If one accepts the reality of brain plasticity, then a child-centred system of education promises to be a more successful approach to develop intelligence and useful abilities than one imposing rigidity that rejects free will. This does not mean *giving way* to learner, but guiding, balancing, challenging, responding and listening. Teachers need to undertake and study current research such as that performed in the Independent Report-Munro Review of Child Protection (2011). This was a '*Child-Centred System-Report Setting Out Reform Proposals to Enable Professionals to Make the Best Judgements to Help Children, Young People and Families.*'

5.4 Parental Perspectives and Involvement

Kirsty Henderson is the mother of two children with additional needs. She is currently undertaking a nursing degree and co-runs a website for SEND advice. Her son (four) is autistic, and her daughter (eight) has Attention Deficit Hyperactivity Disorder (ADHD) and Sensory Processing Disorder (SPD) and is currently on the pathway for an autism diagnosis. She believes all children regardless of SEND have different learning styles and abilities. She thinks it is crucial to quickly identify the needs of each child and get support. She points out that early proactive interventions are vital, clear targets and routines. She thinks all teachers should undertake placements in a special school and attend SEND-related courses during and after training. Henderson says that when a child's needs are not being met it quickly affects their happiness and well-being impacting their education and mental health. See Special Data Thread Endnote.[13]

Henderson sent me an email and in it she writes, *"When labelling children in a school setting, it must be done carefully as to not create a divide. With the discretion of the parent and child I believe there are potential benefits to other students knowing a child's diagnosis. However, to use 'labels' freely in front of other students there must first be better understanding and inclusivity within schools, for example lessons on neurodiversity and disability to avoid bullying and creating an us/them. From a parent perspective I have found that labels have enabled my children to the support they need, although I do believe this should not be the case and a child should receive the appropriate support based on their needs not diagnosis."*

What can be done to help parents like Kirsty and all those other parents concerned about their children's welfare? Sue Gerrard is the parent of a child diagnosed with an autism spectrum disorder and is therefore an expert with lived experience of the subjects she has studied. She also trained as a biologist, psychologist and primary teacher and so has taught all kinds of children. She has researched autism and 'mental disorders' and looked at how scientists form their theoretical models. She says that although psychologists have discovered some aspects of child development, most have failed to fully appreciate that the relationship between genes, environment and behaviour is changeable and complicated. This relationship very much affects the way as a society we see people and can find ways to develop everybody's learning.

Gerrard finds that after 50 years of research into how genetics may affect brain function, only two explanations offered by psychodynamic theorists for 'abnormal/atypical' human behaviour are still generally accepted in educational settings. These are that either there is a medical explanation or the parents must be responsible, with the occasional acknowledgement in a particular case that environmental and 'incidental' childhood experiences may have been beyond parental control (Bowlby, 1988). It seems that on the ground when there is no medical diagnosis, it is simply assumed by everyone (including parents themselves) that the parents are at fault. See Attachment Theory Video Endnote.[14]

This psychodynamic model has been absorbed into our collective unconscious and is still misinforming our child health, education and social care practices. Gerrard says this is because of the way we humans tend to think. Although we are capable of using logical, rational thought, we have evolved as a species that relies heavily on 'rule of thumb' (heuristics) and inbuilt habits, tendencies like default ideas and biases. She cites Herbert Simon (1916–2001), an American economist, political scientist and cognitive psychologist. She points out that rough and ready ways of responding to our environment help us cope with challenges to immediate survival, but are not long term. These strategies to improve lives are overdue and would likely save time and money. The dismissive attitude towards families and teachers by politicians in the United Kingdom is compounded by reluctance by local education authorities to pay for specialist assessment services to help all children develop within the mainstream. See Blog Article Endnote.[15]

Gerrard sees an association between interpretations of psychodynamic psychology and the pseudoscientific movement of eugenics. This movement tried to justify prejudices, 'othering' and hatred of marginalised people and has been used to justify genocide by dictators. There is no justification either economically (in the long term) or morally for these views. Genetic variations are robustly transmitted between generations and are inherited or arise spontaneously. Some are common and others rare. They affect all levels of society. Gerrard points out it is genetic variation that maximises a species' chances of surviving environmental change

(Darwin 1809–1882, 2009). It is inappropriate and often not meaningful to talk about 'normal' and 'abnormal' in descriptive terms (i.e. common and uncommon, typical and atypical). It is also unrealistic, unethical and ultimately detrimental to society to say people should be like this but not like that. She says that parent behaviour and inherited characteristics are only two of many factors that interact in complex ways to influence a child's development.

Other factors involved in learning development are diet, toxins, allergens or infections that disrupt physiological development. It is a societal problem because in addition to parents, other people's behaviour is involved, for example, peers, teachers, neighbours and wider community. Gerrard goes into this in another long and informative article where she points out the number, formation and location of neurons is genetically determined, unless affected by external medical factors. A baby has a full complement of neurons at birth (some neurogenesis even occurs in some areas of the brain in adults) and interconnections between neurons are formed by synapses. See Chap. 3. Brain development is the result of interactions between four factors. These are genetic; epigenetic caused by gene expression impacted by environment; environmental (including both nutrition and the behaviour of others) and behavioural effect (the impact of the child's own behaviour). See Blog Article Endnote.[16]

These factors vary in their impact upon the brains of individuals at different ages and under different circumstances. These issues Gerrard lists have an impact on a child's development and they affect everybody in society in one way or another going forward. The problem is that governments and their advisors tend not to see the long-term value of financing every child towards overcoming these problems and becoming an individually happy, autonomous and productive citizen. They are ignorant about education at a professional level and are not prepared, though they may fund them, to allow researching experts to take decisions and resolve problems. This seems short-sighted considering that the bulk of their current and future electorate from students to parents, grandparents, teachers and employers are education stakeholders.

Jules Daulby, special needs expert, government advisor and head of department in a special school, has written extensively about the lasting impact of the failure of mainstream education to help children with 'learning difficulties.' Gerrard's experience as a parent and teacher and Daulby's research as a special needs teacher give the kind of insights and analysis that can inform a better education system that is looking for solutions for all children. Daulby points out that the narrow attitude of politicians to what constitutes achievement actually does a great disservice to most children in the United Kingdom. They have tended to make GCSE exams harder believing that high grades accurately predict success and standards of 'intelligence.' Decision-makers have never acknowledged the full range of children's and parents' intelligence, skills and abilities as necessary components of a successful society. They have not valued non-academic occupations, wrongly believing a university education is vital to all. See Endnote.[17]

Further education of other kinds, work placements and apprenticeships have been undervalued. Normative exam systems have a suppressive effect on talented mainstream children as well as those with special educational needs. Government often lacks self-criticism and does not listen to research on the ground. Teachers have not been supported or trained to undertake this and in the UK there is no government body co-ordinating their analysis or enacting their recommendations. Governments often invest in teacher training and then do not listen to expert teachers' advice. Recent exam policies have widened the gap between typical students and those disadvantaged by difficulties. Many of the difficulties exist on a spectrum only addressed in extreme cases. The remainder of the school population may be experiencing these problems, but mask their difficulties and 'appear normal.'

Most teachers are not experienced and trained to support diverse needs. Exams becoming less inclusive and varied may be a contributory factor towards the young people's mental health difficulties in recent years. The competitive school system has tended to reduce young people's resilience and is having a damaging impact on quite a large section of society. However, a new public understanding and awareness about mental health, has arisen due to problems that have been highlighted and discussed during the pandemic. It has been realised by parents and children themselves that they need help and support in order to navigate the challenges of modern society.

Many teachers have commented about the system showing they are also concerned from a parental perspective. Harmeet Sahota, senior assistant head teacher and digital consultant, explains why he hopes his daughter underperforms in her statutory assessment tests (SATs). He talks about the competitive culture propagated by algorithms on social media and says one in eight children have a diagnosable mental health disorder and one in six people aged 16–24 have symptoms of depression or an anxiety. He has discovered that half of all mental health problems manifest by the age of

14 and suspects that some of this is a 'knock-on' effect from pressure put on schools to achieve. Exam pressure is one of the reasons students say affects their mental health. He wants his daughter to do well and develop a positive, confident attitude, achieving her own goals. See Blog Article Endnote.[18]

As well as getting the highest grades possible for her, he wants her to receive guidance to encourage self-motivation. Sahota says his daughter is already anxious. He is afraid she might become a child with a mental health disorder. If she underperforms in her SATs, she will not then get pressure to meet the targets set by the results. British schools have raised awareness of mental health problems, but he finds it ironic they can't see they contribute to the problem. His website says government reforms (2014) have had a huge impact on curriculum design for schools. The emphasis on performance tables has increased English Baccalaureate (EBacc) subject exam entries in many schools resulting in a reduction in creative arts, and vocational subjects such as motor mechanics, hairdressing and construction.

Sahota believes the current exam system has had a bad impact on students' well-being and exam anxiety would reduce by decreasing the number of written exams. There are a whole range of subjects useful in adult life, which have been abandoned, that develop the brain. For example, horticulture and floristry, biodiversity, husbandry, nature study, outdoor pursuits, computer engineering (as well as coding), plumbing, cooking, hospitality, child and healthcare, budgeting, woodworking, dressmaking, all kinds of skilled crafting, community services and civic studies around environmental issues, charity work and democracy. All of the above can contribute to a happier life. Some of them create an interest leading to academic as well as other careers. Non-academic learning contributes to intelligence—remember the little girl shoe sorting and Barbara Arrowsmith-Young as an adult who 'changed' her brain by repeatedly using clock flash cards. See Chap. 1. She found she had stimulated her brain to function in other useful ways than just reading the time (2013).

Parent, business leader and educationalist David Harkin describes his hopes for his two young sons, Toby and Harry, who will be back in mainstream after the pandemic. He wants them to be able to be fundamentally happy and fulfilled, find their passions in life, be committed to their choices and add value to whatever they decide to act upon in any future career. Harkin thinks the reason 80% of people in Britain are dissatisfied with their day job is that they did not pursue their passion (or interest) at school.

Happiness is beneficial for our well-being and also benefits careers and the economy tending towards productivity and quality of work. Harkin hopes that when Toby is 18 in 2034 he will be able to explain what he is passionate about. See Blog Article Endnote.[19]

Harkin hopes that all schools will be offering resources with a 'wider scope of entry into passionate pursuits.' By the time a student leaves school, they should not only be clear on their passions, but also believe they can find a career linked to one of them. He wants acknowledgement that students learn from interactions and activities outside school—including the internet. He believes parents and teachers should pay attention to internet content to make people more accountable for the content they publish. He wants schools to be more agile, looking at challenges from different perspectives, experimenting with methods and content to create exciting and memorable lessons for students. He believes teachers are capable of doing this and should be given more preparation time. He thinks a new form of hybrid education could be useful.

He suggests a physical school experience and a 'virtual' extended education addressing 'soft skill' extracurricular skills encompassing a sense of belonging and community in a creative way. He believes 'agile school setups' could be organised following a business model with a chief innovation officer appointed to lead the way. Harkin wants schools to evolve over the next 14 years to make the educational experience 'vastly different' so his sons get a greater chance of happiness. Teachers are not averse to working with parents—many are parents themselves—but they need time, training and resources. One model which might be researched and adapted is that of the Steiner/Waldorf school. Teachers in Steiner schools visit families to act as a 'bridge' between home and school, a practice that is rare in state schools.

Another teacher in the blog below explains that in these schools the teacher-parent relationship is considered crucial for learning. Instead of meeting a new cohort of students every year, he has up to 30 families in total to connect with and eight years to build a trusted working relationship and to foster understanding. Although there might be considerations about being 'stuck' with a particular teacher or group, he thinks this works well. He has also had experience of being part of a vertical tutor group in mainstream where each tutor group had a few pupils from each year group within it so that he took a 'family' of students consistently over several years. He tells us being a tutor in this way has given him some of the most rewarding moments of his career. See Blog Article Endnote.[20]

5.5 Teaching and Assessing All Kinds of Children to Achieve Their Personal Best

There are several logical consequences that follow as a result of the attitude that everyone should be 'typical' or 'normal.' These are that:

- Additional skills, talents, gains and achievements may not be looked for, identified, supported and valued.
- Children who do not conform to norms may be considered as fundamentally 'different' and labelled as 'odd' including those who are academically gifted. They may experience feelings of mental isolation even while typically masking their difficulties in social situations. See Article Endnote.[21]
- Pupils of every ability who do not meet set expectations may be sidelined and even 'able' children can be bored and their learning underdeveloped when they remain under-challenged.
- Learners may become discouraged, depressed and/or disaffected with school because their personal qualities are not acknowledged and celebrated.
- All of the above points mean the benefits of brain plasticity go unsupported and budding opportunities for learning development go unobserved, cut off, discouraged and/or lost.
- Society becomes 'normalised,' accustomed to and comfortable with lower achievements and failure from those it marginalises and excludes.
- Citizens may tend to become less tolerant towards others, less diversely intelligent and less contented, accomplished, 'productive' and creative.

The truth is that the human race is filled with diversity and every human being is similar in fundamental ways, but our adaptable brain plasticity also makes each of us unique. It follows therefore that teachers researching children's learning who support learning differences are likely to benefit more children, and are more likely to value and encourage them when they 'fail.' This includes those who are good at exams and those with cognitive learning challenges caused by accident of birth or genetic make-up still not fully understood. Children who are diagnosed 'high functioning autistic' like Greta Thunberg may have exceptional abilities. Similarly those who are visually or hearing impaired or otherwise medically challenged. See Article Endnote.[22]

Different teachers are specialist experts in their different teaching subjects and roles. However, my research suggests schools should create subject-based teaching and learning councils of teachers who research and solve problems so the progress of different children in various classrooms is constantly being 'assessed' in different subjects and successful practices shared. Something similar to the phrase 'Every Child Matters' or Finland's fundamental principle of 'equality' just might provide a 'self-correcting' educational focus if properly re-conceptualised and applied. In Britain the 'Every Child Matters' initiative affected all children from 0 to 19 and was embedded in the Children Act legislation (2004).

As in America, the phrase was adopted by the United Kingdom when it set out a 'vision of radical reform' for children, young people and families. However, it has not been carried forward as a principle across all of the social services in sufficient depth. In the United Kingdom this social services reform law resulted from a report after the tragic death of Victoria Climbié, a child murder victim, in 2000. It was shared across government services (health, education, social services and probation), enforced by an integrated inspection framework called the Common Assessment Framework. It was meant to ensure professional social agencies services worked together effectively and shared information to prevent record keeping and communication problems. See Report Endnote.[23]

The aim was to ensure all agencies shared information about the children so they and their families were supported. The framework aimed to improve educational achievement, reduce deprivation, "ill health, teenage pregnancy, abuse and neglect, crime, and antisocial behaviour." The system was updated and rationalised to promote 'efficiency' and is now called the 'Early Help Assessment' to determine whether and in what way a child and their family need help and support. Each sector was inspired by and implemented the principle of 'Every Child Matters' but was constrained by central government policies with time-consuming data collection and accountability. There was a lack of political perception, leadership and targeted financial resources to make all these social agencies work efficiently. See Article Endnote.[24]

Teachers often liked the idea of 'Every Child Matters,' but implementation was patchy. Agents of change, including teachers function best within a self-regulatory collaborative research culture, with training and ongoing professional development so that they can create change on the ground. This is a process which should be researched and managed so that the same principles of sharing findings, listening, reflecting and analysing apply to management driven by outcomes on the ground. Experienced managers and politicians should be able to relate to and empathise with children and families. Good governance involves believing everyone should get health care, achieve economic well-being and make a positive contribution to society. Unfortunately the British government has believed simple data collection in education without qualitative analysis will guarantee such caring professionalism.

Jules Daulby, SEND expert and English teacher, says England's exam system sets up children to fail in mainstream schools. She explains the range of examinations is based less and less upon constructive thinking and reflection, original ideas, problem-solving, making meaning and explaining. Examinations have not been about demonstrating originality, depth and understanding. They have been asking for a decreasing range of literate behaviours requiring short term memorisation. Exams have been increasingly about a speedy 'standalone' performance of a required narrow list of 'basic' knowledge at a set level of difficulty. Daulby says that when Mary Warnock (1978) produced a report leading to legislation to include SEND children in mainstream schools, there was no commitment to research by teachers. She points out that the subsequent effect of top-down directives, enforced narrowing down of the exam system and non-compliance with the country's Equality Act (2010, 2015) has had a damaging effect on non-academic pupils.

Issues like responding to failure by exclusion and unethical schools off-rolling/discouraging students from attending because they may get lower exam results have made things worse. Although exams and tests are useful, even necessary, they are not adequate on their own. The problem is that learning is complex. Interest, motivation and curiosity are key to memorisation, retrieval, deployment, productive thought and action. We need more, not less, ways to teach and assess learning because neuroplasticity tells us that people are intelligent in diverse ways. The brain develops best when it has confidence, opportunity, an enabling environment and support. It has become more difficult for neurodiverse learners including mainstream pupils to pass exams that are personally useful to them. The system has not been learner centred, but has been aimed at producing end of year numerical data (a league table score) by which the school is judged, rated and funded.

When children have 'failed' in the system they have often been allocated less resources, less well-trained teachers and poorer school buildings when they actually have more needs. These children need a variety of courses and exams that teach core knowledge and skills in adaptable ways. Courses that are set, moderated and examined by well-trained professional teachers. Teachers who are subject knowledge rich and on their own professional development learning journey. In a supportive system teachers can set, moderate and examine each other's work through discussion, careful monitoring of coursework, formative assessment and individual skill assignments (ISAs) that are variously both practical and 'academic.' Learners should be allowed to pursue an interest within the set syllabus at the discretion of the teacher, provided such an initiative can be justified and shown to be reasonable. Retakes should be allowed under considered circumstances. See Endnotes [25,26].

Children who require a reader or scribe should be marked according to their demonstration of understanding. They should be judged on original ideas, memory and meaning not marked down automatically purely on spelling, punctuation and grammar (SPAG) marks. Complex language, reading, writing, memory and speed of processing, which cause many difficulties for learners in mainstream schools, should not be viewed as the sole measure of intelligence. It is possible to achieve a reasonable level of mechanical reading and factual memorising, but this does not necessarily prove potential, real interest, subject knowledge and comprehension. Many learners struggle with retrieving information to order but this does not necessarily mean they are unintelligent or lack knowledge and ability. For example, English Literature exams that refuse to allow books in for reference. Physical Education exams that disallow physical activity coursework and demand a 60% written exam. See Blog Article Endnote.[27]

Exams that favour high academic literacy skills penalise the achievements of talented young people who have knowledge in a particular subject but lower literacy skills. Maths able learners are penalised because they need high comprehension skills. Words in maths questions can make exam questions inaccessible for students who have relatively good maths abilities useful in certain careers. The Education Endowment Fund reported that literacy levels have affected pupils' science scores. Recently there has been a move by science teachers to make content knowledge and practical demonstrations more central. In science it may not always be appropriate to award marks for the 'quality of written communication,' for example, phonetic spelling can be accepted even if incorrect. Teachers complain that some Biology exams rely too heavily on advanced comprehension skills and yet we need good laboratory technicians as the pandemic has proved.

Subject knowledge should be available to those who have learning difficulties affecting literacy and language. Daulby asks how many great scientists, poets and mathematicians are lost because they are not helped to get around their dyslexia and developmental language disorders. She points out that the majority of these students are in mainstream schools. Teachers themselves become demotivated and disaffected when they see their 'learning difficulty' students suffering further disadvantage as they face being excluded by an inappropriate exam system. In the United Kingdom it has often been unequal and unjust and the system does not encourage teachers to intervene and develop the working practices. This is why teacher satisfaction, employability and retention have been low.

5.6 Teacher Autonomy and Collaborative Research

Nicole Dempsey, an experienced SENCO at an academy school, recommends schools look at methods and good resources working well in other schools and see what new ideas and trends are emerging. They should share and borrow ideas, adapting, mixing and matching, to make approaches suit their own settings. In high schools it is not good enough to exclude SEND students from modern foreign languages, art and music and so on. Many subjects can be designed to include them and additional different provision should be equal quality (given a learner's ability) with minimal disruption to their education. Educational provision should benefit every child and not be curtailed for someone else's convenience. It should be evidence based, advised by an outside expert like an educational psychologist and regularly reviewed.

She reminds us that 'no one teacher knows everything about every possible type of additional need that might be present in their classroom.' In her job she finds getting to know parents and carers, as well as the children themselves and doing 'a bit of research about specific needs' and soliciting advice from knowledgeable colleagues will 'get you to where you need to be quicker.' She tells us to believe in ourselves as subject specialists and pro-activists who have chosen our vocation. 'SEND students need and deserve to be with us.' These are the fundamental aims in teaching that are to some extent appropriate in all educational situations. We are not doing our job unless we are willing to learn more about our learners and their environmental context. Humans are complicated social animals and different kinds of empathic interactions are required to suit different situations and learning cohorts. It is important to interact and listen very carefully to students at all levels, request and reflect on feedback. See Blog Article Endnote.[28]

This is the teaching and learning process that stimulates cognition and neural growth. Subject knowledge is an important varied and complex component in a teacher's 'tool bag.' The same core knowledge can be taught in different ways and expanded in different directions to appeal to learners. Teaching is an art and learning programmes can be constructed with various methods and formats. The complex world of education flourishes through justice, equality of opportunity, agreement, respect for differences, listening and learning from each other. We need to consider contexts and events, generate new jointly engineered ideas and reach wiser conclusions. Why is this taking so long to be incorporated into practice? It seems that experts' traditional, deconstructive analytic training and fixed view about our physical capabilities has become hardwired into their own perceptions. This is detrimental where it creates self-fulfilling outcomes.

This is particularly the case when children have a different physical make-up, clinical condition or have survived different types of stressful events, trauma and deprivation. Learners need to be helped to develop by building on those things which resonate—things to which their senses and emotions naturally respond. The brain will follow and embed those things that stimulate emotional connections and memories. The professionals' job is to understand the end goal of the work, whatever that may be. However, neuroplasticity shows us within each learner's brain, there are variations and different ways those goals can be achieved. Further the biological network building process requires varying amounts of repetition of tasks presented in similar and different ways. The two common teaching approaches of '*stimulating and engaging curiosity and interest*' and '*insisting on the use of repetitive tasks*' have always worked. It is the choices and presentation in action that help them work.

Teaching is a highly skilled job that involves directing and producing a learning experience likely to engage most of the class. Analysing and learning how to respond to different abilities and needs is a necessary teacher learning process in itself. It is a sophisticated professional skill that operates and adjusts appropriately in action—a concurrent and fluid one that develops through experience and practice. The work done by Rachel Lofthouse, professor of teacher education at Leeds Beckett University School of Education in the United Kingdom, confirms that the process described above is true at all levels in education. She runs the Teacher

Research and Practice Centre with a focus on coaching and mentoring and the international 'CollectivED Mentoring and Coaching Hub.'

Lofthouse researches how and what teachers learn as they explore innovative pedagogies and design curriculum and how we can support them to put that learning into practice. Her interest was inspired, when teaching in a high school Geography teacher community; she collaborated to bring the then new national curriculum to life (1988). The process involved deconstructing, reconstructing and interpreting ways to teach the skills and attributes children needed to learn within the new rules. She has gone on to help establish and develop a research-based culture in her university school of education. She wants to help teachers find out how to learn their own lessons from researching their practices, innovating and developing pedagogy to have more impact. It is an iterative action research process using coaching, mentoring, dialogue, creative and critical thinking, lesson study, professional enquiry and inter-professional practice. See Podcast Interview Endnote.[29]

She works with student teachers, school-based mentors, teacher coaches, postgraduate students and school leaders. This includes various specialist practitioners, including those involved in community work, speech and language coaching in multicultural early years and primary settings. The whole process and analysis appears to operate as a complex collaborative narrative network (Hawkins, 2017). She says that in teaching, research should always be related to practice preferably within a collaborative community. She mentions in passing the General Teaching Council for England (GTCE), which supported professional lives of teachers and was abolished by the government in 2012. Evidently teachers still felt the need for such an organisation in spite of the lack of understanding from their political employers.

The Chartered College of Teaching, originally founded in 1846 stepped in to fill the void. Established on an updated basis in 2017, it is now the recognised professional body for the teaching profession in the United Kingdom. As well as supporting teachers continuing professional development it operates as a national research and practice hub. Its work supports the national and international early career framework which requires working teachers to be coached and mentored by other teachers. They are committed to helping each person in the profession to be the best teacher they can possibly be. Such a body supports teachers to deliver better teaching through access to research, ongoing development of their knowledge about effective practice and to take part in ongoing debate.

The Chartered College of Teaching is needed in order to foster a spirit of collaboration in the profession towards self-improvement. It provides a platform for teachers' voices to be heard. Lofthouse suggests coaching and mentoring have evolved because of the repositioning of school-based teacher training standards and assessment. As a result it is being realised by the profession that some practices need o be constructed differently. She says the whole field of coaching and mentoring is made up of a spectrum of many colours and practices. It represents a complex field. Teaching expertise involves adapting and working out how to make quality interventions in different situations. Teachers need to believe their unique contribution matters and that they can make a difference in order to bring their best selves to the job.

The deficit model which has been prevalent has had a detrimental impact on them. Mentoring and coaching skills have developed and adapted in different ways as this evolving education sector has started to use its insight, experience and wisdom to support teachers' practice. The differences and definitions of coaching and mentoring are fuzzy. Two main themes are professional person-centred counselling techniques supporting and promoting a person's own professional development and the passing on of knowledge, advice, practical approaches and expertise in specific tasks and roles (Hawkins, 2017). See Endnote.[30]

The 'mentor' and 'coach' title and role is interpreted and understood by custom and usage by different professionals in different settings. They overlap and blend together in practice. They are both about helping and empowering teachers to learn, rather than instructing them to perform in an unthinking process. They focus on people's different strengths and personalities, encouraging self-awareness and building on personal strengths and interests. People function more efficiently when they network and share practice as part of a learning community. They can share good ideas and values and learn from each other. Lofthouse says, "We need an education system that sustains good practice over time," and to "re-imagine teaching cultures that build on past good practice." Teacher and leader efficacy improves through mentoring and coaching creating a more sustainable system going forward. She feels teacher learning, teacher retention, schools and teachers' lives are currently vulnerable. See Report Endnote.[31]

Lofthouse informs us that, in her opinion, the education system is becoming stripped of its memory because so many *teachers are leaving*. Over the last 20 and 25 years sustained practice is in the process of being lost. This point of view is evidenced by The VITAE Report (2006), commissioned by the Department for Education and Schools (DfES), to investigate factors contributing to teacher effectiveness in different career phases in a range of school settings. It was undertaken by the University of Nottingham and the University of London, Institute of Education. They were interested in teacher effectiveness and pupil attainment over time, how recent major changes to the National Curriculum might have impacted the system and how to minimise bureaucratic demands upon teachers. This is significant today when we look back and consider how we came to be in our current educational situation.

A dialogic body of experienced educators performs an important function. It not only provides a forum but an effective 'clearing house' for discussion and expert advice about new and popular research pedagogies, their deployment and effectiveness in different settings. For example, an article in the Chartered College of Teachers' in-house magazine by Daniel Muijs approves new insights into the 'science of learning' in classrooms and schools. However, he warns against the danger that often scientific ideas may be taken up, misunderstood, mutated and turned into fashionable fads. He tells us that schools and teachers across the country are using concepts like cognitive load theory, perhaps in the way that 'learning styles' (and 'positive mindset') were sometimes misinterpreted. Currently metacognition and self-regulation illustrate this point. He says there is indeed a strong, growing body of research showing that the ability to self-regulate and deploy metacognitive strategies helps pupils to be effective learners.

However, he does not approve of such terms being used irresponsibly in schools, because they can be misunderstood and misused to support unhelpful generic approaches such as 'separate lessons on learning to learn,' or that a teacher's role is merely to guide pupils to develop their own self-regulation. Muijs explains that to him self-regulation is perhaps most usefully understood as consisting of three areas of psychological functioning: cognition, metacognition and motivation. He says that self-regulation is about the learners' own awareness of their strengths and weaknesses. It is about how learners motivate themselves to engage in learning, the strategies and tactics they can use to enhance the learning process. In learning, cognition refers to information-processing strategies when doing a particular task, for example, paying attention and practising; metacognition refers to strategies to control and regulate (and perhaps to analyse). Motivation is related to interest, curiosity and beliefs about oneself related to a task. See Article Endnote.[32]

Muijs goes on to focus on how teachers may be able to improve pupils' metacognition as a useful intervention to attempt, provided they are 'clear on what it is.' *This article is evidence of expert qualitative advice which can help to instruct and focus teachers around how to improve practice.* Teacher challenges are to some extent similar and in other ways different in their various cultural situations with cohorts of learners around the world. In 2019 Cat Scutt and Sarah Harrison edited a report for the Chartered College which talks about some of the international trends, opportunities and challenges in teachers continuing professional development (CPD). They researched through connections through an organisation that links 66 international schools in 29 countries offering personalised learning to about 64,000 students between 2 and 18 years old (Nord Anglia Education). They collected ideas and case studies from teachers to show how professional learning is being developed in practice, its impact on students and evaluation by teachers.

Andy Puttock, group education director of the schools' organisation facilitating collecting the report data, believes there are some universal principles underpinning good professional learning that this report illustrates. He says these involve a "virtuous cycle of performance review and development with feedback, identification of learning goals and true involvement by colleagues in identifying their own needs. He does not believe in hasty external opinion as used in the 'outdated concept of performance management." He favours professional learning conversations, mentoring, coaching and guidance, provided it is done with rigour and the aim of improving and maintaining high standards in professional practice. He admires the Chartered College of Teaching because he says they celebrate, support and connect teachers to take pride in their profession and help to bridge the gap between practice and research, equipping teachers with knowledge and confidence to make decisions about teaching. There has certainly been an increase of interest from teachers in expressing their views on social media since it came into being in its current form. See Report and Webinar Endnotes [33,34].

5.7 Conclusion

As previously stated, we have plenty of historical narrative and environmental evidence of our ability to create nurturing communities and cultures that work. Sadly, human societies also create adverse social conditions that produce unpleasant, cruel, unhappy, deprived and unproductive social environments and national systems. It is time to work collaboratively around the world to create local and national environments and curricula systems that work for everybody's long-term benefit. This chapter has looked at some examples and it seems that some teachers have been capable of quietly doing this, often in isolation, by researching, reflecting, analysing, inferring and implementing within their own time frame and contexts all around the world.

Understanding our own brain plasticity, with its propensity for extension and deterioration, in many of the complex aspects of our lives might help us to do this even better and on a wider scale. Neil Hawkes was a former head teacher in the United Kingdom who worked with his school community to devise and implement an ethical and transformational approach. This is an innovative approach which can be created in different schools. The practical educational philosophy he has developed since then is known as Values-based Education (VbE). Stephanie Giles, former pupil at the first VbE-school, gives a convincing narrative explanation.

Originally at the time Hawkes wanted to enhance and transform his school's culture. This involved forming a core plan, building awareness, sharing and listening to perceptions about learning. His vision was about giving children an 'ethical vocabulary' based on values such as respect, humility, tolerance and justice. He noted that it also profoundly affected the adults in the community as they modelled these values. They were empowered and the approach also empowered children to become self-leaders. They tended to develop an active 'moral compass' that affected behaviour, their critical thinking and the quality of their schoolwork. See Article Endnote.[35]

Notes

1. Daniel Martínez, Lara & Rafa Cano Méndez 2018 An animated film "Alike: In a busy life, Copi is a father who tries to teach the right way to his son, Paste. But… what is the correct path?" La Fiesta, Poducciones Cinematigraficas, S. A. https://www.youtube.com/watch?v=kQjtK32mGJQ&feature=youtu.be. Accessed 31/01/2021.
2. Sue Cowley 'Life in the sandpit' 17 January 2018 https://suecowley.wordpress.com/2018/01/17/life-in-the-sandpit/. Accessed 24/05/2018.
3. M. Suzanne Zeedyk (2006) 'From intersubjectivity to subjectivity: the transformative roles of emotional intimacy and imitation' Research Article,

Infant and Child Development Journal: Wiley. https://onlinelibrary.wiley.com/doi/abs/10.1002/icd.457. Accessed 08/02/2019.

4. Pam Jarvis, Senior Lecturer, Department of Children, Young People and Families, Leeds Trinity University 'It's against human nature to send two-year-olds to school' 10 February 2015 https://theconversation.com/its-against-human-nature-to-send-two-year-olds-to-school-37180. Accessed 03/06/2018.
5. 'The Reggio Emilia approach to early years education' https://www.daynurseries.co.uk/advice/the-reggio-emilia-approach-to-early-years-education/. Accessed 30/01/2021.
6. Nick Corston @nickcorston 'We are all Artists' – A Wag Film about STEAM Co. www.steamco.org.uk Video- https://youtu.be/etuYGxomHWg. Accessed 10/03/2021.
7. **SEND Education Acronyms**

SEND	Special Educational Needs & Disability
ADD	Attention Deficit Disorder
ADHD	Attention Deficit Hyperactivity Disorder
ASD	Autism Spectrum Disorders
AST	Advanced Skills Teacher
AT	Attainment Target (in National Curriculum)
BAC	Behaviour and Attendance Collaborative
BESD	Behavioural, Emotional and Social Difficulty
BSL	British Sign Language
BSP	Behaviour Support Plan
BSS	Behaviour Support Service
CAT	Cognitive Ability Test
CL	Curriculum Leader
CPD	Continuing Professional Development
DCPO	Designated Child Protection Officer

General Education Acronyms

BSA	Basic Skills Agency
DBS	Disclosure Barring Service
DfE	Department for Education
EAL	English as an Additional Language
EDU	Education Development Unit

https://www.athona.com/education/send/information/send-education-acronyms/. Accessed 11/02/2021.

8. Charles Sabel, AnnaLee Saxenian, Reijo Miettinen, Peer Hull Kristensen, and Jarkko Hautamäki. 'Individualized Service Provision in the New Welfare State: Lessons from Special Education in Finland' October 2011 Sitra Report Prepared for SITRA and Ministry of Employment and the Economy, 62

   Sitra Studies, the Finnish Innovation Fund: Helsinki 2011 Finland ISSN 1796–7112 ISBN 978–951–563-824-3. https://www.sitra.fi/en/. Accessed 09/02/2021.
9. Elly Chapple 'Concentrating to hear….#FlipTheNarrative' 3 March 2020. https://candoella.com/concentrating-to-hear-flipthenarrative/. Accessed 11/02/2021
10. Elly Chapple 'Diversity is the key to our survival: The Shoeness of a Shoe' TEDxNorwichED 26 August 2019 https://www.youtube.com/watch?time_continue=6&v=DXBdiGUQ8Lw Accessed 27/08/2019
11. Elly Chapple 'The Deafblind world: The 'Shoeness' of a Shoe' 8 February 2019. https://specialneedsjungle.com/the-deafblind-world-the-shoeness-of-a-shoe/ Accessed 08/02/2019
12. Barbara Miles 'The Importance of Hands & Touch for Children who are Deafblind' The Perkins School for the Blind, Massachusetts USA https://www.perkins.org. Accessed 08/02/2019.
13. 'How it feels to be a new parent of an autistic child' by Kirsty Henderson on Twitter 23 July 2020, 20 tweets, 5 min read https://threadreaderapp.com/thread/1286399737332146176.html Accessed 30/07/2020

   I've read some negative comments regarding the parents of autistic children lately. As a parent of an autistic child & many others who advocate and do everything humanly possible for theirs here, can I explain in a thread what it is like to be a parent from the start?

   1/ You have a beautiful baby who misses all their milestones. You spend months thinking you are a terrible parent and that everything is your fault. Dreading taking them to checks in case the health visitor thinks you are as awful of a parent as you feel.

   2/ They don't laugh, smile, look at you, mumble the words mummy, want to eat, cuddle, be touched or engage in play with you. You genuinely start to think your own child doesn't like you very much.

   3/ So you take them to the GP. They do a referral it's comes back as not enough evidence…this will happen over and over again. In the meantime you are left to deal with the unknown and following your own thoughts that your child may be autistic while also thinking you are crazy.

   4/ You send them to pre-school and hope that socialising with the other children will help. It doesn't. It's awful. You are called to collect them early every day to arrive to a screaming distraught child who have 9/10 injured themselves in their distress.

5/ Still no help from anywhere so you become desperate and find a private autism pre-school and pay £15 a hour to send them there not knowing if they are even autistic because NO ONE will have the conversation with parents about this or say the word autism but will drop a hint...

6/ Finally you get to the diagnosis appointment. After hearing and eye tests, SALT and everything in between. My son luckily had only just turned 3 but others will fight and fight for years for a diagnosis for their child which causes so much emotional distress.

7/ We attend endless parenting courses aimed at parents who have children with SEND. Fitting these in while working/caring/in education/ running a family and a home. You read every book on autism and research online for hours.

8/ Every single time you go out someone will make a comment or stare. It can takes months to feel brave enough to use the disabled badge or disabled toilet.

9/ Your relationship with everyone around you breaks down, with your partner, your friends, your family no one understands, you have no one to talk to and you are left isolated.

10/Then we hit education. Reduced timetables, meltdowns, trashed classrooms and bullying. Receiving phone calls asking you to collect your child and hearing them screaming having a meltdown in the background, you can't get their fast enough to console them.

11/ EHCPs & ILPs. The paperwork and hours spent fighting for these are gruelling. Reading what your child cannot do. How they are 4 but working at 6-12 m & how much help they need when all you want to do is scream how amazing they are & how they know every dinosaur name.

12/ You can't go anywhere, do anything. We watch them go through so much finding their way in this busy, loud, sensory overloading world and it's hard to see. We just want to wrap them in cotton wool forever.

13/ Some of our children do not have a voice yet, so we use ours for them. It doesn't mean we are speaking for them but as parents we are pretty good at being in tune with them and knowing what is best...They can literally mumble 'Nsjdjdnbdjsnd!' and we know they want a cracker.

14/ We spend hours in therapy with them, appointments and endless meetings. We wake up at the crack of dawn to take them to SEN activities to ensure they have some fun, because obviously businesses can only offer autism sessions at 8.30 am on a Sunday morning.

15/ Parents of autistic children are at higher risk of suffering from PTSD, anxiety and depression.

16/ We truly want what's best for them. Maybe we will use the wrong wording & make mistakes sometimes but we are learning too. Take people at their intent & ours is for them to grow up in a world with better understanding and not to face the difficulties we have seen others face.
17/ I can't speak for other parents. This is just a snippet of my experience thus far. But the parents I have met along the way are some of the most incredible, kindest, knowledgeable people I've ever met and who I would turn to before health care professionals for advice.
18/ I wrote this to just give you an idea of what goes on in the background and to help give some you a better understanding of us as so many of you give us of you. ❤
19/ Thank you for everyone's lovely comments, DM's and RT's, I have read them all! I'm so pleased that so many of you could relate to my thread and it was so well received. ❤ #autism

Personal thoughts:

All children regardless of SEND have different learning styles and abilities. I believe it is crucial to quickly identify the needs of each child and be proactive in providing the correct support. Children with SEND require extra support and encouragement with clear targets and routines. I believe that teachers undertaking placements in special schools and attending SEND-related courses will help to give them a deeper understanding of a child's needs and their clinical diagnosis. When a child's needs are not being met it quickly affects their happiness and well-being in turn impacting their education and mental health.

When labelling children in a school setting, it must be done carefully as to not create a divide. With the discretion of the parent and child I believe there are potential benefits to other students knowing a child's diagnosis. However, to use 'labels' freely in front of other students there must first be better understanding and inclusivity within schools, for example, lessons on neurodiversity and disability to avoid bullying and creating an us/them. From a parent perspective I have found that labels have enabled my children to the support they need, although I do believe this should not be the case and a child should receive the appropriate support based on their needs, not diagnosis.

Kirsty Henderson is a mother of two children with additional needs, she is currently undertaking a children's nursing degree and co-runs a website for SEND-related advice. Her son (four) is autistic, and her daughter (eight) has Attention Deficit Hyperactivity Disorder (ADHD) and Sensory Processing Disorder (SPD) and is on the pathway for an autism diagnosis.

14. Attachment Theory: https://www.youtube.com/watch?v=WjOowWxOXCg
15. Sue Gerrard 'Why parents get the blame' 14 May 2012. https://whatisautismanyway.wordpress.com/2012/05/14/why-parents-get-the-blame/. Accessed 20/11/2018.
16. Sue Gerrard 'Policy makers on the brain' 1 July 2012 https://movingonfrombowlby.wordpress.com/2012/07/01/124/. Accessed 20/11/2018.
17. Jules Daulby '40 years on from the Warnock Report and England's exam system is setting up children with SEND to fail in mainstream schools' 10 May 2018. https://www.driveryouthtrust.com/40-years-on-from-the-warnock-report-and-englands-exam-system-is-setting-up-children-with-send-to-fail-in-mainstream-schools/. Accessed 14/7/2018.
18. Harmeet Sahota 'Why I wanted my daughter to underperform in her SAT's' February 2021 https://www.curriculumwide.co.uk/blog. Accessed 20/02/2021.
19. David Harkin 'When Toby turns 18 – My hopes for the future of education' 5 February 2021 https://fed.education/david-harkin-when-toby-turns-18-my-hopes-for-the-future-of-education/. Accessed 22/02/2021.
20. The Hospitable Wanderer: Educational Meanderings: 'The roles and qualities of a Waldorf class teacher.' https://hospitablewanderer.com/2017/12/11/the-roles-and-qualities-of-a-waldorf-class-teacher/. Accessed 08/03/2018.
21. Peter Gray PhD (2011) 'What Einstein, Twain, & Forty Eight Others Said About School: "I was at the foot of my class."—Thomas Edison' https://www.psychologytoday.com/gb/blog/freedom-learn/201107/what-einstein-twain-forty-eight-others-said-about-school. Accessed 10/03/2021.
22. Stephen Shore (2020) 'In Greta, A Moment – And A Champion – For Autism' 24/04/2020 https://www.innovateli.com/in-greta-a-moment-and-a-champion-for-autism/. Accessed 21/02/2021.
23. Independent report Lord Laming 'The Victoria Climbie Inquiry Report' (2003) https://www.gov.uk/government/publications/the-victoria-climbie-inquiry-report-of-an-inquiry-by-lord-laming. Accessed 28/02/2021.
24. Annie Woods 'Every Child Matters: This policy strategy is a shared programme of change to improve outcomes for all children and young people resulting from the Laming Report (2003) into the murder of Victoria Climbié' Quick Reference Overview http://www.everychildmatters.gov.uk/aims/ Explains the strategy and planned outcomes. https://www.oxfordreference.com/view/10.1093/oi/authority.20110803095802724. Accessed 28/02/2021.
25. The Warnock Report (1978) 'Special Educational Needs: Report of the Committee of Enquiry into the Education of Handicapped Children and

Young People' London: Her Majesty's Stationery. Scotland.http://www.educationengland.org.uk/documents/warnock/warnock1978.html. Accessed 03/03/2021.

26. Headteacher Update Article (2013) 'What does the Equality Act mean for schools?' https://www.headteacher-update.com/best-practice-article/what-does-the-equality-act-mean-for-schools/66681/. Accessed 03/03/2021.
27. Jules Daulby '40 years on from the Warnock Report and England's exam system is setting up children with SEND to fail in mainstream schools' 10 May 2018. https://www.driveryouthtrust.com/40-years-on-from-the-warnock-report-and-englands-exam-system-is-setting-up-children-with-send-to-fail-in-mainstream-schools/. Accessed 14/7/2018
28. Nicole Dempsey, Assistant Principal Dixons Trinity Academy, UK 'Equality, Inclusion & education, equality, equity, inclusion, schools, SEN, SENCo, SEND, teaching' 5 September 2020 https://inco14.wordpress.com/2020/09/05/grassroots-inclusion/. Accessed 24/02/2021.
29. Rachel Lofthouse talks to Mark Bocker FCCT (2019) 'Coaching, Mentoring – what's the difference? Rachel Lofthouse explains' Podcast. https://pivotaleducation.com/coaching-mentoring-whats-the-difference-rachel-lofthouse-explains-pp247/. Accessed 30/01/2020.
30. Appropriately applied, I found the following principles helpful to keep in mind when mentoring and/or coaching and evaluating:

    - Maintaining an open mind—being willing to learn from participants' contributions whatever they are, even when the meaning is not immediately obvious.
    - Giving space and listening to thoughts developing as they are expressed.
    - Not feeling I had to understand everything.
    - Suppressing my own urge to share at length (a danger when mentees are sympathetic listeners).
    - Relinquishing 'control' of the discussion.
    - Introducing a topic only as appropriate to the mentee.
    - Backing off when my own avenue of thought was not considered relevant by the mentee.
    - Constructive self-criticism can be supported through mentoring, by acceptance of self-criticism and by pointing out positives.
    - Teachers' reflecting on pupil experience improve their own perspectives about their teaching approaches, planning and assessment.
    - Sharing one's own emotional problems and fallibility to empathic, trusted colleagues, tends to increases empathy and understanding for pupils, helping to prevent hasty judgements and stereotyping.

- The researcher sharing notes, checking out conclusions with the mentee, and accepting amendments and corrections promotes authenticity and increases the data quality.

Hawkins, J. A. (2017) Feelings and Emotion-Based Learning; A New Theory New York: Palgrave Macmillan (Chapter 7).

31. The VITAE Report (2006) Christopher Day [1], Gordon Stobart [2], Pam Sammons [1], Alison Kington [1], Qing Gu [1], Rebecca Smees [2] and Tamjid Mujtaba [2] 'The VITAE Project Report Variations in Teachers' Work, Lives and Effectiveness' [1] School of Education, University of Nottingham [2] London Institute of Education Nottingham © Crown Copyright 2006 ISBN 184478 728 1 DfES Publications www.dfespublications.gov.uk. Accessed 10/06/2018.
32. Daniel Muijs (2020) 'Cognition, learning and educational-research' https://impact.chartered.college/article/cognition-learning-educational-research/. Accessed 06/02/2020.
33. Cat Scutt and Sarah Harrison (eds) Chartered College of Teaching Report (2019) 'Teacher CPD: International trends, opportunities and challenges' https://chartered.college/publications/. Accessed 04/03/2021.
34. SchoolRubric (2021) 'The Sunlight Through the Rain (What's Your School Story?): A tribute to some of these educators and messages of hope that have inspired us' https://www.youtube.com/watch?v=dXlnX7fv8PY. Accessed 09/03/2021.
35. Stephanie Giles; Neil Hawkes (2021) 'What is Values-based Education? Does it work? Stephanie Giles speaks out' https://neilhawkes.org/tenets-of-vbe/. Accessed 10/03/2021.

References

A Difficult Parental Decision: The Arrowsmith Program or NotNancy Gedge. (2019). *Disabled Children Are People.* https://notsoordinarydiary.wordpress.com/. Accessed 14 Oct 2019.

Abbott, J. (2009). *Over Schooled But Under Educated: How the Crisis in Education Is Jeopardizing Our Adolescents.* Continuum.

Allen, R., & Sims, S. (2018). *The Teacher Gap.* David Fulton, Routledge.

An autistic mother. (2020). Many SEND Children Are Not Even Offered the Chance to Show Their Potential: Are We Still AIMING HIGH for Disabled Children/Young People? https://wecanaccess.com/are-we-still-aiming-high-for-our-disabled-children/. Accessed 01 Sept 2020.

Anderson, J. (2018). *A Wall Built to Keep People Out': The Cruel, Bureaucratic Maze of Children's Services: In a System Cut to the Bone, Gaining Access to the Support We Had Been Promised for Our Daughter's Special Educational Needs*

Was an Exhausting, Soul-Sapping Battle. https://www.theguardian.com/society/2018/nov/01/cruel-bureaucratic-maze-childrens-services-special-educational-needs?CMP=Share_AndroidApp_Tweet. Accessed 04 Nov 2018.

Nathan Archer, Early Years Researcher. (2020). *Dictated by Data? Nathan Archer Explores the Idea That an Outcome-Driven, Accountability Culture Is Shaping a Limited and Limiting Research Agenda. He Calls for Richer, More Diverse Perspectives.* https://www.earlyyearseducator.co.uk/features/article/dictated-by-data#.XkP-Rf_Mnx4. Accessed 14 Feb 2020.

Mike Armiger. (2018). *Mental Health – A Shared Language-Teachers Should Have Mental Health Training – Teachers Should Be Trained in Counselling-Teachers Should Understand the Signs of Mental Illness-Teachers Should Be Given the Tools and Language to Have a Conversation About a Pupil's Levels of Distress, Pain or Overwhelm.* https://mikearmiger.wordpress.com/2018/01/07/mental-health-a-shared-language/. Accessed 01 Jan 2019.

Arrowsmith-Young, B. (2013). *The Woman Who Changed Her Brain: How We Can Shape Our Minds and Other Tales of Cognitive Transformation.* Random House.

Jonny Bailey. (2019). *Assistant Head for Inclusion at a Primary School in London. SEND for help! The SEND System Is Broken.* https://educatingjon.wordpress.com/2019/02/22/send-for-help/. Accessed 04 May 2019.

Barbara Arrowsmith-Young. (1996). *Definition of a Learning Dysfunction.* https://arrowsmithschool.org/definition-of-a-learning-dysfunction/. Accessed 21 Apr 2019.

Bignell, S. (2018). *Autism, Asperger's and ADHD: What You Need to Know. A Guide for Parents, Students and Other Professionals.* My Child Services Publishing.

Ian Birrell. (2018, 22 December). I Have Seen So Many Horrors...But This Is Barbarism: Ian Birrell on the 'cruel and destructive' Conditions Children with Learning Disabilities Have Been Kept in Around Britain. *The Mail on Sunday.* https://www.dailymail.co.uk/news/article-6523805/IAN-BIRRELL-seen-horrors-barbarism.html. Accessed 31 Dec 2018.

Black, P., & Wiliam, D. (1998). *Inside the Black Box: Raising Standards Through Classroom Assessment.* King's College School of Education.

Bold Beginnings Report. (2017). Bold Beginnings: The Reception Curriculum in a Sample of Good and Outstanding Primary Schools. *Ofsted.* https://assets.publishing.service.gov.uk/government/uploads/system/uploads/attachment_data/file/663560/28933_Ofsted_-_Early_Years_Curriculum_Report_-_Accessible.pdf. Accessed 15 Feb 2021.

Bowlby, J. (1988). *A Secure Base: Parent-Child Attachment and Healthy Human Development* (Tavistock Professional Book). Routledge.

Cagliari, P., et al. (Eds.). (2016). *Loris Malaguzzi and the Schools of Reggio Emilia: A Selection of His Writings and Speeches, 1945–1993 (Contesting Early Childhood)*. Routledge.

CANDOELLA Blog Elly Chapple. https://candoella.com

CANDOELLA Elly Chapple. https://candoella.com Twitter @elly_chapple

Caroline Bond, Placement Director & Carla Tomlinson, Trainee Educational Psychologist – MIE. (2019, January 15). *Behind the Mask: Recognising and Supporting Autistic Girls in Secondary Schools.* http://blogging2.humanities.manchester.ac.uk/miebee/behind-the-mask-recognising-and-supporting-autistic-girls-in-secondary-schools/. Accessed 15 Jan 2019.

Elly Chapple and Ella Chapple (2018). *It's a Story of Hope, Persistence and Tenacity and It's Rooted in Authenticity Despite Overwhelming Odds.* https://candoella.com/our-story/. Accessed 10 Jan 2019.

Graham Chatterley. (2019, May 8). SEMH School Leader – Has Kindly Provided This Axcis Guest Post Which Offers a Heartfelt Insight into the Ups and Downs of Being a Parent to a Profoundly Autistic Child. My Takeaways from the Autism Roller Coaster (Guest Post). http://testblog.axcis.co.uk.grid-hosted.co.uk/2019/05/my-takeaways-from-the-autism-roller-coaster%ef%bb%bf-guest-post/. Accessed 04 Dec 2019.

Cohen, I. L. (2007). A Neural Network Model of Autism: Implications for Theory and Treatment. In D. Mareschal, S. Sirois, G. Westermann, & M. Johnson (Eds.), *Neuroconstructivism: Perspectives and Prospects* (Vol. 2). Oxford University Press.

CollectivED. (2020). *Working Papers from CollectivED: The Centre for Mentoring, Coaching and Professional UK Learning.* Leeds Beckett University: Carnegie School of Education. https://www.leedsbeckett.ac.uk/-/media/files/schools/school-of-education/collectived-issue-10.pdf?la=en. Accessed 05 Mar 2021.

Cordingley, P., Higgins, S., Greany, T., et al. (2015). *Developing Great Teaching: Lessons from the International Reviews into Effective Professional Development.* Teacher Development Trust. Available at: http://TDTrust.org/dgt. Accessed 15 June 2019

Cowley, S. (2018). *The Ultimate Guide to Differentiation: Achieving Excellence for All.* Bloomsbury.

Cozolino, L. (2013). *The Social Neuroscience of Education: Optimizing Attachment & Learning in the Classroom.* W.W. Norton & Company, Inc.

Curriculum Wide Harmeet Sahota. www.curriculumwide.co.uk. Twitter @ HarmeetSahota

Darwin, C. (1860). *On the Origin of Species by Means of Natural Selection, or the Preservation of Favoured Races in the Struggle for Life* (2nd ed.). John Murray. Retrieved 9 January 2009.

Department for Education. (2012, December 27). *Pupils with SCLN (Speech Language and Communication Needs) in Mainstream Schools – These Reports Present the Findings from a 3-Year Study of Pupils with Language Impairments and Autism Spectrum Disorders in the Education System*. Department for Education. https://www.gov.uk/government/publications/profiles-of-need-and-provision-for-children-with-language-impairments-and-autism-spectrum-disorders-inmainstream-schools-a-prospective-study. Accessed 15 Sept 2021.

Department for Education. (2016). *Standard for Teachers' Professional Development*. Available at: https://www.gov.uk/government/publications/standard-for-teachers-professional-development. Accessed 30 Sept 2019.

Down Syndrome Website. *Misconceptions vs. Reality*. https://www.globaldown-syndrome.org/about-down-syndrome/misconceptions-vs-reality/ Accessed 21 Mar 2019.

Eaton, H. (2018). *A Difficult Parental Decision, the Arrowsmith Program or Not*. http://howardeaton.com/2018/02/a-difficult-parental-decision-the-arrowsmith-program-or-not/. Accessed 25 Feb 2020.

EDUCATING JON. Twitter: @JonnyBailey87 https://educatingjon.wordpress.com/author/jonnybailey111/

Education Endowment Foundation. Working with Parents to Support Children's Learning: Four Recommendations on Working with Parents to Support Their Child's Learning. https://educationendowmentfoundation.org.uk/tools/guidance-reports/working-with-parents-to-support-childrens-learning/. Accessed 20 Feb 2021.

Elvy. *Female, Disabled and Mixed Race Person*. https://elvysblog.com/

Farberov, S. (2012). The Remarkable Tale of a Woman Who 'Fixed' Her Own BRAIN After Spending Nearly 30 Years Unable to Read, Write or Tell Time. https://www.dailymail.co.uk/news/article-2200343/Barbara-Arrowsmith-Young-The-remarkable-tale-woman-fixed-BRAIN.html. Accessed 24 July 2019.

Florance, C. (2004). *A Boy Beyond Reach*. Simon & Schuster UK Ltd..

Gareth Morewood. Website http://www.gdmorewood.com @gdmorewood.

Garvey, P. (2019). *Talk for Teaching: Completely Rethinking Professional Development in Schools*. John Catt Educational Ltd.

Gilbert, I. (2015, 2018). *There Is Another Way: The Second Big Book of Independent Thinking*. Crown House.

Harvey Goldstein, Gemma Moss, Pamela Sammons, Gwen Sinnott and Gordon Stobart (2018). 'A Baseline Without Basis: the Validity and Utility of the Proposed Reception Baseline Assessment in England' British Educational Research Association Full Report.: https://www.bera.ac.uk/researchers-resources/publications/a-baseline-without-basis. Accessed 15 Feb 2021.

Hawkins, J. A. (2017). *Feelings and Emotion-based Learning; A New Theory.* Palgrave Macmillan.

Goodall, J., & Weston, K. (2018). *Engaging Parents: 100 Ideas for Primary Teachers.* Bloomsbury.

Joanna Grace. *Seeking to Contribute to a World Where Everyone Is Understood In spite of Difference.* Website http://www.thesensoryprojects.co.uk/

Grassroots Inclusion. https://inco14.wordpress.com/ @NDempseyTA.

Hale, I. (2018). *Asperger's, Autism and You: Is the Epidemic Real? Are Vaccines the Cause?* Institute for Education, Research and Scholarships.

Hattie, J., & Yates, G. (2014). *Visible Learning and the Science of How We Learn.* Routledge.

Hayward, F. H. (2020). *The Educational Ideas of Pestalozzi and Frobel.* Alpha Editions.

Highlighting Strategies for Student Success. https://edsource.org/topic/special-education

Huxtable, M. (Ed.). (2019). Foreword 'Jack Whitehead and Practitioner Living Theories' University of Cumbria. *Educational Journal of Living Theories, 12*(2), i–viii. http://ejolts.net/drupal/node/345. Accessed 10/12/2019

Innovate My School. *Dr John Patterson.* http://www.innovatemyschool.com/ideas/how-visually-impaired-kids-can-help-education-right-now. @DrJohnAPatters1 @StVincentsL12

Interview with Dr Ian Hale. (2019) and Prof. Newton Lee by Chloe Estelle Carrier on the New Book *Asperger's, Autism & You: Is the Epidemic Real and Are Vaccines the Cause?* Institute for Education, Research, and Scholarships. https://youtu.be/Xq0Fw2GFanE. Accessed 21 Feb 2021.

Jarvis, P. (2020) Attachment Theory, Cortisol and Care for the Under-threes in the Twenty-first Century: Constructing Evidence-Informed Policy. *Early Years.* 10.1080/09575146.2020.1764507. https://www.tandfonline.com/doi/full/10.1080/09575146.2020.1764507. Accessed 23 Aug 2021.

(a) Jarvis, P., (b) Newman, S., & (c) Swiniarski, L. (2013). On 'becoming social': The Importance of Collaborative Free Play in Childhood. (a) Leeds Trinity University, Leeds, UK; (b) Leeds Metropolitan University, Leeds, UK; (c) Salem State University, Salem MA, USA Taylor & Francis in *International Journal of Play* on 12 Feb 2014. Available online: http://wwww.tandfonline.com/10.1080/21594937.2013.863440. Accessed 09 June 2018.

Jarvis, P., et al. (2017). *Early Years Pioneers in Context: Their Lives, Lasting Influence and Impact on Practice Today.* Routledge.

Pam Jarvis. (2017). Is Baseline Missing the Bigger Picture? *Times Educational Supplement.* https://www.tes.com/author/pam-jarvis. Accessed 01 Mar 2018.

Martin Jensen. (2019). *Pedagogisk Leder på Friggfeltet barnehage | Kursholder i Røde Kors 'How Do I Address Conflicts in a Professional Manner?'* March 15, 2018. https://www.linkedin.com/pulse/how-do-i-address-conflicts-professional-manner-martin-s%C3%B8rby-jensen/. Accessed 23 Sept 2019.

Kell, E. (2018). *How to Survive in Teaching without Imploding, Exploding or Walking Away*. London.

Kennedy, I. (2010). *Getting It Right for Children and Young People: Overcoming Cultural Barriers in the NHS so as to Meet Their Needs*. HMSO.

Pooky Knightsmith. (2021). Mental Health Educator, Speaker, Author & Advisor. Website https://www.pookyknightsmith.com/. Accessed 21 Feb 2021.

Pooky Knightsmith, Autism, Anorexia & Me – Pooky Explains About Being Autistic to Gareth Morewood. (2019, February 18). https://www.youtube.com/watch?time_continue=24&v=PqrAgjV8OnA .Accessed 21 Feb 2021.

Lampert, M. (2001). *Teaching Problems and the Problems of Teaching (math)*. Yale University Press.

Andrea Libutti. (2019). 'Autism: 7 Reasons Why YOU Have A Choice.' https://www.andrealibutti.com/2019/04/24/autism-7-reasons-why-you-have-a-choice/. Accessed 25 Apr 2019.

Luqman Michel Twitter. @Luqmanmichel.

Luqman's Dyslexia Blog. (2010). *Dyslexia – A Different Perspective: My Experiences Teaching Children with Dyslexia. Dyslexia – An English Language Problem*. https://www.dyslexiafriend.com/. Accessed 23 Feb 2021.

Lynn McCann Is an Autism Specialist Teacher, Trainer and Author. https://reachoutasc.com/

MacEachron. (2018). *Video 'What You Need to Know About Neurodiversity*. https://nowthisnews.com/videos/news/what-you-need-to-know-about-neurodiversity?utm_content=What+You+Need+to+Know+About+Neurodiversity&utm_source=twitter&utm_medium=social-media. Accessed 25 Feb 2021.

Mainstream Sen. https://mainstreamsen.wordpress.com/

McCann, L. (2017). *How to … Support Children with Autism Spectrum Condition in Secondary School*. LDA.

McGill, R. M. (2017). *Mark. Plan. Teach*. Bloomsbury Education.

Michel, L. (2018). *Shut Down Kids: How to Prevent Kids from Shutting Down*. Notion Press.

Barbara Miles. Reflections on Deafblindness: Hands & Touch. http://www.perkinselearning.org/videos/webcast/reflections-deafblindness-hands-touch. Accessed 05 Dec 2018.

Montessori education. https://en.wikipedia.org/wiki. Accessed 10 Feb 2021.

Montessori, M. (1991). *The Montessori Method*. Schocken Books Inc.

Montessori, M. (1965). *Dr. Montessori's Own Handbook: A Short Guide to Her Ideas and Materials*. Schocken Books Inc.

More Than A Score. (2018). *Baseline Testing in Reception*. https://youtu.be/OxwjEPyKW0s. Accessed 01 Mar 2021.

Gareth D. Morewood. *Autism, Education & Emotional Regulation*. Website www.gdmorewood.com

Mariette Mulvenna. (2019). *Action on Hearing Loss 'What It Really Feels Like Growing Up Deaf.'* https://www.belfasttelegraph.co.uk/life/health/what-it-really-feels-like-growing-up-deaf-38085521.html. Accessed 07 May 2019.

Munro, E. (2011). *Independent Report-Munro Review of Child Protection: A Child-Centred System-Report Setting Out Reform Proposals to Enable Professionals to Make the Best Judgements to Help Children, Young People and Families.* Published 10 May 2011. Department for Education. https://www.gov.uk/government/publications/munro-review-of-child-protection-final-report-a-child-centred-system. Accessed 15 Sept 2021.

National Center on Deaf-Blindness. https://www.nationaldb.org/

Graham Newell (2012) .'Going Beyond CPD: To develop Outstanding Teaching and Learning.' https://www.irisconnect.com/us/wp-content/uploads/sites/3//2014/08/going-beyond-UK-IRIS-Connect.pdf

Patterson, J. A.. (2020, November). Principal, St. Vincent's School for Sensory Impairment. *Impact Journal, Chartered College of Teaching.* https://impact.chartered.college/article/lighting-pathways-investing-visually-impaired-creativity/. Accessed 12 Dec 2020.

Report. Special Educational Needs and Disability Code of Practice: 0 to 25 years' 2015 United Kingdom: Her Majesty's Stationary Office download. www.gov.uk/government/publications. Reference: DFE-00205-2013. Accessed 15 Feb 2021.

John Roberts. (2020). *SEND omitted from some new Ofsted reports: Charity director raises concern after analysis shows 16 new school inspection reports make no mention of SEND pupils.* https://www.tes.com/news/first-look-ofsteds-new-inspection-reports. Accessed 25 Feb 2020.

Rogers, C., & Freiberg, H. J. (1983). *Freedom to Learn* (Rev. ed.). Prentice Hall International.

Rogers, C. R. (1951). *Client-Centred Therapy.* Constable.

Rogers, C. R. (1961). *On Becoming a Person: A Therapist's View of Psychotherapy.* Constable.

Tom Rogers. (2019). *There Are a Handful of Research Findings That Have Made Tom Rogers Rethink His Practice as a Teacher.* He Presents Them Here https://www.tes.com/news/eight-key-pieces-research-teachers. Accessed 24 Dec 2019.

Michael Rosen. Children's Author and Educational Expert 'Michael Rosen Talks in Depth About Dyslexia: What It Is, How to Understand It and Useful Advice for Parents and Teachers, with Expert Professor Maggie Snowling CBE. President of St John's College. https://www.bbc.co.uk/sounds/play/m0002gr1. Accessed 25 Feb 2019.

David Roy, Lecturer in Education, University of Newcastle & David Armstrong, Senior Lecturer in Special and Inclusive Education. (2019, January 15). What You Need to Know If Your Child with a Disability Is Starting School Soon: Parents May Need to Play the Advocate for Their Child, Especially If Their

Child Has a Disability Which Affects Their Ability to Communicate. RMIT University, 7 *The Conversation* Australia @dmsroy. https://theconversation.com/what-you-need-to-know-if-your-child-with-a-disability-is-starting-school-soon-107275. Accessed 25 Feb 2019.

Sherrington, T. (2019). *Rosenshine's Principles in Action*. John Catt Educational.

Shore, S. (2001). *Beyond the Wall: Personal Experiences with Autism and Asperger Syndrome*. Autism Asperger Publishing Co.

STARLIGHTMcKENZIE. https://starlightmckenzie.wordpress.com @ StarlightMcKenz

Stenhouse, L. (1975). *An Introduction to Curriculum Research and Development*. Heinemann Educational.

Street, M. (2020). Benefits and Limitations of Eliciting the Well-Being Views of Two-Four Year Olds Living in a Low-Income Area in England. *The Child Indicators Research*. https://doi.org/10.1007/s12187-020-09773-1

Sutton Trust. (2011). *Improving the Impact of Teachers on Pupil Achievement in the UK – Interim Findings*. Available at: https://www.suttontrust.com/wp-content/uploads/2011/09/2teachersimpact-report-final.pdf. Accessed 30 Sept 2019.

Katharine Swindells. (2020, January 25). *Teaching Intelligence: How to Support Visually Impaired Students: Subjects Traditionally Closed to Visually Impaired Learners Are Being Opened Up by New Tech – But Universities Must Provide Policies, Training and Resources to Match*. https://www.timeshighereducation.com/teaching/teaching-intelligence-how-support-visually-impaired-students. Accessed 04 Mar 2020.

Taylor, T. (2016). *A Beginner's Guide to Mantle of the Expert: A Transformative Approach to Education*. Singular Publishing.

Team Coaching Charmaine Roche. https://www.youtube.com/watch?v=vkUU56RtUDc&feature=youtu.be; https://www.lifeflowbalance.co.uk/why-i-coach-teams-team-coaching-insights-vlog-series/

Teather, S. (2011). *Support and Aspiration: A New Approach to Special Educational Needs and Disability – A Consultation*. HMSO.

Tharby, A. (2018). *How to Explain Absolutely Anything to Absolutely Anyone: The Art and Science of Teacher Explanation*. Crown House.

The National Autistic Society. (2020, September). *New Police Autism Guide Released*. http://www.npaa.org.uk/police-autism-guide/. Accessed 21 Feb 2021.

The Perkins School for the Blind. https://www.perkins.org

The Special Needs Jungle. https://www.specialneedsjungle.com/

The Whole School SEND Consortium & National Association for Special Educational Needs (nasen). https://www.sendgateway.org.uk/resources

Thinkpix Blog. (2019, February 9). *Lollygagger or Learning Difference? How Do Schools Create 'lollygaggers'?* https://thinkpix.blog/2019/02/09/lollygagger-or-learning-difference/. Accessed 15 Feb 2019.

Thinkpix Blog. (2020, November 14). *Intervention and the Medicalisation of SEND*. Twitter @thinkpix_suze. https://thinkpix.blogintervention-and-the-medicalisation-of-send/. Accessed 10 Dec 2020.

UK Department for Education. (2018). *The School Snapshot Survey: Summer 2018*. Available at: https://assets.publishing.service.gov.uk/government/uploads/system/uploads/attachment_data/file/773842/Summer_2018_SSS_Final_Report.pdf. Accessed 30 Sept 2019.

Watson-Druee, N. (2015). *From Impossible to Possible: the Leaders Code*. Markham.

Sally Weale, Education correspondent. (2019). *Special-needs Children Lose Out on £1.2bn of Support, Says Union Teachers Say Government Funding Has Failed to Keep Pace with Soaring Demand*. https://www.theguardian.com/education/2019/apr/15/special-needs-children-lose-out-on-12bn-of-support-says-union?CMP=share_btn_tw. Accessed 15 Apr 2019.

WeCanAccess. https://wecanaccess.com

Karen Wespieser. (2019). *It's Not Fair That Getting an EHC Plan Could Depend on Your Postcode*. Director of Operations, Driver Youth Trust https://schoolsweek.co.uk/its-not-fair-that-getting-an-ehc-plan-could-depend-on-your-postcode/?unapproved=148217&moderation-hash=77e1ab8d44fd4cc92acbb90aebb62919#comment-148217. Accessed 30 May 2019.

What Are Speech, Language and Communication Needs (SLCN)? https://www.afasic.org.uk/about-talking/what-are-speech-language-and-communication-needs-slcn/. Accessed 15 Sept 2021.

Willingham, D. T. (2009). *Why Don't Students Like School? A Cognitive Scientist Answers Questions About How the Mind Works and What It Means for the Classroom*. Hoboken NJ.

Working papers from CollectivED. (2018, November). The Hub for Mentoring and Coaching, A Research and Practice Centre at Carnegie School of Education Issue 6. CollectivED@leedsbeckett.ac.uk @CollectivED1.

Jill Young. (2019, March 5). *Book review: 'An Introduction to Autism for Adoptive and Foster Families*. https://www.nafp.co.uk/blog/book-review-introduction-autism-adoptive-and-foster-families. Accessed 21 Mar 2019.

CHAPTER 6

Behaviour, Inclusion and Mental Well-Being

6.1 Introduction

Lisa Feldman Barrett (2017) says there should be a change in the way we view the brain. We have been teaching children in a cultural climate of dissonance, discord and double values for too long. For example, when we make and model arbitrary judgements about situations and other people without knowing anything about them. Her conclusion that the brain is constructive and predictive in nature has been reached after many years of working as a neuroscientist. My own research has come to parallel conclusions starting from an intuitive teacher base and investigating how learners are affected by their life experiences in various ways. Unfortunately our culture has been subliminally influenced by the popularity of using 'deconstruction' negatively based on assumption in psychodynamic arguments about human psychology.

This traditional scientific approach has led to the categorisation of human behaviour influenced by populist prejudices. Categorisation, although it is of initial useful as a cognitive gadget, cannot deal with human changeability and complexity. For example, Sigmund Freud (1856–1939) made some useful observations, but assumed all his patients could be put into fixed categories of mental health difficulty (2010). He

Special thanks to Dr Christine Challen 'Happiness' for A poem from 'My Mind's House', 20 August 2019.

J. A. Hawkins, *Brain Plasticity and Learning*,
https://doi.org/10.1007/978-3-030-83530-9_6

rationalised their conditions into a set of 'abnormal' disorders which could be diagnosed and delineated within the medical models he observed and 'invented.' So popular were his theories and observations that most people accepted and believed in these kinds of restricted diagnoses. Unfortunately these ideas and tropes still tend to influence the public view of education.

There is truth there but it makes no sense to regard the brain as necessarily fixed and typical or atypical. It is diverse, variable and plastic in nature. The psychologist Carl Jung (1875–1961) was more open-minded and, in view of his era, took a surprisingly insightful and pragmatic approach. He listened to his patients and attempted to understand each person's sometimes extraordinary experience. He was ahead of his time, sympathetic and open-minded, in spite of a rigid religious upbringing. He listened to people and considering the times in which he lived, he showed an extraordinary degree of insight and empathy. This may be why his work still has relevance for psychotherapists today (1995). National governments are still influenced by traditional psychological approaches. They see education as an important area of justification. They tend to view it as representative of their country's status in the world.

Historically uninformed external social and political attitudes and pressures have filtered through and in turn influenced some teaching practices. Even today some traditional ideas are influential. It takes time to change mindsets around accepting the full diversity and value of human populations and cultures. However, a change of attitude is evident in much of the data I have collected. This is evident in individual teachers', psychologists' and parents' blogs, articles, books and academic research. The issues of pollution, global warming, environmental threats, industrial globalisation, population health and welfare are causing concern, forcing a search for solutions and a concentrating of minds upon how education might be used to create a better future that *includes* everybody. Some issues around making education fit for this purpose are discussed in this chapter.

6.2 Teaching Approaches That Encourage Inclusion

Empathy helps us to understand 'common' mistakes and guide learners through specific subject difficulties as we plan our teaching. This is often because we learned how to solve these difficulties ourselves when learning particular concepts and ideas. However, empathy is harder to feel when the reason for the difficulty is completely hidden from us by a total lack of experience and knowledge about that particular problem. We cannot imagine it. Bruce Perry found this when he found himself faced with seriously damaged children and realised that he must develop his own skills of empathic enquiry in order to help them. We will look into this problem in this chapter (Szalavitz & Perry, 2010; Perry & Szalavitz, 2006).

An ordinary everyday example of a lack of empathy might be a 'naturally' numerate Maths teacher who may go too fast for students who are lacking in or haven't built particular neural capacities for calculation. In Math you can, of course, achieve a correct result by various methods, but some are easier for some people than others. If you have no difficulty then it is hard to understand the problems for others. A less able teacher may learn to be a better one because of an ability to break down difficulty into recognisable stages. It is a skill that can be developed by listening and open-minded inquiry. If we can accept that other people may conceptualise differently then we may be able to offer different ways to do so. We can model this behaviour in school.

The existence of brain plasticity informs us that if we can stimulate, intrigue and interest learners in the 'right' way, they may have learning capacity available. In the classroom this is our challenge rather than theirs. Sometimes this may be through teachers presenting brilliantly, but often it may be through asking for and listening to learner feedback, creating an empathic relationship based on interested observation and kindness. All ages and abilities of learners benefit from one to one, peer and social support. Empathy creates and opens up connection and enables mental support and development. This has been recently evidenced by the isolation and mental health problems experienced during the worldwide pandemic.

The reverse of empathy through unjust comment and abuse is also evidenced by the effects of trolling on social media and highlighted by the positive aspects of empathy by online peer support and counselling. We can encourage empathy by helping learners to 'experience' other people's lives by engaging in stories, describing feelings and situations using imagination. As teachers, journalists, film-makers and creative practitioners we can take learners through the 'second hand' experience of stories in literature and drama classes. We can make them 'real' and visceral experiences by stimulating the senses and emotions, engaging and helping people to make the physical electrochemical connections that encourage inferential, sympathetic and imaginative thinking (Zacks, 2015). Neuroscientist and film buff Jeffrey M. Zacks explains this in his book *Flicker*. See Article Endnote.[1]

We know now brain plasticity can have a downside on the internet. It can be used as a deliberate unethical process where conspiracy theories and financial scamming are involved. That this works can be understood by the effect of the repetition of 'fake news' by bad actors in the modern media. Such a strong 'mental reaction' must be countered by all kinds of teachers and politicians with good intentions in order to prevent the exploitation of this negative neural conditioning. Such a force must be underpinned with values and regulation if the world is to be 'improved.' We can have more empathy for victims and understand why people affected by conspiracy theories feel and behave the way they do. Brain plasticity explains how shallow and basic algorithms online using behavioural psychology 'conditioning' have been able to impact our lives (Skinner et al., 1957).

Education can employ and deploy the use of discussion (dialogic) and inferential logic in order to counter this kind of 'neural brain washing.' In order to do this we need some awareness of our own susceptibility. Corey Block is an American teacher of English in New York who felt a strong call to action around politics in 2017. His normal practice is to make his English teaching relevant to the times he and his pupils are living through. He realised however, it would be unethical to impose his own political opinions on his students in spite of the strength of his feelings. He was tempted to teach by directly addressing political injustice, but pulled back because in order to help pupils to think for themselves he accepts the need to also show professionalism and restraint. See Article Endnote.[2]

He felt sharing his own political views was not part of his job, even though he was pressured by colleagues. He says that although teaching consistently is not as 'cathartic as protesting,' over time it can be 'even more subversive'—in the long run. This has always been a possibility and reminded me of another iconic book *Teaching as a Subversive Activity* by Neil Postman and Charles Weingartner (1969). He had been tempted to change the curriculum to discuss and debate current events. However, he decided to stick to his plan and teach: *Night* by Elie and Marion Wiesel (2008); *The House on Mango Street* by Sandra Cisneros (2004); *American Born Chinese* by Gene Luen Yang and Lark Pien (2006). These are all books that encourage pupils to make connections with disadvantaged children's lives. Even though they are modern stories, they are not brand new. However, he knows them well and there is a teaching advantage with ease in style of presentation to be gained.

He points out that these three books are diverse but all have human experiences in common—ones that his pupils, many of whom are immigrants, can probably relate to. This is the skill of the language teacher, but even so perhaps immigrant children should experience more stories in their native languages. If this is done well it might very well help to build a bridge to the secondary language being taught. See Sect. 6.6. He notes that these stories relate to teenage angst in different environments. He says that stories entertain and unite people especially if they provoke sympathy, empathy and understanding. They take the reader out of their comfort zone. They help us care about other people and understand their feelings in different situations.

Like many teachers he is a thoughtful and a practical philosopher working to inspire empathy and demonstrating it himself. Block understands how to help his pupils develop and make mental connections and he listens when they relate the books they are reading to their own experience of 'Real' or 'Fake News.' He wants them to feel compassion and perhaps choose themselves to write about human issues, controversy and conflict. He gives them the 'experience of reading and being part of a classroom community.' He points out that we humans *are* all stories and believes this is the way 'to help to shape our collective story for the better' because the memory of a book lasts in the mind for longer than, for example, a slogan on a protest sign or an internet meme.

New experiences can be created through exposure to situations outside a learner's everyday life. For example, through exposure to music, drama, social projects, group endeavours and different kinds of tribal celebrations, for example, sports like the Olympics. Empathising is about 'feeling' other people's feelings as if they were our own and we can gain extra experiences and information in this way. When we empathise we come to see another person's perspective and may understand it even when we disagree. Sometimes, when it turns out they know a different truth to our own, we benefit from some new information and it can help to hone and rearrange our own ideas. We can thus generate and produce new thoughts separately and together by combining ideas. This is as true in maths and science classes just as in the humanities or any other subject. A good actor and scriptwriting teacher Hywel Roberts (2012) is an expert in engaging learners of all ages by intriguing them with stories. He explains how in his book *Oops! Helping Children to Learn Accidentally.*

Teachers have their own understanding of empathy and have often set out to create sympathetic, enabling environments within their classrooms and schools. A teacher's own personality contributes to the feeling and the ambience adds to student experience. Within a happy school different personalities blend and interact to create a varied social experience for children. Elizabeth Peyton has been teaching in a small urban middle school for refugee and immigrant kids for 12 years where 99% of students receive free school lunches. She says she's tried a lot of classroom management and pupil behavioural approaches—for example, (1) picking-up on the 'small stuff,' (2) establishing strict, detailed behaviour routines she had difficulty remembering herself and (3) positive reward systems. See Article Endnote.[3]

She knows these can work for other teachers but her own favourite effective strategy is to always default to compassion. She asks them if they are hungry and tired and 'what is the matter' and tells them the group have missed their presence when they are late—by sympathising first she finds she discovers the truth much quicker in most cases. She says that if you start with compassion and empathise, you are more likely to find out the cause, because some students like adults face invisible challenges they don't know how to solve. Even if you cannot fix the problem you can help the student to verbalise and give them support and sympathy. Peyton says compassion builds relationships with students and she finds erring on the side of understanding works better than being overly harsh.

She says this approach can help prevent the 'school-to-prison pipeline' and the needlessly punitive discipline policies that punish vulnerable students. She admits that default to compassion will not solve all classroom management problems, but it will certainly help, because if you treat a student like a decent person they are more likely to act like one. She has been able to get help to students—like the girl whose single parent had been arrested leaving her alone at home, hungry and afraid—no wonder she was not doing her homework! Some students may see a teacher showing compassion as weakness, but at least they will know their teacher cares. She finds also that following up on explanations generally reveals deceptive behaviour before it becomes embedded.

The logic of Peyton's approach described above is supported by the work of Peter Fonagy, head of psychology and language sciences at University College London and a proponent of attachment-aware teaching. He says attachment is important, not just for vulnerable pupils who have experienced trauma, but to every child. This process is one which all children and young people and even adults go through when they go into a new school or job environment. He regards attachment as a simple biological process, which affects us when we are nervous or fearful or in an unfamiliar situation. This makes sense if one considers the anxiety caused by attempting to learn something new that appears at first sight difficult. If we have the support of trusted teachers and colleagues it is easier to be persistent and do the necessary work. See Article Endnote.[4]

6.3 Positive Behaviour Changing Solutions

The ability to empathise and interact in order to enable learning in others is a learned skill in itself. Now that it is endorsed by the concrete scientific evidence of practical neuroscientists it should be properly acknowledged as a type of medical and teaching intervention. Across and between the fields of medicine and education there are already many kinds of specific 'alternative' interventional therapies which can now be scientifically validated, rejected and/or adjusted appropriately to suit patients, clients and learners. Regulatory bodies can and should be updating their information to train and support teachers, social practitioners and therapists to research and reflect about if and how new neuroscience discoveries might improve and justify their work (Chatterley, 2020).

Lori Desautels, a former special education teacher, is currently an education professor (Desautels & McKnight, 2019). She co-teaches and researches with teachers using brain and trauma-informed approaches to teaching. She investigates ways to prepare the brain to learn, by calming stress response systems and finding strategies that help teachers be more attuned to learners' changing brain states. She teaches teachers about how nervous response systems of traumatised people can skew, subvert and block communication, attention and retention. In doing this every teacher can develop their own individual therapeutic style and expertise with the guiding support and supervision of a mentor or coach. See Article Endnote.[5]

Desautels (2020) points out that Western education has encouraged social agencies, schools, teachers and parents to bring up children through the lens of Behaviourism. These ideas have been embedded in the society along with common ways to address particular problems. *The focus has been on narrow compliance and control, judged by behaviour observed according to appearances.* She says that zero tolerance educational policies designed by privileged school leaders and social groups have caused misunderstandings. Teachers in these systems are often forced to abandon empathy and deny the effects of past experience on children. Teachers are not generally trained sufficiently in ways to respond sympathetically to children with problems.

There are many reasons for showing and demonstrating empathy and compassion towards learners—not least it encourages them to think more intelligently about the world and other people. Students are more likely to engage with other people's ideas and this is a good basis for developing inferential thinking. In fact inappropriate programmes using behavioural conditioning tend to discourage sophisticated thinking. Resentment, confusion and rebellion are unlikely to cease just because we tell students they are wrong. Children's natural process of logical connection will continue to operate beneath the radar even if it is incorrect. It is easy to discount other people's thought processes, but coercion is counterproductive in teaching and the very opposite of empathy. In some cases it constitutes even abuse and neglect.

Sometimes 'inappropriate' thoughts and feelings require verbalising and acknowledging to make them register before a person works them out and sees why they are wrong. Other thoughts may be inappropriate to be voiced in certain situations. We see this currently on social media platforms. However, it should be explained that it is the way we act upon our thoughts that matters most in all aspects of life. People who acknowledge their mistakes generally deserve to be allowed to change their behaviour and move on. We need to listen and validate children's feelings, even if we cannot immediately make sense of what they say or do. We can also model positive acceptance by our general demeanour. Geoffrey James teaches and practices a technique called solutions-focused coaching (2016). Years ago James worked in a pupil referral unit in the United Kingdom. His job involved working with children who were failing to achieve at school. He now goes into schools and 'sorts' problem behaviour out using one-to-one interviews.

While studying for his PhD he learned about 'solution-focused thinking'—a therapy approach developed by Harvey Ratner (2012). He has developed that approach so that with training, it can be used appropriately in schools. The approach is so non-invasive and non-controversial it can even be demonstrated to staff when he talks to a fresh child and under his expert guidance, it still works extraordinarily well. I have watched James at work and it is interesting to see how he has trained himself to be very adaptable to a child's responses and the questions of a teacher he is training. He gives the positive regard recommended by Carl Rogers & Freiberg (1983), the founder of person-centred counselling by picking up immediately on anything that can be viewed as a positive within a child's or adult's answers to his 'open' and 'positively focused' questions.

James does not deconstruct or conduct a negative analysis of 'what went wrong.' He has interpreted and developed this approach in a workbook in simplified form so that teachers can develop their own expertise (2019). It gives them a supportive framework of questions to fall back on and explanatory case studies to think about. He says that he realises that often 'children are doing their best even when it might not look like it.'

This may be simply because they are working with and responding to the experiences and information they have and are misinterpreting them. They may be unable to see a way out of situations. When they have been sent to see him and just at the point where the children expect to be criticised, he gets them in a good frame of mind by asking them, what they like best.

James goes on to point out any good aspects and strengths within their responses using his own expertly tailored methods and intuition. He asks them at a suitable point what will help to make their situation at school better. The problem is usually revealed by default. One teacher commented that learning this new positive teaching approach was like learning a new language. It is the opposite of 'finding out what is wrong.' He stresses that his approach requires thoughtful interpretation by an ethical teacher, but generally provides a system that is safe for the child because they decide how they will solve problems. It is appropriate whether you consider yourself a 'traditional' teacher or a 'progressive' or a mixture of both.

The workbook and method does not require a teacher to be trained in psychoanalysis. It demonstrates how to use a basic set of questions in a practical positive way. James points out that the method tends to aid metacognition. It might also encourage more mental flexibility, adjustment, confidence, self-control and self-management. We can imagine it as if a dormant positive cognitive net or map lights up with an endorphin hit that helps push a habitual negative reflex response to the 'back' of the mind. This is an example of the benefits of researching aspects of neuroplasticity and the secret is often in the situational nuance and detail.

Like James many social research practitioners are gradually helping to improve our understanding of autonomic physiology. Although still in the early stages of research, it might explain much about why, in the past, such work has often been successful. Psychiatrist Stephen Porges proposed the Polyvagal Theory in 1994, believing that the physiological state of the human autonomic nervous system affects behaviour. The theory and its implications are currently being researched and considered in education and social welfare settings (2011, 2017). For example, he has been developing music-based interventions that can improve language processing, reduce hearing sensitivities and improve spontaneous social engagement.

Stephen W. Porges has developed a procedure called the 'Safe and Sound Protocol' which is currently being trialled. See Video Presentation Endnote.[6] He believes that neurophysiological states affect attachment, communication, emotion and self-regulation. He says that 'feeling safe' has a transformative effect on behaviour. Picking up on this, Desautels, who works in the same university in Indiana, is currently teaching teachers applied educational neuroscience and the effects of trauma on the brain. She is researching how an understanding of these matters can help develop strategies and practices in the classroom to help the brain to learn while dampening down stress response systems.

Desautels teaches educators about Polyvagal Theory and how to regulate our minds and bodies by finding strategies to 'improve' the balance of our mental and physical 'states.' She says that learning requires physiological state regulation so human brains can access and integrate cognitive and mental abilities to navigate life's experiences. She says in the United States today, teachers are used to a traditional 'law and order' mentality measured by observed behaviour. This is now out of date and inefficient in the light of new neuroscience and clinical psychology research. We now know that our nervous systems and physiological states create these behaviours and it is impractical to spend a lot of time labelling and classifying them. They are too complex and transitory. Themes within them may be useful but circumstances change.

6.4 Childhood Experiences Affect Teenage and Adult Mental Development

In 2019 Mary Meredith started a study group amongst colleagues on Twitter to read and discuss Bruce Perry's book *The Boy Who Was Raised as a Dog: And Other Stories from a Psychiatrist's Notebook* (2006). She produced a series of blog articles and brought her considerable experience as a local education authority Inclusion Officer in the United Kingdom to the task of understanding the implications for her work with teachers and schools. In considering the book and her own work, she was forcefully struck by the huge underlying influence and effect of human social interaction upon the way children behave. Two things struck her as particularly useful: the idea of enlisting peer support 'within a school's provision for pastoral care' and the benefits of listening to the family in order to gain understanding of each child's situation. See Blog Article Endnote.[7]

She found particularly significant one of Perry's cases (Peter) where he discovered and unravelled this healthy baby boy's life story. He looked at Peter from Peter's point of view, given his babyhood experiences. He imagined the baby boy's situation in a clinical and inhumane orphanage. After listening carefully to the adoptive parents' story, he was able to work out the reasons for Peter's behaviour. He explained some of the little boy's story to his peers at school and asking them to help him. He sensed that other small children might be able to understand. They became keen helpers. Perry is adept at making different imaginatively targeted interventions using his professional judgement. For example, the ones he made with and around Peter helped to counteract developmental gaps in his experience 'across emotional, social, motor and cognitive domains.'

Meredith notes that there had been a 'downward spiral of rejection, confusion and frustration' and that each were fundamentally reversed to create a 'cascade of positive reinforcement' that 'fed' Peter's neural development. He eventually recovered and went to high school, doing well both academically and socially. After many years of dealing with cases of ordinary and extraordinary instances of child trauma, Perry has analysed the necessary conditions that are necessary for healthy intellectual child development in his neurosequential model. He says this is not a fixed brain model map but rather a useful process approach model. Meredith views these ideas as of fundamental importance in her own work. Her group of educationalists and practitioners have decided to continue their discussions around these ideas going forward. She recommends some resources in the endnotes below.

Perry is continuing to stress the importance of complex social interaction in his analysis of his work and developing the application of his neurosequential model, which Meredith shows and discusses in her article. Meredith says that in a sense the 'golden threads' of the model are Perry's three R's—Regulate, Relate and Reason—and that as well as sequential they were 'woven together' in Peter's case. This observation may be key to understanding how our brains and bodies gain different apperceptions about our world. These *sequential conditions* are the ones we need to enable our brains to cope with, learn to adapt to and survive in the world. We might frame learning in this way as positive neural connective proliferation. Meredith considers each condition and looks for practical resources and interventions that promote each of them by relating them to resources in her own field of inclusion support. See Advisory Website Endnote.[8]

Regulating Neural Regulation involves the brainstem and midbrain, the sensory motor brain where stress responses may need soothing and calming. For example, the impulse to fight, flee or freeze. As previously mentioned, the best way to help is by being present in a calm and centred way because of the way we human mammals reflect and mirror with our neurobiology each other's states of being. Meredith says that children can be helped to develop ways to 'regulate, relate and reason' themselves and that some of these can be taught to all children in helping themselves and their friends. See Endnote.[9]

Relating It refers to the emotional relational brain or limbic brain when the child forms an attuned connection with another person. It is a deep social instinct based in our mammalian history of group membership for survival. It is a sensitive and perhaps to some extent sensitised relational ability, which a teacher or caring adult can exercise themselves and encourage by listening attentively and validating with acceptance. It is a skill that can be practised and developed, which involves working with different people in various situations and learning different appropriate levels of thoughtfulness and empathy. Meredith recommends an article by Kim Golding giving advice on how to approach this process with children who present challenging behaviours in adoption or foster care needing relationship-based parenting or 'connection before correction.' See Article Endnote.[10]

Reasoning It involves the brain cortex which operates best when we are calm, connected, focused and ready to learn. We cannot calmly progress and learn when we feel dysregulated and disconnected from the reality of the task, our own grounded physical reality, our environment and/or other relevant people. We are able to work out, accept and set limits on behaviour, when we are feeling accepted, calm, confident and safe. Ideally home and school community should provide these conditions for a child. See Szalavitz & Perry (2010). Such treatment builds up a resilience and a confidence which can resonate through life. Those who never experience such acceptance often struggle to cope and adjust to future situations in life. Meredith looks at schools and finds them very much lacking. She says our approach requires a radical change of attitude.

All kinds of individual vibrant learning experiences—through loving to frightening emotional events can have a lasting positive and/or negative impact on children's lives. For example, children who have been nurtured by some adults, have also suffered emotional trauma and/or physical abuse from others and even quite often suffered both from the same adult. One might think of it as a form of torture. Bad experiences are known as adverse childhood experiences (ACEs) and they are always complicated. Parents had similar childhoods and are repeating behaviour patterns. Lenore

Walker coined the phrase that describes it—the 'cycle of abuse'—and it takes many forms (1979). Troubled children are suffering—they are often anxious and aggressive. Punishment, deprivation and force re-traumatise 'damaged' children and tend to reinforce their difficulties.

We need to teach vulnerable children to work within the school rules within a framework of loving attention and positive regard even when they let us down—in order to help them create positive neural connections. Meredith thinks a 'reliable 3 R respecting plan' is useful on these occasions. We may or may not be able to change circumstances depending on each unique situation, but patient, loving, consistent care does support traumatised children and adults at any age. This is true even when carers, teachers or therapists are not aware of the full context. For example, Jaz Ampaw-Farr (2021) tells us how teachers at school helped her to build resilience and cope with an extremely violent, tragic and chaotic dysregulated home life they were mostly unaware of. She thanks her teachers and explains how she has succeeded in life in spite of that traumatic upbringing. See Video Endnote.[11]

Yousif Davis, an MSc in Health Psychology, has worked with States of Mind, a group of 12 community schools in London. While doing a placement with them he researched their mental health education programme for teenage students. He says that mental health and well-being support in schools does not generally meet young people's needs, but that this programme works by helping to empower them to solve their own problems. He says if they are given a chance to tell you what they need along with some training and resources they are capable of finding their own solutions. See Article Endnote.[12]

6.5 Teaching That Is Trauma-Informed and Adaptable

Graham Chatterley (2020), former Social-Emotional-Mental-Health (SEMH) lead tutor, turned training provider and parent of a child with special educational needs explains that no child can be expected to explain their own behaviour in adult terms. They are inexperienced in life and in a process of learning how to become an adult—sometimes society asks too much of them. We often cannot necessarily understand a child's point of view unless we understand what has happened to cause their behaviour. The fact we think they are behaving inappropriately does not mean they are doing it to annoy us. Restorative practice is all about seeing from the eyes of the person your actions have affected. Some schools are getting better at this, spending more time teaching children and teachers to see the perspectives of others.

This is also something we all need to do as adults and teachers. We need to alter our perceptions and change our own logic which has similarly been moulded by our beliefs and experiences. Christine Challen has written a poem about happiness, she is a scientist and a further education teacher. She and her children have experienced personal trauma and like many teachers her experience has helped her to develop her own understanding of herself and others. Her academic paper about the benefits of 'being, becoming and thriving,' learning and recovering through art and literature is based on her own recovery as well as her teaching experience. See Academic Paper Endnote[13] and Special Data Contribution.[14]

All of us are liable to suffer trauma at some time in our lives and there are many different uniquely various situations that cause problems. Desautels (2020) says there are many layers of different kinds of adversity and trauma within the lives of children, families and communities. Currently evidence suggests a surge in anxiety and depression among children and adolescents. Reports in the United States by the Center for Disease Control and Prevention (CDC) say there are an estimated one in five US children experiencing a behavioural, emotional or mental disorder exacerbated by the world pandemic. They include attention deficit hyperactivity disorder, anxiety, disruptive behaviour and depression.

She says the body and brain development of children and youth is shaped by their perceptions of human interaction going on around them. Children who are traumatised, fearful and angry are not able to hear instructions. They respond instinctively to tone of voice and cannot think clearly. Teachers and social practitioners need to learn to 'read' emotional states, empathise, offer protection, redirect and respond appropriately with warmth and positivity. Teachers need training to understand about emotion contagion because counter-aggression is a human biological instinct—an autonomic response that is activated when we feel at risk.

Our autonomic responses affect everyone alike, including teachers and children. They trigger dissociation and fight or flight nervous system pathways with a likelihood of cutting off empathy for others. Porges (2011) explains that humans communicate through dyadic reciprocal reactions. These kinds of reactions are 'beneath the radar' in everyday life as we read each other's body language and use physiognomic perception to judge each other. This is why traditional discipline does not always enable learning, particularly if people are distressed. Learning works best when everyone involved is calm and feels safe. Teachers need to be trained in how to bring an awareness of their own ability to use co-regulation to their teaching and use it in the classroom. It is a skill that can be learned and developed through practice. Co-regulation involves the teacher asking themselves questions about the situation. For example, how and where can learners feel safe? What do they need to know to set their minds at rest? Are they frustrated by a task? What do I need to understand? How can I reinforce and complement what is going well?

Co-regulation may involve body language, non-verbal communication and a calm presence in order to induce a sense of relaxed calm. Words are used to offer validation through gentle questioning, telling a story or reframing an experience. Teachers need to know that most troubled youth and children have nervous systems that are wired in a survival brain and body state that is programmed to defend and protect against attack. Teachers need to learn how to develop trust and create an appropriate atmosphere and sense of safety. It can be done by giving space and time for nervous systems to settle and by skilled redirection. Young people who 'misbehave' often have racial and special educational needs inequalities and genuine differences in ability and experience. In order to be able to take in new information and to learn they need to feel calm, confident and supported in school. Porges' Polyvagal Theory (2011) appears to fit in with the work, therapeutic interventions and independent analyses of traumatised children by Bruce Perry (Perry & Szalavitz, 2006).

At the beginning of Perry's career as a psychologist he learned to work with marginalised children. They were not only misunderstood within their communities but by professionals who suffered from what he calls '*child illiteracy*.' These people worked in judicial, educational and clinical jobs that even included child protection. As he tried to help a variety of different children, all of whom had different personalities and life experiences, he listened and observed carefully. Over the years, as he worked to help them, he carefully reflected, researched and reflected again upon each child's case. In the process he found out a great deal about human mental health at any age (Szalavitz & Perry, 2010). See Website Endnote.[15]

Perry found out some universal truths about the way our bodies and minds use neuroplasticity in order to survive. It turns out his work as a psychologist with traumatised children reveals a great deal about how the plastic human brain responds to experiences throughout adult life. This is discussed in his next book, *What Happened to You? Conversations on Trauma, Resilience, and Healing* co-authored with Oprah Winfrey (Winfrey & Perry, 2021). It is worth considering the implications of his work for ourselves with the possible intention of researching his theories with other age groups. If this theory is of practical use, can be seen to describe some fundamental truths and makes sense to enough people, it may become more widely evidenced and substantiated in other areas of social research.

Activist Armand King (2019) explains his story; he says he 'never felt peaceful as a child' because his parents were always fighting. They argued constantly and he grew up trapped in a cycle of verbal and mental abuse. When his parents split up, he was 12 years of age. When he went to find his father to get money for food he said, "Pretend I'm not your Dad!" King ended up in a gang, 'messed up' and involved in criminal activities like selling drugs. For a while, he says making money felt cool to him and he was influenced when he saw prosperous older black men who seemed happy as pimps. He became a pimp himself at 17. He knows now that 60% of his community age group were involved in this lifestyle.

King eventually got 32 months in federal prison far away in Oregon. One day he was told to "Call your Mom—your brother died." He knows many of his peers lost their lives to drugs, murder and madness. He says that now he can't sleep—he just works because 'too many people need help.' "The kids who killed my brother are away in prison." King wants people to empathise with his mission and learn about why he is so inspired to do this work. He says although his life was chosen for him and not by him, he wants to dedicate his future to help stop the cycle of death and incarceration that has plagued his community. He has asked San Diego State University to consider making a film about his life and he has also published a book (2019).

King hopes his story can give hope and guide other young men and women like him out of the destructive lifestyles they are born into. These problems are universal human problems, which manifest themselves in different ways. They are human problems affected by cultural factors that are not just about particular kinds of racial groups. For example, there are groups who are involved in sex trafficking all over the world, for example in Western and Eastern Europe and Asia. It is not just a 'black problem' it is a human problem. Human beings are similar creatures in fundamental ways, even when we think we look different. At the same time people who may look the same to us are also unique in hidden ways, because everyone's personal physical make-up, experience and history is unique. This is the existential human paradox created by our neuroplasticity. See Video Endnote.[16]

6.6 Cultural Diversity, Differences and Similarities

It is often difficult for humans of different cultural and ethnic backgrounds to accept each other's experiences. We assume our own are the same as everyone else's. We think other people choose to be different when they have perhaps often had little or no choice. In order to gain authentic information it is important to our intelligence that we can feel empathy for other people's points of view. Empathic connections develop thinking and learning, contribute to and increase knowledge and intelligence. For example, we all learn many useful things from diverse peoples and cultures—learning different facts and ways to view the world increases our mental agility and can make the world richer, more interesting and more enjoyable.

At the same time, examples and social interactions from role models we see as like us can also encourage and help us to make these kinds of empathic connections. In many cases survivors of poor and criminal lifestyles are the best people to understand these particular social problems. They can inform remedies and make interventions. Armand King's story shows this and that there is always hope even when your upbringing has been extremely traumatic (2019). He says the fight against sex trafficking needs the perspective of someone like him in order to succeed. He dedicates his life to 'helping others help others' and has written a book to teach them. He is an expert on this 'criminal lifestyle culture.'

Adults who have experienced traumas often think there is nothing that can improve their lives, but people do recover. Some of them find a way to recover on their own and others are fortunate enough to get the right kind of support. Interventions can be made by coaches and mentors in a range of contexts. The fact of brain plasticity can now explain this and why much depends on personal experience and the environments we encounter. Bryan Stevenson (2020) found himself, as a lawyer in a very similar position to Dr Bruce Perry when he became involved in defending prisoners on death row. He found that many of them were there, because of upbringing and bad luck and were victimised in the criminal justice system in the United States through absolutely no fault of their own. His book *Just Mercy: A Story of Justice and Redemption* tells about the cases he defended.

Sadly the privations and violence of slavery against them under colonialism have also often denied groups of people the truth about their own heritage. This video is about Tamara Lanier's enslaved great, great, great grandfather Rentry and his daughter Delia. They bear testimony to a group of people whose lives and images were disrespected by people of 'white' privilege and whose descendants ignorantly assume they are still entitled to continue to do so. All humans have a fundamental need to know who they are, where they come from and to belong to the society in which they live. Everyone needs social respect. The colour of our skin is and should be irrelevant to our human value. We need to consider how we can work to develop healthy groups and societies that operate as benevolent families. These should be inclusive and value difference, forging new traditions that include everyone's useful contributions. See Video Endnote.[17]

Toria Bono, who is a teacher, interviewed Paul Hume on one of her website podcasts. In these podcasts she asks a variety of other teachers about their personal motivations and experiences which led them into teaching. Hume gives out friendly teaching advice on Twitter and is known as Teacher Paul. She wanted to find out about his journey to becoming a teacher. Paul is thoughtful when asked who he is and says that being 'Teacher Paul' is only part of his life. He is surprised about his popularity in the teaching community on Twitter. Paul comes from a town in Scotland, where he loved school and enjoyed all the subjects. He thought about doing law, Russian and marine biology at the age of 16 or 17 when he was trying to decide what to study at university. Paul eventually 'followed his friends' and went to university in Manchester, England, at the age of 20 to study architecture and then switched to complete an art degree in fashion. See Podcast Endnote.[18]

After university, Paul worked in the food service industry and then got a job in a call centre eventually becoming a supervisor. After a while, he and his partner who is also called Paul moved to London. At first they both worked as teaching assistants in a variety of London schools before Paul decided to become an early year's teacher and after a number of school moves as a supply teacher he became a deputy head. At the end of their talk Bono and Hulme discuss the benefits of admitting mistakes and learning to mentor others in order to achieve 'becoming' the people they are now. Nine years after starting university, he became a teacher and had been a teacher for 13 years at the time of the podcast interview.

Human beings live not just by and within a spectrum of ability that is predetermined. As illustrated above, our life journeys vary enormously. We have a range of inherited traits and abilities that can be activated (potential) and we live in a whirlpool of different complex experiences that influence us towards or against such processes. We are all unique and fluid, flexible and adaptable within our capacity to adapt neuro-plastically in given moments. We develop and remain dormant neurologically in different ways each day under different environmental conditions. It is important that we question fixed assumptions and perceptions and remain open to learning. We need to share each other's knowledge in order to survive. It would benefit us all greatly if we could develop both ourselves and future generations to be more aware of the desirability of becoming better custodians of all our peoples, social groups and cultures throughout our world going forward.

Carmel O'Hagan, a modern foreign languages teacher (MFL) in the United Kingdom, has written a blog article about language and identity. She asks us: who are we if not our name? She says that as a linguist she believes in the power of language to solve many of the world's problems. She says that as a little girl growing up in Wales she wanted a different name to her Irish one, but fortunately in her schools there were lots of children with 'unusual' names. They were all the children of immigrants—Irish, Scottish, Italian, Caribbean, Indian, Polish, English, German, French, Lithuanian, Czech, South African and more. She says that although the children may not all have liked their names they knew their names were their identity. She has been annoyed by hearing teachers on Twitter joking about "names that are always trouble." She thinks they might be stereotyping classes of people and mocking parents' rights and choices. See Blog Article Endnote.[19]

O'Hagan thinks such talk might easily become racist and that educators should know better. She fell to pondering about the names in her own family. It occurred to her they were all chosen carefully with joy by proud parents. She makes a long list and dares anyone to mock them. She asks us not to make assumptions and says there are so many ways we 'label and define children that are just so wrong.' She thinks it is uncaring and unprofessional and she recommends some books about language names and identity for anyone who 'might want to think further about the power of language.' For example, *The Speckled People* by Hugo Hamilton; *Travellers in the Third Reich* by Julia Boyd; *The House by the Lake* by Thomas Harding and *Homme de Couleur* by Léopold Sédar Senghor available in French, German, Spanish, Italian and English.

Dean Hobbs (2016) finds that bilingual children competent in their first languages demonstrate a greater ability to learn other foreign languages due to their metalinguistic skills. Third language learning helps them to improve their competency in the first two languages. Even so immigrant children in the United States do not typically study third languages because educational leaders are unaware of the benefits. Monolingual educators are unaware of the advantages of previously learning a language. Each language that learners study serves to reinforce the concepts, skills and intuitions assimilated in previous language learning cultivation. Unfortunately minority dialects often appear as errors on national standardised tests. The fact of higher immigrant dropout rates indicates students feel demoralised by a lack of appreciation of their abilities.

Other research demonstrated that teachers in English scored students higher who where first language English speakers. Teachers awarded lower grades to second language English speaking children even if they performed the tasks set. Learners appeared to lose motivation when they realised their efforts were not appreciated. As well as being taught, Dean Hobbs (2016) found children need to be assessed in their first languages as well as in English to gain accurate evaluations in knowledge growth and linguistic skill development. Foreign language learning mitigates against ethnocentrism, racism and xenophobia. Multilingualism offers students larger interaction repertoires with a greater number of people and multilingual opportunities contribute to important democratic values of equality, tolerance and mutual respect.

Carmel O'Hagan says in 2018 the UK government announced new funding for her subject while quietly withdrawing funding from the national Centre for Information on Language Teaching and Research (CILT). She had found it was useful in her long teaching career in making the links between academic research and action research done by teachers in schools and classroom practice. It was the national resource library with thousands of precious texts, perhaps now in storage in a library at Cambridge University. She would like to know where and what is happening to the secondary websites with thousands of resources that supported MFL teaching in secondary schools and Post 16 years. She wonders what has happened to the Primary Languages Training Zone with thousands of film clips and continuing professional development (CPD) materials supporting primary languages teachers. See Blog Article Endnote.[20]

Many are not specialist linguists and were using these resources to merge their primary pedagogy with sound language teaching. Hundreds of well-researched, well-crafted and easily usable materials designed by teachers for teachers existed in the online library to support language programmes in French, German and Spanish to support primary teachers foreign language skills. It included a phonology module to teach non-specialist teachers and thus their pupils the accurate pronunciation in French, German and Spanish. It served as an online, virtual home for many state and independent language teachers, with links to and between ITE, HE and languages. It included a comprehensive national and regional 'training trainers' programme so that schools would not need to spend money on consultants. It appeared to make good financial sense and was a resource for the country's future and yet it is gone apparently without consultation.

A particular world problem is that monolingual students may miss critical thinking and developmental opportunities. To not acknowledge first languages of ethnic minorities deprives monolingual students of the opportunity to learn from multilingual classmates. Monolingual students and teachers are unaware of the advantages inherent in multicultural knowledge embedded in blended identities of these students. Research shows that bilingual students outperform monolingual students on practical and metalinguistic intuitive tasks. Many teachers are unaware of the bilingual and multicultural advantage for all pupils of acknowledging all of the student cultures in a class. Appreciating the skills and experience of minority learners enhances their own and other students' academic performance.

Human learning is a much more complex business than previously understood by educational experts particularly academically based ones. Emotional intelligence is an active and necessary force in brain plasticity. The motivations, physical sensations, mental feelings and reactions of the therapy practitioner and child are an important part of any remedial process. For example, Sarah McDermott tells the story of how her foster child 'asked to be adopted' by drawing himself into a family photo. Learning is also very much involved with parents', learners', therapists' and teachers' personal insights and relationships. Listening to the parents' and learners voices' and observing socially transactional behaviour takes on an enhanced level of importance. It challenges our intelligence and helps us to interpret situations developing and using our insight and inference skills. See Article Endnote.[21]

Amy Gaunt and Alice Stott (2019) explain how teaching and learning can be transformed and stimulated by talk. They point out it is the basis of all reading, writing and communicating through our evolving social and technological networks. Teachers can enable learning by providing themes and structures in a much more focused and targeted way, but with the proviso that new ideas, improvisations and original thought can be valued.

This requires a great deal of professional skill, but is done routinely by language and drama teachers who are able to set up and manage innovative thinking activities while using well developed teaching methods and skills. One-to-one tuition is also a useful targeting approach and tool for facilitating learning. Teaching the isolated facts without context that are needed to be memorised for an exam in large classes cannot *always* trigger or provide the kind of rich stimulation which our plastic brains need and thrive upon.

The professionals' job is to be experienced in understanding the end goal of the work, whatever that may be. However, neuroplasticity shows us that, within each individual brain, there are variations and different ways that goal can be achieved. Further the biological building process requires varying amounts of repetition of tasks presented in similar and different ways. This explains why the two common teaching approaches of 'stimulating and engaging curiosity and interest' and 'insisting on the use of repetitive tasks' both work. In any given lesson the teacher's job is to direct and to produce a learning experience that is likely to engage most of the class. Sorting out and learning how to respond to different abilities and needs is a necessary learning process in itself and to some extent a concurrent one in teaching.

The human race is generally quite forgetful about the past unless it is educated to value it at a young age. There are white people whose ancestors were slave traders, but there are also privileged humans like some white educated middle classes whose families have only been in that position for a relatively short time. Many of us come from families who worked their way up. Many of us have very short memories or families that did not pass on their history. Some of our families were lucky enough to live in countries with governments that believed in human rights. Even though our ancestors fought for those rights and they were hard won, we may have difficulty valuing those democracies and educating our children to appreciate them. Looking at people's experiences in this section it is evident that humans lead diverse lives but also have many common needs.

6.7 Leading an Inclusive, Developing and Supportive Learning Community

We need educational leaders who:

- believe that every child matters and deserves a happy education and can make sure the staff, children and parents know this so that the school grows an autonomous 'equality' ethos (that is, one that gives agency as appropriate to everybody to solve problems).

- are constantly working to understand their school community, how it exists in its own world and how it relates to the community outside.
- constantly work to create, build up and make accountable a system that enables everyone to learn to the best of their ability and personal potential.
- understand that while sharing these responsibilities with staff, it is the leader's ultimate responsibility to enable a system that delivers a 'good' education to everybody.
- listen to people, including the children, taking a coaching and mentoring role when necessary in every area of the organisation.
- are able to inspire confidence by delegating people, enabling them to do their own research and use their own initiative to solve problems within the parameters of their job description.
- Encourage, make time and continuing professional development resources available for staff to develop their own knowledge and career pathways.
- request explanations, justifications, suggestions and 'sharing' contributions from staff in order to learn about new developments, maintain consistency, morale and the quality of the teaching.

A SENCO (Special Educational Needs Coordinator) is a teacher who is responsible for special educational needs in school. Every school in the United Kingdom is obliged to employ a SENCO to ensure all students with learning disabilities get the support they need at school. The evidence from parents and psychologists suggests the system is not operating properly in the United Kingdom and provision is often inadequate. Nicole Dempsey, experienced SENCO and assistant principal of an academy school in the United Kingdom, makes the following points. She writes that the biggest barriers to inclusion in mainstream schools are caused by assuming SEND students are quite unlike other people. True inclusion would be a system designed to meet the full range of needs as part of the main offer. Here social acceptance is key. She thinks that having inclusion as a separate 'bolted on provision' creates a feeling of separation within classes and schools. See Blog Article Endnote.[22]

She finds that SEND students are disadvantaged by the current approach because strategies and systems that work well in one school, or even a few schools, don't necessarily work in all schools. Much of what happens in schools is due to habit and custom. We assume that traditional practices are correct, because we don't stop to question if they are appropriate. Dempsey challenges teachers to ask themselves if SEND pupils are being treated equally and actually receiving the education they need. How much better could it be if their provision was adapted and how would that look. Truly dedicated and professional teachers know 'what a high-quality education looks like' and try to achieve this for most students.

If the children who challenge because of difference and/or special needs are to be included they must also receive the best education possible after questions have been asked and adjustments made by professional teachers. Anything less is simply not education. She believes meeting the needs of SEND students should not be just an add-on because those children should be included in a teacher's thinking when they plan for any class, subject, year group or whole school. This is about changing fundamental attitudes to teaching on the part of society so that teachers are given 'permission' to develop themselves as agents of change. She does not believe that teachers can learn to be inclusive simply by reading about methods in books. Their job is about improving their professional practice.

We are all neurotypical and neurodiverse in different ways and teachers have to learn to deal with all sorts of children on a daily basis. Dempsey recommends collaborative planning by classroom teachers, department heads or senior leaders planning and designing lessons, schemes of work, and routines for students as a whole. In effect she appears to be telling us that mainstream teachers know that SEND students are going to be in mainstream just as they are in society. There are certainly students who need additional and different in order to succeed, but she believes it is possible to integrate, research and collaborate to meet the needs of a wider range of students. Separate provision is not the same thing as additional provision, but it needs organising within school, with resources and outside expertise and support.

She asks why this is taking so long to be incorporated into practice. It seems that experts' traditional, deconstructive analytic training and fixed views about human capabilities has become hardwired into their own perceptions. Across the board this is detrimental where it creates self-fulfilling outcomes. This is particularly the case when children have a different physical make-up, clinical condition or have survived different types of stressful events, trauma and deprivation. They need to be helped to develop by building on those things which resonate—things that their senses and emotions naturally respond to because the brain tends to follow and embed those things it connects with. This requires planning, responsive assessment and interactive responsive teaching. It is about making meaning in the process, even if resolved in retrospect as learning development moves forward.

A placement in a SEND setting for all teachers in training would help them to learn skills and put their work with vulnerable students into context. Dempsey advocates that all teachers read the SEND Code of Practice and note the statutory requirements about their role and their position in the network of systems in place to support vulnerable learners. Dempsey recommends all teachers should read and use it to better meet the needs of students. They should know about educational and adult outcomes for SEND students so that they can improve their prospects. She feels that a teacher's role as an inclusive educator is important. It is crucial that they get it right and continue to improve provision into the future. See Webpage Endnote.[23]

Teachers can do the job provided they are given the training and resources needed to organise and create the necessary systems for this to happen. Jackie Ward, a former deputy head/SENCO at a Pupil Referral Unit (PRU) in the United Kingdom, would certainly agree (2019). She currently goes into schools to give training and advice. Mary Myatt (2018), an expert upon leadership and the curriculum looks at 'essentialism' as an approach to teaching (McKeown, 2021; Greg McKeown, 2014). She thinks teachers should be able to ask 'why' they are being asked to work in particular ways. This is not about being rude but about gaining clarity in their work. For example, we should ask ourselves (and our leaders) why we are being asked to submit detailed plans for lessons and inputting particular sets of data; in asking 'why' we are seeking justification for processes and systems and Myatt suggests that 'some of these are important, but others are not' or have become redundant. See Blog Article Endnote.[24]

Myatt says that everyone in any organisation should be entitled to ask 'why?' In schools these include trainee and newly qualified teachers, teaching support assistants, experienced classroom teachers, governors and leaders. We need to ask, 'what is the problem we need to solve right now?' She believes this can shift the focus to activities and practices that can really make a difference. We should shift from trying to see how to make established routines work to thinking about how to solve current problems. If everyone stops to ask why something is needed, we will find that some things are unnecessary and can concentrate on what needs to be achieved—and it will have to be justified. This can produce clarity and release energy to move work forward.

We can become overwhelmed when overloaded with redundant activities and can end up in a state of learned helplessness, feeling unable to make a difference. Cutting back on unnecessary routines can release new insights and improve outcomes. In another article Myatt says the contemporary sceptical viewpoint recognises problems but without necessarily taking a 'cynical defeatist' stance. This is useful teaching advice especially in the current general and future post-pandemic teaching climate in the United Kingdom. I note, however, that if one looks up the roots of the sceptic and cynic philosophies, there were schools of thought under each title that had connecting themes that can be seen generally as 'positive' and 'negative' as areas of belief. It depends on the concepts held *by the people involved* and the meanings made relevant to specific situations.

Reading across the various 'cynic positions' from a teaching leadership perspective, one might extract advice to think about, be unsurprised by and compensate for possible self-serving aspects of human nature, while appealing to people's more altruistic side. This might be used in a different but 'similar' way to Peyton's 'default to compassion' student management approach (referred to in this chapter at Sect. 6.2). For example, asking the staff to commit to 'value-based education' tends to reveal failures and encourage self-correction—again by default (Hawkes & Hawkes, 2018; Hawkes, 2014). On the whole, Myatt tells us, it is good to take a sceptical position because one avoids 'trite exhortations'; can show an awareness of possible problems; take up a clear-sighted position and challenge staff by asking questions about whether things are strong enough, good enough and relevant enough for current circumstances.

Myatt gives us a useful interpretation of a sceptical approach. She informs us that it turns questioning into an art form that is based on a Socratic line of enquiry. The outcomes are not predetermined or not set out in advance but are engaged in, in order to find a 'better way of going about business.' The elements of our practice that remain after such an inquiry are likely there because they have proved their worth and not because they have become a habit and are just assumed to be right. A sceptical mindset involves asking 'why' in a spirit of being prepared to be surprised and having the self-confidence to not know the answers or solutions straightaway. In this, worrying about image, appearance and approval has to be set to one side in order to consider fresh perspectives and pursue clarity. It is an approach that encourages us to find better alternatives. See Blog Article.[25]

Myatt cites Matthew Syed in his book *Rebel Ideas* (2019). He recommends that it is important to keep an open mind, looking for possibilities for what might be improved. Problems and their solutions are often in plain sight and we need to pay closer attention to cultural aspects of our lives. We should listen to other people and develop collective wisdom together in dealing with our problematic situations. Such groups consist of individuals who are able to share their thoughts respectfully but purposefully in open dialogical discussion. A range of views stimulates attention and productive thought. Following on from this and in another article, Myatt discusses the fact that in education we tend to 'privilege writing above everything else' and discount high-quality talk. This is important because we tend to assume in education that written outcomes matter more than talk. See Blog Article Endnote.[26]

She argues that 'high-quality' talk and listening underpin reading and writing. This is very important all across education and is especially vital to early years—a major argument against formal written testing at too young an age. Myatt says that listening and speaking, unlike reading, get little attention in the national curriculum even though it improves literacy standards and reading comprehension. She cites a summary of reports by the United Kingdom Education Endowment Foundation which says oral language interventions are highly effective in producing higher learning outcomes provided they are carefully implemented to introduce new and extended spoken vocabulary. They work best when related to current content studied in school involving meaningful learning activity, especially involving dialogic and collaborative talk that promotes metacognition. Reports Website Endnote.[27]

On average, pupils who participated in oral language interventions made about five months' added progress in a year and some studies showed a larger effect with younger and disadvantaged child backgrounds (up to six months). These approaches can include targeted reading aloud, book discussion, story improvisation, structured questioning and purposeful, curriculum-focused, dialogue and interaction. Oracy skills should have equal status to reading and writing in school because they promote learning in all subjects and settings. Teachers should be challenged to find ways to facilitate oracy in purposeful and meaningful ways—to model listening and speaking and to value and celebrate the spoken word. Students of all ages should be expected to memorise poetry and prose, explain their ideas, discuss, debate, engage in Socratic dialogue and recognise others' points of view (Cambridge Primary Review Trust, 2020).

6.8 Conclusion

What can we learn about the human issues that have created our problems that education might help us to address? How can all of our future generations be helped to lead a happy life? It seems we have already been looking for answers by investigating our own brains. Some of us are actually becoming more enlightened about ourselves and our abilities and many of us have become great 'teachers' in the broadest sense inside and outside of education. Our ability to communicate great ideas through information technology has reached sophisticated capabilities. Unfortunately these technologies are also able to channel all the worst aspects of humanity's greatest flaws and weaknesses.

We are facing a crisis around how to manage our own human nature on multiple fronts. I suggest that our happiness rests on determining how to promote the best aspects of human nature to the benefit of the greatest number of people. This is not about control, but about the development of intelligent, flexible understanding. The good news is that we have all the ingredients of knowledge and action we need to achieve this goal, provided we proceed wisely. Empathy and the ability to see the logic of where someone is 'coming from' is very important in this and so is good teaching. It is important we all produce and engage with wide-ranging and intelligent research data to gain improved outlooks, in order to solve wider world problems.

Notes

1. Mary Myatt (2019) 'Using stories in the curriculum' https://marymyatt.com/blog/2019-12-14/using-stories-in-the-curriculum#disqus_thread Accessed 15/12/2019.
2. Corey Block (2017) 'Why Compassion (and Great Stories) Trumps Current Events in the Classroom' https://www.weareteachers.com/stories-not-current-events/ Accessed 11/03.2021.
3. Elizabeth Peyton (2019) 'The Secret to Classroom Management—No Matter Where You Teach' https://www.weareteachers.com/classroom-management-title-i/?fbclid=IwAR30UVUK5sO Accessed 19/01/2020.
4. Grainne Hallahan (2019) 'Why attachment-aware teaching matters for every child Attachment is important for every child in your class – not just the vulnerable pupils, says Professor Peter Fonagy' https://www.tes.com/news/why-attachment-aware-teaching-matters-every-child Accessed 14/10/2019.

5. Lori Desautels (2021), Professor in the College of Education, Butler University, Indiana 'Bad Behavior or Nervous System Response: A New Lens for Discipline'. https://endseclusion.org/2021/02/20/bad-behavior-or-nervous-system-response/ Accessed 15/03/2021.
6. Guy Stephens (2021) 'Revolutionizing our understanding of autonomic physiology: An interview with Stephen W. Porges, PhD' https://www.youtube.com/watch?v=ZBA1RaxEU3c&t=73s Accessed 15/03/2021.
7. Mary Meredith (2020) 'Towards a pastoral pedagogy fit for the pandemic age – a final reflection on The Boy who was Raised as a Dog' https://marymered.com/2020/05/16/towards-a-pastoral-pedagogy-fit-for-the-pandemic-age-a-final-reflection-on-the-boy-who-was-raised-as-a-dog/ Accessed 17/03/2021.
8. National Center on Domestic Violence, Trauma and Mental Health 'Exercises for Grounding, Emotional Regulation & Relaxation for children and their parents' http://www.nationalcenterdvtraumamh.org/wp-content/uploads/2012/01/Exercises-for-Grounding-Emotional-Regulation-Relaxation-Final.pdf Accessed 18/03/2021.
9. Karen Treisman Website Safe Hands Thinking Minds http://www.safehandsthinkingminds.co.uk/covid-anxiety-stress-resources-links/ Accessed 18/03/2021.
10. Kim Golding (2015) 'Connection Before Correction: Supporting Parents to Meet the Challenges of Parenting Children who have been Traumatised within their Early Parenting Environments' 2015/05/04 Volume 40(2):1–8 Children Australia Journal DOI: https://doi.org/10.1017/cha.2015.9 https://www.researchgate.net/publication/276500716_Connection_Before_Correction_Supporting_Parents_to_Meet_the_Challenges_of_Parenting_Children_who_have_been_Traumatised_within_their_Early_Parenting_Environments Accessed 22/03/2021.
11. Jaz Ampaw-Farr (2017) 'The power of everyday heroes' TEDxNorwichEDhttps://youtu.be/q3xoZXSW5yc Accessed 18/03/2021.
12. Yousif Davis (2020) 'Evaluating Evidence-Based Practise In Your Workplace – The Teaching'. https://www.statesofmind.org/journal/2020/09/07/yousif-research-paper.html Accessed 19/03/2021.
13. Challen, C. (2020). 'The importance of art/poetry and philosophy in educational leadership, well-being and engagement of STEM subjects'. Journal of Higher Education Policy and Leadership Studies, 1(3), 41–54. DOI: https://doi.org/10.29252/johepa.1.3.41 Accessed 17/03/2021.
14. Special Data Contribution

'Happiness' A poem from 'My Mind's House' by Dr Christine Challen 20 August 2019

Dedicated to Chris Martin who introduced me to Rumi, to my children who are a constant source of amazement and love. To those at Harbour who help victims of domestic abuse and family and friends who have supported us on this journey. I hope that this helps those who suffer trauma to have the courage, strength and hope that you can rebuild and recover in time.

Happiness
How do we define,
A state of emotion,
Dependent of mind
But really it's more
Perception and Presence
Positive reflections through
Time, peace and solitude
Healing wounded scars over,
Mindfulness moments, a
Long tangled journey
To slowly go.
Emotions like a kaleidoscope
Both grief and sad
Positive and gratitude
Negative and bad
All form to create colours
Opportunities to be had
In each we embrace
Experiences to call on
For happiness to hold
Through perceptions and
Feelings moods to unfold

So key to happiness
Collide emotions of grief
Creating warmth and energy
Uplifting your mood
Remember struggles are temporary
Not a permanent state
But a useful reminder
To be grateful for the good
Happiness is evolving
It spins without and within
Depends on the present
And a positive take
On the past

15. Bruce Perry (2021) 'The Neurosequential Model' Website https://www.neurosequential.com/ourteam and Video https://www.youtube.com/watch?v=_3is_3XHKKs Accessed 17/03/2021.
16. Armand King (2021) 'Using pain to Pave a Way' https://youtu.be/wQY9Jz-mi1A Accessed 22/03/2021.
17. Tamara Lanier 'The World Is Watching: Woman Suing Harvard for Photos of Enslaved Ancestors Says History Is At Stake' 29 March 2019 Democracy Now! https://www.youtube.com/watch?v=-PMgWraTJz4 Accessed 23/03/2021.
18. Toria Bono (2021) Tiny Voice Talks Podcasts 'Becoming, with Paul Hume': In this episode, Toria talks to Paul about 'Becoming: a teacher, and a leader.' @teacherpaul1978 @MenTeachPrimary. https://podcasts.apple.com/us/podcast/becoming-with-paul-hume/id1526852152?i=1000492531394 Accessed 19/03/2021.
19. Carmel O'Hagan (2020) Language and identity – who are we if not our name? https://carmelohagan.wordpress.com/author/carmelohagan/ Accessed 26/02/2020.
20. Carmel O'Hagan (2018) 'New funding for MFL or a political stranglehold on teaching and learning? So is this good news?' 6 August 2018 https://carmelohagan.wordpress.com/author/carmelohagan/ Accessed 28/03/2021.
21. Sarah McDermott (2020) Adopted Children-'Our foster child asked us to adopt him – by drawing himself on to a family photo' BBC News https://www.bbc.co.uk/news/stories-54071599 Accessed 06/10/2020.
22. Nicole Dempsey, Assistant Principal Dixons Trinity Academy, UK 'Equality, Inclusion &, education, equality, equity, inclusion, schools, SEN, SENCo, SEND, teaching' 5 September 2020 https://inco14.wordpress.com/2020/09/05/grassroots-inclusion/ Accessed 24/02/2021.
23. Statutory guidance: SEND code of practice: 0 to 25 years (2014) United Kingdom (England): Department for Education and Department of Health and Social Care-updated 30 April 2020 https://www.gov.uk/government/publications/send-code-of-practice-0-to-2 5 Accessed 14/06/2021.
24. Mary Myatt (2021a) Culture-Essentialism 'The main thing is to keep the main thing the main thing.' https://www.marymyatt.com/blog/essentialism Accessed 26/03/2021.
25. Mary Myatt (2021b) 'Culture-Cynic or skeptic? In a democratic society scepticism is a virtue in history as well as in philosophy – Napoleon Bonaparte' https://www.marymyatt.com/blog/cynic-or-skeptic Accessed 26/03/2021.

26. Mary Myatt (2021c) 'Culture-Walking the talk-"Writing floats on a sea of talk" – James Britton'. https://www.marymyatt.com/blog/walking-the-talk Accessed 26/03/2021.
27. The Education Endowment Foundation (2019, 2018, 2016) 'Oral language interventions Moderate impact for very low cost, based on extensive evidence' https://educationendowmentfoundation.org.uk/evidence-summaries/teaching-learning-toolkit/oral-language-interventions/ Accessed 29/03/2021.

References

A Portrait of Care was a collaboration between Dr Aoife O'Higgins, Rosie and The University of Southampton Widening Participation department. It Was an Online Exhibition That Ran During National Care Leavers Week (NCLW) Via Instagram Using Self-Portraiture as a Way to Combat the Negative Stereotypes People Have About Children in Care – Twitter: @CareExp_Culture Press Release: https://careleaversinfiction.wordpres... https://youtu.be/b_ZZhS3qYcE. Accessed 10 Mar 2021.

AcEd Advocating Creativity in education Kevin Hewitson https://4c3d.wordpress.com/ & http://www.ace-d.co.uk/

Adrian Bethune. (2018). *Four Ways to Grow Kind Kids: Teaching Children Prosocial Behavior Can Benefit the Whole Family*. https://www.livehappy.com/practice/four-ways-grow-kind-kids. Accessed 30 Aug 2018.

Ampaw-Farr, J. (2021). *Because of You, This Is Me*. Crown House Publishing.

Armand King. (2021). *Using Pain to Pave a Way*. https://youtu.be/wQY9Jz-mi1A. Accessed 22 Mar 2021.

Bhopal, K. (2018). *White Privilege: The Myth of a Post-Racial Society*. Policy Press.

British Cohort Study '*50 years of the 1970 British Cohort Study: 50 stories in 50 weeks This April the 1970 British Cohort Study and our study participants turned 50!*' https://cls.ucl.ac.uk/bcs7050stories/. Accessed 23/12/2018.

Bruce Perry. (2021). *The Neurosequential Model*. Website https://www.neurosequential.com/ourteam and Video https://www.youtube.com/watch?v=_3is_3XHKKs. Accessed 17 Mar 2021.

Cambridge Primary Review Trust & York University. (2020). *Dialogic Teaching: Improving Children's Learning by Improving the Quality of Classroom Talk: The Programme Uses Video Review, Print Materials and In-School Mentoring to Support Teachers' Practice Across English, Maths and Science Lessons* (EEF Summary, Key Stage 2, 4985 Pupils in 76 Schools) Independent Evaluator Sheffield Hallam University Education Endowment Foundation. https://educationendowmentfoundation.org.uk/projects-and-evaluation/projects/dialogic-teaching/. Accessed 18 Sept 2021.

Carmel O'Hagan. (2018, August 6). *New Funding for MFL or a Political Stranglehold on Teaching and Learning? So is This Good News?*. https://carmelohagan.wordpress.com/author/carmelohagan/. Accessed 28 Mar 2021.

Carmel O'Hagan. (2020). *Language and Identity – Who Are We if Not Our Name?* https://carmelohagan.wordpress.com/author/carmelohagan/. Accessed 26 Feb 2020.

Challen, C. (2020). The Importance of Art/Poetry and Philosophy in Educational Leadership, Well-Being and Engagement of STEM Subjects. *Journal of Higher Education Policy and Leadership Studies, 1*(3), 41–54. https://doi.org/10.29252/johepal.1.3.41. Accessed 17 Mar 2021.

Corey Block. (2017). *Why Compassion (and Great Stories) Trumps Current Events in the Classroom*. https://www.weareteachers.com/stories-not-current-events/. Accessed 11 Mar 2021.

de Thierry, B. (2019). *The Simple Guide to Attachment Difficulties in Children: What They Are and How to Help*. Jessica Kingsley.

Dean Hobbs, R. (2016). *The Global Child: How Experts Would Change Education. Research-Based Acquisition of Languages*. Lap Lambert Academic Publishing.

Dee Michell and Rosie Canning Talk to Lisa Cherry. (2021). Children and Young People in Care: A New Digital Archive Called Care Experience and Culture. A Portrait of Care Was an Online Exhibition that Ran During National Care Leavers Week (NCLW) via Instagram Using Self-Portraiture as a Way to Combat the Negative Stereotypes People Have About Children in Care – Twitter: @CareExp_Culture https://youtu.be/b_ZZhS3qYcE. Accessed 10 Mar 2021.

Desautels, L. (2020). *Connections Over Compliance: Rewiring Our Perceptions of Discipline*. Wyatt-MacKenzie Publishing.

Desautels, L., & McKnight, M. (2019). *Eyes Are Never Quiet: Listening Beneath the Behaviors of Our Most Troubled Students*. Wyatt Mackenzie Publishing.

Diversity. (2020). *Diversity Take to the Stage with POWERFUL Black Lives Matter Performance | Semi-Finals | BGT 2020 a Group of Performers Who Express a Vision of the World*. https://www.youtube.com/watch?time_continue=1&v=kzFNKFitHjw&feature=emb_logo

Dix, P. (2017). *When the Adults Change, Everything Changes: Seismic Shifts in School Behaviour*. Independent Thinking Press

Dr Sonia Hutton-Taylor. (2018, 21 October). *From the Ashes of Burnout....* https://www.linkedin.com/pulse/from-ashes-burnout-sonia-hutton-taylor/. Accessed 29 Aug 2019.

Ed Finch. (2019). *Is Teacher Instinct Becoming Extinct?* https://www.tes.com/magazine/article/teacher-instinct-becoming-extinct. Accessed 20 Oct 2019.

Elizabeth Peyton. (2019). *The Secret to Classroom Management—No Matter Where You Teach*. https://www.weareteachers.com/classroom-management-title-i/?fbclid=IwAR30UVUK5sO. Accessed 19 Jan 2020.

Emma Turner. (2019). *Support Staff: Some Don't Realise How Special They Are: Meet Mrs Rush, the Nursery Nurse Who Reassures and Supports with a Wink, a Twinkle in Her Eye and a Wide Grin.* https://www.tes.com/news/support-staff-some-dont-realise-how-special-they-are. Accessed 22 May 2019.

Exercises for Grounding Emotional Regulation-Relaxation. http://www.nationalcenterdvtraumamh.org/wp-content/uploads/2012/01/Exercises-for-Grounding-Emotional-Regulation-Relaxation-Final.pdf. Accessed 18 Mar 2021.

Feldman Barrett, L. (2017). *How Emotions Are Made; the Secret Life of the Brain.* Pan Macmillan.

Gaunt, A., & Stott, A. (2019). *Transform Teaching and Learning through Talk: The Oracy Imperative.* Rowman & Littlefield Publishing Group Inc.

Gedge, N. (2016). *Inclusion for Primary School Teachers.* Bloomsbury.

Gilbert, I. (2015). *There Is Another Way: The Second Big Book of Independent Thinking.* Independent Thinking Press

Glazzard, J., & Bostwick, R. (2018). *Positive Mental Health: A Whole School Approach.* Critical Publishing.

Graham Chatterley. (2020). *Building Positive Behaviour: Returning to Learning Using a Sequential Approach.* Email gchatterleyroasa@gmail.com self-published: Amazon Kindle Direct Publishing.

Grainne Hallahan. (2019). *Why Attachment-Aware Teaching Matters for Every Child Attachment Is Important for Every Child in Your Class – Not Just the Vulnerable Pupils, Says Professor Peter Fonagy.* https://www.tes.com/news/why-attachment-aware-teaching-matters-every-child. Accessed 14 Oct 2019.

Grassroots-Inclusion-Nicole Dempsey https://inco14.wordpress.com/2020/09/05/grassroots-inclusion/

Greg McKeown. (2014). *Essentialism: The Disciplined Pursuit of Less.* Virgin Books, Random House Publishing.

Guy Stephens. (2021). *Revolutionizing Our Understanding of Autonomic Physiology: An Interview with Stephen W. Porges, PhD.* https://www.youtube.com/watch?v=ZBA1RaxEU3c&t=73s. Accessed 15 Mar 2021.

Hawkes, N. (2014). *From My Heart: Transforming Lives through Values.* Independent Thinking Press.

Hawkes, N., & Hawkes, J. (2018). *The Inner Curriculum: How to Nourish Wellbeing, Resilience and Self-Leadership.* John Catt Educational.

Heather Lucas. (2020). *FOLLOW THE ARROWS! A Guest Post by @HLucas8 on Healing School Communities.* https://marymered.com/2020/05/27/follow-the-arrows-a-guest-post-by-hlucas8-on-healing-school-communities/. Accessed 18 Mar 2021.

Hejlskov Elven, B., & Wiman, T. (2017). *Sulky, Rowdy, Rude? Why Kids Really Act Out and What to Do About It.* Jessica Kingsley Publishers.

Henden, J. (2017). *What It Takes To Thrive: Techniques For Severe Trauma And Stress.* World Scientific Publishing Company.

Holt, J. C. (1982). *How Children Fail*. Perseus Books.

Human Rights Law Centre-Australia Website https://www.hrlc.org.au/what-we-do

Human Rights Law Centre-Australia. (2019). *Dujuan, Aboriginal Child to Address UN Human Rights Council and Urge Australia to Stop Sending 10 Year Olds to Prison*. https://www.hrlc.org.au/news/2019/9/9/aboriginal-child-to-address-un-human-rights-council. Accessed 09 Sept 2019.

Immordino-Yang, M. H., & Knecht, D. R. (2020). 'Building Meaning Builds Teens' Brains Connecting adolescents' Concrete Work to Big Ideas May Help Shape Their Neural Networks Over Time. *Learning and the Brain*, *77*(8), 36–43. *Educational Leadership Journal*-online. http://www.ascd.org/publications/educational-leadership/may20/vol77/num08/Building-Meaning-Builds-Teens'-Brains.aspx. Accessed 16 Oct 2020.

Jackson, P. Z., & McKergrow, M. (2006). *The Solutions Focus: Making Coaching and Change Simple*. Nicholas Brealey International.

James, G. (2016). *Transforming Behaviour in the Classroom: A Solution-Focused Guide for Teachers*. Sage Publications.

James, G. (2019). *Solutions Focused Coaching: A Workbook for Educators*. By Email drgj@thesolutionsfocusedcoach.com and https://www.thesolutionsfocusedcoach.com/

Jarlath O'Brien. (2018). *I Don't Know Where You Are*. https://jarlathobrien.wordpress.com/2018/03/06/i-dont-know-where-you-are/. Accessed 16 Oct 2020.

Jaz Ampaw-Farr. (2017). *The Power of Everyday Heroes*. TEDxNorwichED. https://youtu.be/q3xoZXSW5yc. Accessed 18 Mar 2021.

Jung, C. G. (1995) (1875–1961). *Memories, Dreams, Reflections: An Autobiography*. Fontana Press.

Karen Treisman. http://www.safehandsthinkingminds.co.uk/covid-anxiety-stress-resources-links/. Accessed 18 Mar 2021.

Kim Golding. (2015). Connection Before Correction: Supporting Parents to Meet the Challenges of Parenting Children Who Have Been Traumatised Within Their Early Parenting Environments. *Children Australia Journal*, *40*(2), 1–8. https://doi.org/10.1017/cha.2015.9. https://www.researchgate.net/publication/276500716_Connection_Before_Correction_Supporting_Parents_to_Meet_the_Challenges_of_Parenting_Children_who_have_been_Traumatised_within_their_Early_Parenting_Environments. Accessed 22 Mar 2021.

King, A. (2019). *Raised in Pimp City: The Uncut Truth About Domestic Human Sex Trafficking*. Ariginal One Publishing.

KJ Williams. (2017). *Importance of Teaching Empathy: How Can Technology Teach Kids Some of the Most Significant Social-Emotional Skills?* https://www.parentherald.com/articles/79384/20161031/importance-teaching-empathy-how-can-technology-teach-kids-some-most-significant-social-emotional-skills.htm. Accessed 15 Jan 2020.

LENABELLINA Musings and Ramblings About Life, Education and Wellbeing. https://lenabellina.wordpress.com/about/

LITTLEMAVIS. (2018). *Privilege and Oxbridge: An Autobiography.* https://littlemavis.wordpress.com/author/littlemavis/ https://littlemavis.wordpress.com/author/littlemavis/. Accessed 23 Apr 2019.

Lori Desautels. (2021). *Professor in the College of Education, Butler University, Indiana 'Bad Behavior or Nervous System Response: A New Lens for Discipline'.* https://endseclusion.org/2021/02/20/bad-behavior-or-nervous-system-response/. Accessed 15 Mar 2021.

Lyn Romeo. (2018, December 19). *Social Pedagogy… Social What?* Categories: Education and Training, Knowledge and Skills https://socialworkwithadults.blog.gov.uk/2018/12/19/social-pedagogy-social-what/. Accessed 14 Dec 2020.

Lytra, V., & Martin, P. (2010). *Sites of Multilingualism: Complementary Schools in Britain Today.* Trentham Books.

Mary Meredith. (2020). *Towards a Pastoral Pedagogy Fit for the Pandemic Age – A Final Reflection on the Boy Who Was Raised as a Dog.* https://marymered.com/2020/05/16/towards-a-pastoral-pedagogy-fit-for-the-pandemic-age-a-final-reflection-on-the-boy-who-was-raised-as-a-dog/. Accessed 7 Mar 2021.

Mary Meredith's Blog, Local Education Authority Inclusion Officer https://marymered.com/ @marymered

Mary Myatt. (2019). *Using Stories in the Curriculum.* https://marymyatt.com/blog/2019-12-14/using-stories-in-the-curriculum#disqus_thread. Accessed 15 Dec 2019.

Mary Myatt. (2021a). *Culture-Essentialism 'The Main Thing is to Keep the Main Thing the Main Thing.'* https://www.marymyatt.com/blog/essentialism. Accessed 26 Mar 2021.

Mary Myatt. (2021b). *Culture-Cynic or Skeptic?* https://www.marymyatt.com/blog/cynic-or-skeptic. Accessed 26 Mar 2021.

Mary Myatt. (2021c). *Culture-Walking the Talk-'Writing Floats on a Sea of Talk' – James Britton.* https://www.marymyatt.com/blog/walking-the-talk. Accessed 26 Mar 2021.

McKeown, G. (2021). *Effortless: Make It Easier to Do What Matters Most.* Virgin Books, Random House Publishing.

Michelle Whitstone. (2021, March 26). *Graduate Student Presentation Critical Pedagogy as Decolonizing Practice.* College of Education 2021 National Congress on Rural Education in Canada. https://www.youtube.com/watch?v=TwQzitMMzSk. Accessed 30 Mar 2021.

Myatt, M. (2018). *The Curriculum: Gallimaufry to Coherence.* John Catt Educational.

Nancy Angevine-Sands 'Immigrant Connections ft. Laura Gardner: Why Family Engagement Matters. Nancy Angevine-Sands (2019)'. It's Much More Than Parent Engagement. It's About Engaging the Entire Family!. 2019 voicEd Radio Canada https://voiced.ca/podcast_episode_post/immigrant-connections-ft-laura-gardner/. Accessed 14 Dec 2020.

NEXUS EDUCATION https://www.nexus-education.com/blogs/

Nic Askew. The Soul of This Human Race on film, in all its shades and from all four corners. Video Portraits Soul· Biographies https://soulbiographies.com/library/

Nicole Dempsey. (2020, September 5). Assistant Principal Dixons Trinity Academy, UK. *Equality, Inclusion &, Education, Equality, Equity, Inclusion, Schools, SEN, SENCo, SEND, Teaching*. https://inco14.wordpress.com/2020/09/05/grassroots-inclusion/. Accessed 24 Feb 2021.

Pam Jarvis. (2019). *English Childhood*. Article (PDF) https://www.mdpi.com/books/pdfview/book/1895. Accessed 17 Dec 2019.

Pascoe, B. (2018). *Dark Emu*. Scribe Publications UK Ltd..

Pearce, S. (2005). *You Wouldn't Understand: White Teachers in Multi-Ethnic Classrooms*. Trentham Books.

Perry, B. D., & Szalavitz, M. (2006). *The Boy Who Was Raised as a Dog: And Other Stories from a Psychiatrist's Notebook*. Hachette Book Group.

Porges, S. W. (2011). *The Polyvagal Theory: Neurophysiological Foundations of Emotions, Attachment, Communication, and Self-Regulation*. Norton Publishing.

Porges, S. W. (2017). *The Pocket Guide to the Polyvagal Theory: The Transformative Power of Feeling Safe*. Norton Publishing.

Postman, N., & Weingartner, C. (1969). *Teaching as a Subversive Activity: A No-Holds Barred Assault on Outdated Teaching Methods-with Dramatic and Practical Proposals on How Education Can Be Made Relevant to Today's World*. Dell Publishing Company.

Ratner, H. (2012). (with Evan George & Chris Iveson-contributors). *Solution Focused Brief Therapy: 100 Key Points and Techniques*. Routledge.

Rita Pierson 2013' Every Kid Needs a champion. A Teacher for 40 Years with a Rousing Call to Educators to Believe in Their Students and Actually Connect with Them on a Real, Human, Personal Level. TEDTalks. https://youtu.be/SFnMTHhKdkw. Accessed 10 July 2019.

Roberts, D. (2017). *Empowering Generation Z: How and Why Leadership Opportunities Can Inspire your Students*. John Catt Educational Limited.

Roberts, H. (2012). *Oops! Helping Children to Learn Accidentally*. Independent Thinking Press.

Rogers, C., & Freiberg, H. J. (1983). *Freedom to Learn*. (Rev. ed.). Prentice Hall International.

RogerTitcombe. (2019). *Is the Mental Health Crisis in English Schools Being Driven by the Changes in Teaching Methods Demanded by Marketisation?*

https://rogertitcombelearningmatters.wordpress.com/2019/03/04/is-the-mental-health-crisis-in-english-schools-being-driven-by-the-changes-in-teaching-methods-demanded-by-marketisation/. Accessed 30 Mar 2021.

Safe Hands Thinking Minds Website Dr Karen Treisman, MBE, Clinical Psychologist. http://www.safehandsthinkingminds.co.uk/covid-anxiety-stress-resources-links/

Sam Carr Lecturer in Education, University of Bath. (2016). *What Other Cultures Can Teach Parents About Raising Their Children*. https://theconversation.com/what-other-cultures-can-teach-parents-about-raising-their-children-61479. Accessed 22 Aug 2019.

Sarah Dove FCCT. (2020). Creator of Embracive Education Phoenix Website on SEND provision @PhoenixEdSarah Founder of @phoenixgrouphq President of @prusapuk Committed to Giving Every Child a Sense of Belonging London www.phoenixeducationconsultancy.com https://www.phoenixgrouphq.com/. Accessed 23 Jan 2020.

Sarah McDermott. (2020). Adopted Children – Our Foster Child Asked us to Adopt Him – By Drawing Himself on to a Family Photo. *BBC News*. https://www.bbc.co.uk/news/stories-54071599. Accessed 06 Oct 2020.

Shalaby, C. (2017). *Troublemakers: Lessons in Freedom from Young Children at School*. New Press.

Sigmund Freud. (2010) (1856–1939). *Sigmund Freud Collected Works: The Psychopathology of Everyday Life, The Theory of Sexuality, Beyond the Pleasure Principle, The Ego and the Id, and The Future of an Illusion*. Pacific Publishing Studio.

Skinner, B. F., Ferster, C. B., Cheney, C. D., & Morse, W. H. (1957). *Schedules of Reinforcement*. The Skinner Foundation.

STALLIE'S BLOG Teacher-in Brussels who expresses her emotions in her writing http://stallie-aslifegoesby.blogspot.com/ @carostal

States of Mind Website Dr Chris Bagley https://www.statesofmind.org/services @hiddendepths

Statutory Guidance: SEND Code of Practice: 0 to 25 years (2014). United Kingdom (England): Department for Education and Department of Health and Social Care-updated 30 April 2020 https://www.gov.uk/government/publications/send-code-of-practice-0-to-25. Accessed 14 June 2021.

Stevenson, B. (2020). *Just Mercy: A Story of Justice and Redemption*. Scribe Publications

Suzanne Yvette Culshaw. (2020). *An Exploration of What it Means to be Struggling as a Secondary Teacher in England*. University of Hertfordshire Research Archive Downloaded fulltext (PDF, 3Mb). https://uhra.herts.ac.uk/handle/2299/22082. Accessed 30 Nov 2020.

Syed, M. (2019). *Rebel Ideas: The Power of Diverse Thinking*. John Murray Press.

Szalavitz, M., & Perry, B. D. (2010). *Why Empathy Is Essential and Endangered: Born for Love*. Harper Collins.

Tamara Lanier. (2019). *The World Is Watching: Woman Suing Harvard for Photos of Enslaved Ancestors Says History Is At Stake*. Mar 29, 2019, Democracy Now! https://www.youtube.com/watch?v=-PMgWraTJz4. Accessed 23 Mar 2021.

TEAPOT TEACHER https://teapotteacher.school.blog/ Twitter @ teacher_teapot

TEAPOT TEACHER. (2019). *And so the Analysis Begins….* https://teapot-teacher.school.blog/2019/08/21/and-so-the-analysis-begins/. Accessed 22 Aug 2019.

The Education Endowment Foundation. (2019, 2018, 2016). *Oral Language Interventions: Moderate Impact for Very Low Cost, Based on Extensive Evidence*. https://educationendowmentfoundation.org.uk/evidence-summaries/teaching-learning-toolkit/oral-language-interventions/. Accessed 29 Mar 2021.

The Hospitable Wanderer: Educational Meanderings: https://hospitablewan-derer.com/ Twitter @MrBayew

The Sacred and Profane-Peter Ford. https://edsacredprofane.wordpress.com/

The Times Educational Supplement Helping your school to succeed. https://www.tes.com/for-schools

The Times Educational Supplement. https://www.tes.com/news/fancy-teaching-podcast-then-look-right-here and https://feeds.captivate.fm/tes-podagogy/. Accessed 01 Mar 2021.

Tomsett, J. (2017). *This Much I Know about Mind Over Matter: Improving Mental Health in our Schools*. Crown House Publishing.

Toria Bono. (2021). *Tiny Voice Talks Podcasts 'Becoming, with Paul Hume: In This episode, Toria Talks to Paul About 'Becoming: A Teacher, and a Leader.'* @teach-erpaul1978 @MenTeachPrimary https://podcasts.apple.com/us/podcast/becoming-with-paul-hume/id1526852152?i=1000492531394. Accessed 19 Mar 2021.

Trevor Noah. (2021, March 29). Celebrating Asian-American Voices & Stories. *The Daily Show*. https://www.youtube.com/watch?v=kWi8wo74BBM. Accessed 30 Jan 2021.

United Nations Engagement Website. Our Vision: Australia Upholds the International Human Rights Standards it Has Promised to Comply with and Champions Human Rights on the World Stage-Holding Australia to Account. https://www.hrlc.org.au/united-nations

University of Hull. *Introduction to University Study: Priority Matrices*. https://libguides.hull.ac.uk/introduction/matrix. Accessed 26 Mar 2021.

Walker, L. E. (1979). *The Battered Woman*. Harper and Row Publishers.

Walker, P. (2013). *Complex PTSD: From Surviving to Thriving: A GUIDE AND MAP FOR RECOVERING FROM CHILDHOOD TRAUMA*. Azure Coyote Publishing.

Ward, J. (2019). *On the Fringes: Preventing Exclusion in Schools Through Inclusive, Child-Centred, Needs-Based Practice*. Crown House.

Will Hazell. (2019). *Trainee Teachers with More Friends Less Likely to Quit 'Personal and Professional Relationships Enhance Resilience', Study Finds*. https://www.tes.com/news/trainee-teachers-more-friends-less-likely-quit. Accessed 11 Apr 2019.

Williams, L. E., & Bargh, J. A. (2008). Experiencing Physical Warmth Promotes Interpersonal Warmth. Published in final edited form as: *Science, 322*(5901), 606–607. https://doi.org/10.1126/science.1162548. PMCID: PMC2737341. https://www.ncbi.nlm.nih.gov/pmc/articles/PMC2737341/. Accessed10 Sept 2019.

Willingham, D. T. (2009). *Why Don't Students Like School? A Cognitive Scientist Answers Questions About How the Mind Works and What it Means for the Classroom*. Wiley.

Winfrey, O., & Perry, B. (2021). *What Happened to You? Conversations on Trauma, Resilience, and Healing*. Bluebird Publishing.

Yousif Davis. (2020). *Evaluating Evidence-Based Practice in Your Workplace – The Teaching*. States of Mind. https://www.statesofmind.org/journal/2020/09/07/yousif-research-paper.html. Accessed 19 Mar 2021.

Zacks, J. M. (2015). *Flicker: Your Brain on Movies*. Oxford University Press.

https://voice21.org/

https://educationendowmentfoundation.org.uk/evidence-summaries/teaching-learning-toolkit/oral-language-interventions/

https://www.school21.org.uk/oracy

Harari, Y. N. (2015). *Sapiens: A Brief History of Humankind*. London: Penguin, Vintage Publications.

https://www.teacher.org/daily/what-is-the-harkness-discussion-why-ive-embraced-this-method-and-how-its-worked-for-me/

http://robinalexander.org.uk/dialogic-teaching/

https://www.gov.uk/government/publications/school-inspection-handbook-eif

https://thinkingtogether.educ.cam.ac.uk/about/

https://www.esu.org/resources/

https://www.mentalhealth.org.uk/learning-disabilities/help-information/learning-disability-statistics-

https://www.encap.org.uk/learning-disability-explained/research-and-statistics

https://www.scope.org.uk/media/disability-facts-figures/

CHAPTER 7

Reassessing Our Ideas About Knowledge

7.1 Introduction

Modern neurocognitive and clinical psychologists are telling us that the human brain is complicated and uniquely individual. They have proved it has the ability, given the right 'conditions,' to transform itself in ways that scientific research never previously thought possible. We are now aware of the brain's neuroplasticity. This has important implications for our essential information, accepted views and understandings about ourselves. We are being offered new learning possibilities, enabling us to challenge, deconstruct and amend some established discourses and 'facts.' Such a fundamental discovery as brain plasticity does not *just* alter the way we look at and understand the living physical brain in the laboratory and operating theatre.

It can help us to shift our perspective to explain and infer more about how we are able to discover, learn, process and create knowledge. Such a crucial change in our view of how we think gives us new information about human intelligence, its potential and propensities, both negative and positive as we reflect. This has both personal and professional implications. Knowledge in the areas of psychology, philosophy, sociology, medicine and education, which existed before can now be looked at differently. We can be *more* not less objective as we research and develop these

Special thanks to Muhammad Shehu Shuaibu for 'My Education Journey' 20 January 2021.

J. A. Hawkins, *Brain Plasticity and Learning*,
https://doi.org/10.1007/978-3-030-83530-9_7

disciplines in new ways. Old knowledge, ideas and evidence data sets can be re-examined and adjusted in ways which may offer us alternative and better explanations.

Old ideas can be looked at in a different way, proved, disproved, adjusted and reimagined. They may be correct in some aspects but are no longer *necessarily* fixed into old frameworks in the same way. They may be open to different interpretations. It is time to reassess our knowledge—our human history, our mistakes and achievements, goals, concepts, beliefs and assumptions—in the light of this new information. We can reconsider the implications of human qualitative conclusions around the existing body of scientific knowledge made up of stable, provable facts and fundamental laws. This has become more and more complex as facts are discovered and redefined. The current definition of scientific knowledge by Encyclopaedia Britannica is necessarily broad and divided into discipline areas and titles. No one person can possibly remember all of it. We have already had to acknowledge the need for a range of ethical collaborative systems in order to manage it, for example, the Wikipedia Website. See Encyclopaedia Britannica Definitions Endnote.[1]

Many scientific models involve the establishment of a body of work that is the result of 'unbiased' observation and 'systematic' experimentation. However, such activity necessarily involves human thought and interaction, therefore to a greater or lesser extent, *feelings and emotions* must be involved. They are present in scientific observation and experimentation methods and evaluations—particularly in processes and uses. They should therefore be similarly and appropriately checked out, examined and re-examined collaboratively from different points of view. This has not always been done and so self-destructive errors abound, for example, in the development of plastics, insecticides and weapons. We now have new opportunities to research and develop authentication procedures in those human aspects of research. Human values, ethics and imagination are important components of successful scientific endeavour.

Traditionally feelings and emotions in social research were noted in passing, but not considered particularly important. We now know they are key to understanding our perceived realities because they enable and *drive* our neuroplasticity. Feelings and motivations help to explain why past knowledge was developed, used, valued, forgotten and preserved. They can give us information, often not previously deemed relevant to science. We can use our full intellectual powers to develop our knowledge further by discovering some general and specific truths about how our brains function. This in turn can change fixed approaches, unlock and make possible the acquisition of new knowledge.

There are social learning implications for our expectations of each other and how we organise the fabric of society to allow for *different* perspectives. We can no longer assume that everyone sees the world as we do. We can now *accept* and *expect* other people to see the world differently even when we agree on points. Other people's emotional responses and analyses of the world become useful as scientific data in complex ways. They have the potential to provide us with new information. Facts may alter and combine when seen from different angles in learning—in a similar way to the ways colours separate, mix and change at different angles as light shines 'through a prism.'

This chapter considers how some human motivations, attitudes, feelings and emotions have affected the body of human knowledge. It looks at some ways in which old perspectives have changed and are currently changing in the influential fields of psychology, philosophy and education. It suggests that a better understanding of brain plasticity may help us to preserve knowledge. It might help us develop our intelligence to learn how to use potentially impactful and important areas of knowledge with more wisdom in the future.

7.2 Knowledge as a Human Resource: Valued, Ignored, Destroyed

The body of available and accessible knowledge that exists in our world is varied and complex. Knowledge is viewed in different ways, but in simple summary it exists as and consists of an amalgam of past and present truths. We pull ideas from the body of knowledge we inherit. We compare this information with new knowledge, we use it and we contribute to it. Although there *are* indisputable facts in our world, much stored knowledge of practical uses represents an ever-changing legacy depending upon our purpose. This external knowledge can enrich, restrict and destroy our lives depending on our perceptions and use—both our own and other people's. As well as external knowledge we also continually absorb and create our own personal internal knowledge contributing to our personal emotional model of the world (Feldman Barrett, 2017). See Chap. 2.

We are ultra-social animals influenced by other people's opinions and our own social relationships in different contexts. We are affected by personal emotions like enthusiasm, prejudice, fear, curiosity caused by events and experiences, our sense of justice, and personality. Our attitudes to knowledge have a profound effect upon how we use and benefit from it in both positive and negative ways. It is evident our education, upbringing and emotional responses can actually restrict our outlook and destroy our ability to take in what exists as factual information for others. We see this in the ways the worldwide environment debate plays out in media outlets across our planet Earth (Thunberg, 2019).

We have accumulated and lost a great deal of real, practical and artistic imaginative knowledge created by human intelligence with its diversity, destructiveness, inventiveness, misapprehensions, adaptability, creativity, transformational and observational skills. We can see this in the literature, artefacts and cultures contributed by our religions, leaders, healers, thinkers, artists, musicians, philosophers, historians, dramatists, poets, mathematicians, architects, engineers and scientists. Knowledge has been passed on orally and in written text, diagrams and symbols in ancient libraries and archaeological remains. See History Websites Endnote.[2]

For example from such civilisations as the following: The First Nation People of North and South America, Australasia (50,000–65,000 years ago) and Mongolia (20,000–85,000 years ago); Mesopotamian (3500 BC–500 BC) with their philosophies, customs, arts, technologies and oral histories; Persian (550 BC–331 BC) with libraries in Syria and Iran; Indus Valley (3300 BC–1900 BC) with Nalanda University and Buddhist library; Egyptian (3150 BC–30 BC) with libraries in Alexandria and Serapeum; Mayan (2600 BC–900 AD) leaving evidence of a complex understanding of Astronomy; Roman (550 BC–465 AD) which adapted and spread knowledge across its empire; Aztec (1345 AD–1521 AD) and Olmec (1500 BC–400 BC) in Mexico; Chinese (1600 BCE–1046 BCE) with a library at Xianyang Palace (sophisticated weapons, paper, silk, chemical, mechanical and hydro-engineering); Greek (2700 BC–479 BC) with its legacy in the modern world of philosophy, democracy, sportsmanship, drama, poetry, sculpture and the arts; Inca (1438 AD–1532 AD) in Ecuador, Peru and Chile in South America with governmental systems, temples and monuments; Malian in Africa with the ancient Muslim library preserved by local people in Ahmed Baba Institute in Timbuktu in Africa; Ethiopian civilisation with the earliest Christian writings (2000 BC onwards).

In Western Europe the ancient pre-Christian Druidic education system taught each student for a period of 12–20 years. It existed for hundreds of years before Christ (BC) and was the university system of its age in Western Europe with a mainly oral tradition which relied on wise elders or druids. It was well established to include a legal, political, religious, historical, musical, artistic and medical body of knowledge. It was destroyed by the invading proponents of Rome who favoured Christianity (Beresford Ellis, 1994). Many first nation people around the world continue to perpetuate in some ways similar cultural education traditions. For example, the Xhosa people of South Africa have an oral storytelling tradition called 'iintsomi,' where stories are reimagined from folklore to suit the audience. Native Australian grandparents still perform their traditional task of 'Dadirri,' telling children about their 40,000 years' culture (Gough 1990; Ungunmerr-Baumann 2002 cited in Jarvis, 2019; Kelly, 2017).

Pam Jarvis (2019) says it is the development of cultures and languages that has given humans the capacity to understand and innovatively adapt to different environments. She quotes Jonathan Gottschall (2012) who said that when the human mind was young and our numbers were few, tens of thousands of years ago, we were telling stories to each other. She takes us through some of the main points made by modern experts in evolutionary psychology around this area of knowledge and cites their books on the link below.

- The sharing of symbolic meaning helps us to 'operate within our world' by internalising language and using its system of signs (Bruner, 1996).
- Humans in different environments understand similar aspects of their world in very different ways (Bruner, 1996).
- Humans create social systems and interactive networks requiring co-operation between people who have never met before and may never meet again (Chase, P. 1999).
- Bruner said understanding through storytelling has played a central role in early child development and is amongst 'the most widely used forms of organizing human experience' (Bruner, 1996).
- Sue Lyle (2000) extended the point by describing human beings as 'storying animals' using stories to make sense of thoughts and events resulting in human beings living in 'a largely story-shaped world' (2000). See Article Endnote[3]

Jarvis agrees with Jerome Bruner (1996) in saying storytelling is a 'bio-cultural paradigm'—an 'evolved human competency' using language as a symbolic method of communication. She says, "*The human primate comes equipped with the potential for enormous elasticity in cognitive and social development, far outreaching any other species on Earth.*" She goes on to her main theme explaining in detail the history and advantages of storytelling to our species and to teaching in particular. This is a particularly well-researched, academically and comprehensively referenced informative article.

Knowledge in the form of great narratives in the oral tradition still exists in many cultures around the world passed on by word of mouth, custom and song. Even though we may not entirely agree with them, they can give us information and pause for thought. Such historical information is now constantly being proved to be substantially correct with better, more authentic explanations and interpretations by archaeologists, scientists and historians, for example, by reconsidering historical records and using accurate scientific methods, carbon dating, DNA testing and satellite photography. There remains a huge area of potential research into all aspects of knowledge due to the traditional 'scientific' Western lack of attention paid to voices, histories and cultural heritages of indigenous peoples.

This is now starting to be addressed as first nation people gain confidence, are able to contribute to scientific research and collaborate in the development of worldwide cultural ideas. Putting ideas from different subjects and disciplines together in a process of cross-fertilisation has frequently generated new inventions. New scientific research and the connectivity of technology is increasing potential for new discoveries and ideas to be created. For example, ancient medicinal cures and life philosophies respecting the natural world environment can inform modern science and politics. For this reason, knowledge which appears obsolete to some should be respected and preserved as potentially useful for the future, even when we do not immediately see its relevance or feel a direct personal connection.

In the social knowledge aspect of life, a recent example of useful 'old' knowledge is Nelson Mandela's approach of 'reconciliation and reparation' in South Africa (1918–2013). This initiative in my opinion owes more to Zulu educational and judicial philosophy than to his Western lawyer training. This is revealed in his autobiography *Long Walk to Freedom* (Mandela, 1995). See Website Article Endnote.[4] In his supporting Data Contribution, Muhammad Shehu Shuaibu tells me that he hopes to complete his doctor of philosophy programme and go on to help his country, Nigeria. His story about his schooling shows how old issues in Africa continue to be ongoing. He does not mention it directly, but the legacy of colonialism and corruption is a back drop to life in Africa where corporate mining companies and politicians have exploited resources and neglected citizens. See Special Data Contribution.[5]

There are important implications for many forgotten and oppressed communities around the world when their history, cultural practices, practical knowledge, philosophies and achievements are rediscovered and acknowledged. For example, Bruce Pascoe reveals that the Aborigine people of Australia were agricultural engineers and wise land conservationists for thousands of years before being invaded and subjugated by Western explorers, immigrants and settlers (Pascoe, 2018). This is helping them and their fellow citizens to acknowledge their past achievements, oppressions and difficulties so that they can speak out and work to achieve their rightful place in the modern world. Instead of losing their cultural heritage many first nation peoples have the ability and knowledge within their culture to enrich multi-ethnic cultures as we learn together. People from indigenous cultures should be given and *take* more opportunities to explain their own culture. It seems extraordinary that they are often ignored by foreign academics and documentary makers even when they know (and understand) such a great deal about their own heritage.

Old practices are being studied, revived, taught and may contribute to solutions to new ecological problems. Take, for example, ways of living that can produce a lower carbon footprint and less invasive and destructive farming and mining methods. Methods of economic development and ways of living are being developed with better scientific human insight, for example, renewable energy. Examples of large-scale human 'objective scientific mistakes' are the American Dustbowl, the Chernobyl nuclear disaster, the mining invasions into Northern Tundra and the burning of the Amazon Forest. Many of these were and are driven by an emotional model of greedy capitalism. However, there are pockets of 'unscientific' human knowledge that have always existed in different cultural communities. See Journal Article Endnote.[6]

The psychologist Carl Jung (1875–1961) conceptualised another universal type of knowledge he found in archetypal human symbols, social meanings and behaviours we share. He coined the term the '*collective unconscious.*' He pointed out that similar forms of social practice, concerns and intuited knowledge have existed repeatedly in every human era. This includes our modern age. People have similar human concerns, recurring themes, archetypal ideas and symbols representing human issues and preoccupations. These include birth, life, death, each other and our ideas about our world which are both conscious and subconscious. Take, for example, the super heroes in films which are present in the characters in children's comic magazines. Their stories either retell or create similar stories in a modern context to those of the old Greek heroes and gods of myth and legend.

Our collective unconscious exists through human themes shared across time and geographical locations. These universal connections are ways of knowing and represent strongly felt emotional expressions of thoughts. They have taken the form of sculptures, ceremonies, customs, artworks, monuments, poetry, music, stories and narrative histories. Depending on the research purpose and subject discipline, their importance is not always fully appreciated as scientific multicultural human data. Beliefs, practices and customs are a real and important influence in human societies. They are scientific in as much as they are time, person and place specific, communicate ideas across time and distance and trigger human behaviours. They affect and cause real events. For example, the afterlife stories believed by Christian missionaries and Muslim activists; the perceived usefulness of plastics and non-renewable fossil fuels which many of us now realise have contributed to climate change.

Our perceptions of reality and the boundary between solid and non-material matter are being challenged and investigated in many areas of science. For example, Einstein's theory of relativity is re-examined in Stephen Hawking's *A Brief History of Time* (1988). The existence of previously unresearched energy may partly explain some older metaphysical beliefs that are increasingly seen as connected to actual scientific research ideas and discoveries. There may even be connections here to some of Carl Jung's perceptions about synchronicity (Jung, 1991). The sciences are updating, finessing and changing their approach to knowledge all the time. Neuroplasticity is now proven to be affected by our motivations, perceptions and predictions as well as 'actions' so that physical brain change viewed through 'imaging' is shown to have been brought about by feelings and thoughts.

There are useful and proven ways to illustrate and describe our envisioned realities, for example, new formulas and imaginary numbers in mathematics and physically in Quantum Electrodynamics (QED) in quantum physics where different types of matter are studied as forms of energy. Feelings and emotions around these enable the imagining and creative processes and can be recorded scientifically as part of development, authentication and evidencing. Historic human perceptions about some ideas and facts can be investigated and communicated in a range of new ways. Imaginative pilot research data can also be practical data—it is after all necessary to first imagine something before we can discover and/or invent it in both material and non-material forms. See Endnotes[7,8]

Looking at modern technology, the way that we gain access, select, envisage and create knowledge and share it affects its potential use. Technology companies have invented storage in the cloud where we are able to achieve reproduction of an enormous quantity of digital texts and images stored in a physically smaller space than previously. However, these are vulnerable to loss for several reasons. For example, they are dependent upon computers, the electricity supply and the cost of power and storage including the physical problems of overheating and security. Potentially they are also more vulnerable to total deletion if back-up methods fail. We are experiencing problems of copyright and data ownership because many systems are dependent upon unregulated digital caretaker companies motivated purely by profit.

In 2019 it was announced that a social media and storage company called 'My Space' had lost all of the content uploaded to its site before 2016 (including millions of songs, photos and videos) which disappeared when they went out of business. They failed to keep their promise to keep their customers' information safe. This raises the problem of who decides what is worth saving and even though information may be backed up by its author whether it will remain accessible since according to this same article it currently costs around $2 to archive 1 gigabyte of data permanently. Currently there is a problem around regulation. For example, the massive amount of electricity thus consumed contributes to climate change. The preservation of knowledge in future will depend upon an extension of legal protections. See Article Endnote.[9]

It will also depend on education towards a human appreciation and respect for all knowledge as a social value, irrespective of obvious personal usefulness to ourselves. This goes against current self-interested fashion trends. As time goes on data breaches and ethical problems are coming to light and legal systems of regulation are requiring government intervention and development to take place. For example, the profiteering mindset of some large pharmaceutical and internet companies around promoting sales is proving a problem for human world health. On the plus side, the internet provides almost instant speedy access to a huge body of knowledge which can be used in traditional and innovative ways to connect well-meaning people and generate new ideas. It provides a constant source of access to great writers of both past and present through websites, search engines and platforms. It provides a forum for new and current ideas to be presented through books, newspaper and blog articles and videos. For example, it has been the main route to fresh information for this book.

Here are some examples of feelings about knowledge, which have caused and are currently causing both negative and beneficial outcomes for humanity:

- Indifference—it is lost, mislaid and destroyed because it was not perceived useful by the people who could have preserved it. Perhaps their reasoning told them it was not important because it was not about them or perhaps they just did not understand its potential value.
- Deceitfulness—it is often deliberately used in particular ways, ignored, changed or distorted to serve commercial interests and those who have 'ownership' of it.
- Self-interest—people who developed knowledge from it but failed to credit and preserve their sources. For example, some of the inventions and explorations of Renaissance Europe were quite possibly based on earlier Chinese maps and engineering manuals (Menzies, 2009).
- Profit—the monetisation of data as a 'product' for sale makes knowledge less accessible and true facts problematic in some cases.
- Self-empowerment and pride—sometimes knowledge is erased in order to replace other peoples' cultures, for example, the books by Julius Caesar about conquered peoples (Caesar, 1982). First nation cultures have been affected all around the world.
- Descriptive confusion—discipline areas of knowledge can overlap and intersect but each tends to have its own technical language and conceptual frameworks.
- Respect—there have been people who went to enormous trouble and even risked their lives to hide and preserve knowledge. For example, the people in Timbuktu in Africa saved many ancient documents from foreign invaders which are still being uncovered today, those who have often saved archaeological treasures in Egypt from thieves so they can be kept on public display.
- Love appreciation and respect for tradition—some Middle Eastern peoples recently gave their lives to protect their ancient history in recent conflicts in Syria, Iraq and Egypt as 'ISIS' religious extremists attempted their destruction.

- Open-mindedness and the excitement of discovery—exploring different points of view can leverage new knowledge and create solutions.
- Compassion and empathy—for example as in schools of medical research and well-run, knowledgeable social care and charitable systems.

7.3 Back to Basics: Popular Philosophies, Ideas and Visions

The process and discipline of philosophy is about finding, questioning and working out the truth about all types of knowledge. This might involve surmising and speculating, about the meaning of all aspects of human life including all of the intellectual disciplines or subjects within the arts and sciences. There are many celebrated academic philosophers that are revered and referred to by those in a position to study them. Ideas based on their thoughts can sometimes influence those in power and educated groups, for example, university lecturers. However, it is those more fashionable, popular and changeable often political philosophical arguments that directly affect everyday social relationships and influence ordinary human learning opportunities in homes and workplaces.[10]

Different social groups on social media, lifestyles, relationships, perceptions, political arguments, choices and behaviours are related to Jung's collective unconscious (1991). They deal with 'social vibes,' group beliefs, loyalties and identities, cults, intellectual ideas, fashions, trends, politics, economics and other human concerns, which mutate in form and fashion through history. The philosophies of ordinary people at all levels of society sometimes create social movements that affect history. For example, the charitable self-organising evening institutes in Victorian times in the United Kingdom where the working classes educated themselves. For example, in trades and crafts from clock and furniture making to engineering, dressmaking, plumbing and horticulture. Co-operative movements, credit unions and working-class unions fighting for workers' rights were and are all based on popular philosophies.

Jonathan Jones (2018) describes how science influenced art. He gives a good overview of the forces at play in the United Kingdom which generated the cultural developments and struggles in the historical period known as the Enlightenment in the seventeenth century. He suggests that the English Bill of Rights (1689), the scientific work of Isaac Newton (1643–1727) and the philosopher John Locke (1632–1704) who wrote two influential works called *An Essay on Human Understanding* (1869) and *Treatise on Government* (1869) generated many of the democratic ideas of this Age of Reason as it became known.

Some philosopher activists who have changed society are the lords who made the English king sign the Magna Carta (1215); the English parliament who passed the English Bill of Rights (1689); James Madison Jnr (1809–1817) who drafted the American Bill of Rights (1791); Susan Anthony, Elizabeth Cady Stanton and Lucy Stone who fought for votes for women in America (1920); Emmeline Pankhurst (1858–1928) and daughters Christabel and Sylvia who campaigned for women's votes in the United Kingdom (1928); Mahatma Gandhi (1869–1948) who peacefully liberated India; Martin Luther King (1957–1968) and Maya Angelou (1928–2014) who progressed the cause of civil rights in the United States and Nelson Mandela (1918–2013) who negotiated equal rights for citizens in South Africa.

Perhaps for our purposes here, Thomas Paine (1737–1809) should also be included. He was a British-American political writer, theorist and activist who had some influence on ideas leading the Declaration of Independence in the United States. He wrote three philosophical works titled Common Sense, Rights of Man and The Age of Reason. These ideas grew and were adapted by other thinkers as they spread around the world. The Enlightenment was created during the development and expansion of learning in different areas of science causing cultural, artistic and social change. Some of its proponents developed a benign and humanistic value system aimed at improving the human condition.

Other modern Western philosophies originated and were influenced by scientific discoveries, but their influence has not been so beneficial—for example, ideas of empire, eugenics (a theory which denies human rights to some social groups), supremacist beliefs extolled by existing ruling classes and the acquisition of wealth and privilege as a goal for life. The different threads of these philosophical thoughts have influenced our understanding of our human lives and destiny spreading throughout the world causing oppression, misery, injustice and unrest. Bogus philosophical arguments and nefarious interpretations also cause negative events—as when Hitler presented his evil justifications for the annihilation of the Jewish, disabled and Roma people and Joseph Stalin (1878–1953) misused Communist philosophy to control, imprison and execute his political opponents.

Positive and negative philosophical arguments, wisdoms and conclusions interconnect, develop and are variously deconstructed and adapted or dismissed as humanity moves through history. The value to everybody's quality of life through the education of ordinary people gradually came to be more and more appreciated, depending very much on government interests and geographical location. These ideas have in recent times receded into the background in modern consumer cultures, but are currently increasingly relevant due to 'self-inflicted' environmental imperatives created by the unforeseen effects of global warming and pollution.

There is a body of non-academic philosophy, driven by pragmatism and perceptive perspectives generally ignored by academia. They may be seen as outliers, but are useful for the different insights, emotional models and perspectives given. These books are worthy of study and reflection in understanding how we got to where we are now. In understanding how our minds may be working—a mental reality check is needed as we face the Earth's manmade environmental crisis. Malcolm Gladwell's book *The Tipping Point: How Little Things Can Make a Big Difference* (2002) showed how susceptible we are to memes, fashion, movements and ideas that become enormously popular, making fortunes, causing bankruptcies, impacting and affecting lives, causing death and destruction in war zones and then fading away.

Huge fashion and political trends can be started by a small communication or human incident that starts a trend and builds an 'empire' or political movement based on emotionally 'cloned' or 'mirrored' behaviour. Elaine Hatfield et al. (1994) wrote a book about 'Emotion Contagion' which shows how susceptible we are to the emotional communicators and manipulators amongst us. Gladwell (2002) mentions Howard Friedman, a psychologist who developed the Affective Communication Test, which attempts to discover how this is achieved. People adept at these skills often become political leaders and celebrities in society. They have been able to use social media to their advantage. It is interesting to note that some of these books were written in the 1990s before the heightened current concerns about internet celebrity and the rise of dictatorships started to cause alarm.

In his book *Enlightenment Now* Steven Pinker does a comprehensive review of what might be the current state of human enlightenment (Pinker, 2018). He examines a vast array of statistical evidence and points out the benefits we have gained in health and lifestyle. He finds that modern people (who are addicted to consumerism lifestyles) particularly in the West do not appreciate (apart from the growing problems of global warming) just how fortunate they are. The young Swedish activist Greta Thunberg is arguing that we face an existential crisis caused by our excessive consumption and must change our behaviour before it leads to an irreversible ecological disaster affecting the whole world (2019). These are modern philosophical arguments, which are affecting us now.

Knowledge can influence humanity in positive and negative ways depending on personal interpretations by ourselves and others. Research into brain plasticity tells us everybody relates personal knowledge to their past and current situation at some level of consciousness informed by events, memories, feelings and emotions, perceptions and our own decisions. Our ideas and responses to knowledge are underpinned by emotional worries such as failure and inadequacy and ambitions such as being able to live happily, our value in society and the ability to earn a living. As well as our upbringing and education, they are also affected by the cultural memes, tropes, fashions and ideas we encounter (Syed, 2019).

7.4 Psychology: Traditional and Hybrid Approaches

Over the years human psychological research divided into many different sub-disciplines as different methods, interpretations and solutions to problems were sought and invented. See Endnote.[11] The fact that they are so many and varied shows us the level of complexity and difficulty we experience in understanding ourselves (Hawkins, 2017). Historically, there have been many different kinds of psychological research with different purposes influenced by the perceptions, individual questions and purposes of the researchers. In particular, traditional psychologists concentrated mainly upon collecting what they considered to be 'factual data' but did not question their own assumptions. They often thought their own views as observers were objective and therefore 'scientific.' They did not triangulate opinions or admit there were alternative points of view.

In psychology placing ourselves in the position of an 'objective' researcher causes problems, because we all have different frames of reference. Learned behaviours, assumptions, prejudices and conclusions are open to question and interpretation because they are based on individual experience and upbringing. We must recognise we are always likely to have more or less limited life experience depending on the situations we are in. Human life experiences and habits affect our thoughts and behaviours in the case of both 'observers' and 'subjects.' As researchers we should be aware the human 'subject's' brain is very unlikely to be fixed and is more likely to be constantly assessing and interpreting the research situation. For example, even the two revered Austrian psychologists, Freud and Jung, themselves may provide us with two retrospective 'feeling and emotion' case studies. They were responsible for popularising the social science of psychology in its infancy and they were undoubtedly affected by the social conventions and scientific methods, discoveries and the 'zeitgeist' or spirit of their times.

They were influenced by and had to conform to the scientific, religious, social and political influences of the day. Freud and Jung brought their own personal take on knowledge to their work and had to present it in ways that could be understood by their society in order to further their careers. We should realise that although knowledge accumulates, modern researchers are similarly influenced both consciously and unconsciously. The Freudian and Jungian schools of psychology arose from these two fathers of psychology and their ideas were often taught as fixed methods together with those of other psychologists who contributed ideas and terms. Knowledge advanced as different solutions under variable and similar conditions were developed. Freud (1856–1939) was a psychologist (and psychiatrist) who was hugely influential in his lifetime, particularly in the United States.

His discoveries about human psychology helped to establish the discipline and many of the terms, processes and concepts he used and 'invented' are still used. Although his method of psychoanalysis, appearing to concentrate more on reason than therapy, has become less common as a diagnostic and clinical practice. His ideas remain an influence in psychology, psychiatry, psychotherapy and across the humanities. Freud treated people with mental health problems according to the medical model. He assumed that people's mental health difficulties could be analysed as fixed and as a medical doctor felt it was his job to find explanations and solutions. However, his tendency to analyse from his own '*objective*' perspective did not necessarily help patients to change, develop and evolve according to their own insights and motivations.

From a modern perspective, Freud and his contemporaries' methods were particularly affected by their *attitude towards women*, because their society generally believed them to be less intelligent than men. However, their work contained some useful insights which have since been built upon and adapted. Although some claims seem inappropriate to us, they were not always completely wrong. Some lines of inquiry can still be explored and tested and some ideas may be compatible with modern science and our current view of the world. Mistakes often lead us to fresh information. For example, the Freudian view of the world was accepted, especially in the United States and many people believed it and to some extent lived their lives accordingly. *We now understand that we humans have this tendency to enact and live according to taught belief systems—a useful piece of information and a warning to our own times.*

Another example might be women and ethnic groups who have accepted and brought their children up to accept their restriction and repression. Freud's methods of psychoanalysis opened up aspects of how these levels of consciousness inform behaviour and can interact. He created a theory of the unconscious with the concepts of the id, ego and super-ego to explain some of the subconscious dynamics behind thoughts and behaviour. Freud developed psychoanalysis with therapeutic techniques such as the use of free association by connecting words spontaneously and transference assuming someone is behaving like someone else or projection when a person assigns their personal 'take' on the world to external situations. Freud discovered the existence of complexes. These are concentrations of strong feelings that can have an overly intrusive influence on our reasoning and behaviour. They may sometimes become obsessive and problematic.

Many of Freud's useful ideas were adapted and developed by others, for example, the inferiority complex (Adler). However, some aspects of Freud's work might be better understood in relation to his upbringing and personal life. He deserves our respect as a founding father of psychology, but perhaps he had issues he could or would not talk about. It may be too late to draw any concrete conclusions, but considering his upbringing might help us to get his research conclusions into perspective and help us make allowances for possible bias (Clark, 1987). His life illustrates the problem of accepting the 'objective' conclusions of social research without considering, questioning and reflecting upon the context and veracity of the researcher's possible subconscious bias.

Freud made a study of infantile sexuality and proposed the existence of the Oedipus complex, which is based on a boy's feelings about his mother. He believed children of both sexes had an unconscious sexual desire for one parent and a likely dislike or feeling of competition with the other parent. He thought that everybody experienced sexual feelings in infancy in this way. He decided his ideas were universal and 'normal' and used them to diagnose his patients' adult behaviour. He published '*Three Essays on the Theory of Sexuality* in 1905, which were at first greeted with strong protest, but were eventually accepted by his colleagues influencing their own work with their patients.

It may be that Freud's ideas and personality were influenced by his own family situation and I wonder if he might have had a tendency towards an 'Oedipus complex' himself of some sort or seen it within his family. This might have been his emotional model. For example, Sigmund's father Jacob Freud, aged 39, married Amalia Nathanson, aged 19, and he was their first born child. In age his 'beautiful' mother was halfway between himself and his father. Sigmund's two bachelor stepbrothers from his father's first marriage were a little older than his mother. Seven children were born to the couple, but Sigmund, the eldest was always his mother's favourite. She called him "my golden Sigi" and he was given a much higher level of education than his sisters including a room to study in with his own desk and an oil lamp in order to be able to read in the dark.

When I first read his biography, I remember joking that (from my modern female perspective) Freud's sisters' might have had 'oil-lamp envy' rather than 'penis envy'—which was one of the ideas/theories he proposed about women. Like many other women of their day they were not considered to require a scholarly education. Sometimes our appreciation of the original context becomes important to our understanding when accepting 'facts.' It is possible that Freud suffered from some transference or 'projection' problems himself and may have known it because he took firm steps to prevent details of his own inner analytic reflections from being published. It is ironic if this was indeed the case and one wonders if Jung (1865–1961) who was his great friend and follower realised this when he went off to pursue his own interests and his own self-analysis (1991). Jung's father was a Christian cleric. He came from a very religious Christian family and yet he went on to study many aspects of other ancient religions.

Jung eventually developed his own ideas building on Freud's *Interpretation of Dreams* (1899) and many other ideas about psychotherapy. Jung said that dreams are a way of connecting with what is going on in our personal subconscious (1995). This ties in with modern research on sleep. See Chap. 1, Sect. 1.5. Jung became another important 'teacher' of analytic psychology. He collaborated for a while with Freud on their joint vision. However, Jung was more open-minded and carried on researching, exploring and developing new ideas. His work has contributed many of these to psychiatry, anthropology, archaeology, literature, philosophy and religious studies. Jung came to disagree with Freud's rather fixed categorising approach to diagnosis. Jung developed ideas about the Psyche, Spirit or Soul and Self and often referred to the self as representing a person's inner feelings, emotions and personality.

He studied complexes such as neurosis. He talked about human consciousness and fields of awareness, dividing them into personal consciousness and the subconscious as well as collective or social consciousness. Jung believed that every human being's goal in life was to achieve personal individuation by learning to balance the psychological aspects of their personality. This is thought of as an ongoing process which Jung felt and believed brought inner harmony and independence. This appears to fit in with Carl Roger's concept of self-actualisation, which is referred to and used as a concept in his person-centred counselling theory (1951).

As mentioned previously Jung coined the concept of the collective consciousness. He was interested in similar myths and motifs in various civilisations and studied human instincts and archetypes (frequently occurring human behaviours, images and ideas). He thought that these were human predispositions to perceive, think, act and feel in particular ways for subconscious and often hidden reasons. He considered them universal and believed they occur regardless of class, creed, culture, location, personal history and historical epoch. For example, mandalas or complex symmetrical circular images found in different cultures. He thought they might possibly be subconscious psychological representations of the world, cosmic symbols or even representations of the human mind.

Jung believed that mental stability and physiological health depends on the human ability to be integrally connected so that the conscious and unconscious work well together 'on parallel lines.' He thought being split apart or dissociated within the mind caused psychological disturbances. Jung noted that personalities displayed different habitual behaviours, for example, extraversion (energetic and sociable behaviour) and introversion (solitary and withdrawn behaviour). During his lifetime he moved away from the rather rigid view of behaviours and labels that clinical psychologists had 'discovered' using the medical model.

Jung was well ahead of his time in many respects and had a flexible approach to his patients, which allowed him to work to some extent collaboratively with them towards solving problems. Since Jung's day clinical psychologists, therapists, social practitioners and teachers have developed these with their own approaches. Good practitioners vary their techniques and have a practical appreciation of the importance of their clients' and students' feelings and emotional feedback. Unfortunately, however, some old attitudes linger uninformed by proof from modern neuroscience that intelligence can develop in some broadly similar ways—and yet be ***dynamic, complex, variable and actively plastic.***

It seems our current ways of working are also still affected in some ways by historical theoretical, cultural, social perspectives and purposes. Meanwhile modern therapists have developed methods adapted to modern clients' 'perceived needs' in different ways. In private educational psychology they assess for giftedness, cognitive profiling for physical 'disability,' social-emotional behaviour problems, and psychoeducational profiling for attention deficit hyperactivity disorder (ADHD), autistic spectrum disorders and dyslexia often in order to gain client's and children extra help and funding. Research and development is needed in order to find ways to support without detrimental labelling.

I have edited a list of some of the types of assessment tests currently offered to those who seek psychological treatment privately. These are based on an article by Giuseppe Accard (2014), a psychologist in practice, explaining his work to prospective clients. He says these can help psychologists work out what makes clients behave in particular ways. He says that such tests go back to Confucius (551–479 BC) in China. Although used then for purposes of appointing officials, they may have been more like examinations of competency, character and knowledge. He says that the unscientific methods used in the eighteenth and nineteenth centuries are now 'updated' and 'standardised.' See Website Article Endnote.[12]

Four types of testing may be done by pre-agreed permission that results in a report. These are done through norm-referenced testing such as personality, giftedness and IQ tests, interviews with a therapist to gain extra information, ongoing observation and other informal assessments during therapy. The four categories of psychological assessments include (1) an initial clinical interview which may involve a discussion of problems, medical and family history and 'other assessments' to 'inform' therapy; (2) a critical IQ test (Intelligence Quotient) to assess intellectual functioning and give a cognitive 'picture of the mind'; (3) a personality assessment to identify traits that may be 'holding you back'; (4) a behavioural assessment which may identify 'out of control' behaviours, which range from just 'annoying' to 'dangerous.'

'Test instruments' may be administered by an assistant or psychometrician and are chosen to give the psychologist insights into the client's problems. These 'tests' may also be done at home and by computer. Accard explains that all of the above are done by measuring against scales of 'normalcy' to measure and diagnose for issues of mental health such as depression, bipolar conditions and levels of functioning in old people and children. He defines a 'normalcy' scale as a general comparison of common human traits and behaviours, strengths and limitations. He evidences the psychologist's difficulty in dealing with *the problem of interpretation*.

The article clearly and honestly states some of the problem areas that may exist. I have edited these below and put words in commas which I believe cannot necessarily be simply or exactly defined. These 'problems' appear to include that:

- It is accepted the psychologist's purpose involves the making of 'subjective assessments.'
- The psychologist will have the ability to 'correctly' extract and interpret 'relevant' information.
- The use of a computer in no way compromises the data.
- Tools and tests will be used to 'predict and manipulate' behaviour.
- Even when 'well-meaning' psychological therapy is offered, there may be a problem of validity and reliability.
- The testing and assessment being done will tend towards helping in the 'resolution' of 'mental health difficulties.'

That the client signs up to 'being manipulated' might potentially cause problems especially in the case of vulnerable adults or minors. These methods require ongoing research, analysis, adaptation and supervision in practice. In cases of good practice this would involve checking, observing and collaborating with patients, families and colleagues to find out and discover their point of view in order to achieve ethical balance. They certainly require some form of independent supervision, checks and balances and regulation. In cases of bad practice a complaint and restorative justice system according to law sometimes already exists and is definitely needed.

Another research method used by social psychologists is the randomised control trial, but its use depends upon a clear and honest assessment and declaration of likely limitations. This can produce interesting negative points, which are useful to discover. For example, Naila Kabeer's recent research article (2019) supports the fact that we cannot assume humans are fixed in their behaviour even though we know they have habits and tendencies. She says the suitability of randomised control trials (RCTs) in social sciences has been widely questioned. She confirms that assessing their effectiveness has proved difficult in practice. She evidences her opinion by referring to recent randomised control trials used in a research programme to alleviate poverty in West Bengal. She points out that when used in isolation, they fail to recognise problems and are skewed by *human agency*, in other words people's tendency to be spontaneous and unpredictable! See Article Endnote.[13]

Of course, this proves and confirms that our psychology *is* complex and not fixed or always predictable and this may help us to look out for and pinpoint variables. Brains have unique aspects and behaviour is often due to unknown variables, which are interwoven with feelings and emotions. This does not mean they cannot be researched. These influences are as scientific as more obvious conclusions, but can cause major setbacks in research authenticity and progress. They may even be unethical if 'subjects' are subjected to inappropriate situations. For example, behaviourists have not always understood or been willing to explore potential problems arising from influencing 'subjects'; in particular, the ethics of creating situations which play on human susceptibilities dishonestly to elicit *particular* uninformed responses (Skinner et al. 1957). This relates to unethical practices carried on by social media platforms today (see sections on Silicon Valley).

Ethics are a major concern in psychological therapy and in all areas of the wider discipline. It is presumed that the independent ongoing mental health of the client is the therapist's aim, but in our monetised, commercial society there is quite likely to be a conflict of interests. The knowledge which is produced in this area is likely to be most effective when an appropriate collaborative disclosure approach is adopted both with the client and with independent colleagues. If intelligent regulation is not enforceable 'knowledge' can be distorted and abuses of vulnerable people may go undetected. This can be conscious and/or unconscious. For example, Tom Chivers (2019) asks if modern psychology's conflict of interest problem is causing a credibility problem. He notes there are currently calls for more transparent disclosure due to a reaction against psychologists who promote/allow their work to be used to influence others with insufficient evidence and/or in an unethical way.[14]

7.5 Current Educational Tenets, Ideas, Problems and New Possibilities

A tenet is a principle, doctrine, practice or belief generally held to be true especially when it is believed by members of an organisation, movement or profession. Historically, there have been many different educational schools of practice and thought, with philosophical viewpoints and frames of reference. Some appear as new approaches and arise from custom and practice and are sometimes adopted by leaders and imposed. In education there have always been tensions and disagreements between proponents of different ideas. There are conflicting values and confusing fundamental issues in education, which cause operational problems. We need to keep improving our checks and balances by creating systems that listen to and are informed by stakeholders like parents, pupils, students and teacher researchers, gaining a range of perspectives.

Political managers and civil servants are responsible by law for education at government level. However, brain plasticity tells us their ideas, assumptions and prejudices are often informed by their own career ambitions, unique upbringing, schools and college experiences. This is true of everybody, but these chosen individuals (even those with previous educational credentials) are in a position to impose their ideas upon the education system. They are remote from the everyday consequence of their actions affecting learners and teachers. Politicians are also influenced directly or indirectly by fashionable political, social, psychological, philosophical and populist thought and the pressures of re-election.

Academic 'research' and popular arguments are often produced and used to justify inflexible government policies—including some teaching practices that actually tend to discourage 'educational' learning if used in the wrong context. In recent times it has become fashionable for Western consumer societies and their politicians to value and promote 'high level' academic subjects over the arts. At the same time the education inspectorate have promoted a narrower curriculum, being required to only value knowledge that can be demonstrated statistically. They have adopted the common practices of the business world converting educational data into a spreadsheet. As previously discussed, this has tended to encourage the commodification of learners and schools undermining educational principles that recommend the individual's human right to a broad and rich education.

Schools have been forced towards accepting more inflexible input and output methods developed by behavioural psychology. They were spread around the world by market forces of capitalisation originally from the United States (Skinner et al. 1957). This trend gained impetus when incorporated into the internet marketing strategy of Silicon Valley entrepreneurs using the World Wide Web. They linked all human activity on their websites to the collection of data, selling and advertising it for economic profit. See Chap. 8. The habit of relating and valuing all aspects of life by referring to financial gain as a measure of worthiness has become pervasive in our culture.

In the United Kingdom, Roger Titcombe (2015) is a former secondary school head teacher and college principal of many years. He is passionate about the value of a learner-centred approach to education through careful open-minded planning and research at all levels in the system. This is an approach which worked for him as he worked in comprehensive schools from 1971 to 2003. He worked in the state school system and has considerable knowledge and practical experience of the fairly autonomous school management system at that time. When academy schools were introduced they were given a great deal more independence than state schools in the management of their systems and curricula. These schools were set up to operate like businesses. They tended towards ideas around *the pupil as a product* for which sponsorship and payment is received. Predictably there were failures and successes.

Titcombe (2019) has shown how this commercial outlook has limited and reconfigured the education system—being a factor in its considerable deterioration in some cases. This is not least because such an approach has also resulted in the neglect of 'less able' and 'more able' children within a normative system and a deterioration in mental health across the country. After undertaking extensive research into exam standards, he finds that these have fallen in real terms. He makes a convincing case showing how education outcomes have been devalued for young people through the 'gaming' of exam results over the last 20 years to produce the band of 'average' attainment results viewed by the government as satisfactory. There appears to be a likely long-term effect in terms of dumbing down the variety and depth of knowledge across the nation. See Newspaper and Blog Article Endnotes.[15,16]

Some traditional locally governed schools have followed academies in becoming more competitive and some less academic pupils have been overlooked, discarded and neglected as not 'profitable' in terms of results.

Academy schools work well when those in charge have an inclusive and knowledgeable educational approach, but not when corporate school image and exam outcomes take precedence over human values. In the United Kingdom this attitude has been encouraged by publicly available schools' league tables judged by government inspection. *The political system has become indifferent to both intrinsic and tacit knowledge in assessing and regulating schools.* Over the last 20 years, there has been an increasing tendency for teachers to be required and trained to teach and assess mainly instrumental knowledge.

Teachers have been forced towards 'teaching to the test,' using formulaic educational approaches to produce norm-referenced statistical results. Tensions have risen and teachers have become more stressed and less effective. They have been disempowered and their own brain plasticity in developing their insight, creativity and judgement in teaching children has not been sufficiently engaged and valued. Many have left the profession as a result. The situation of teacher retention in the United Kingdom has become so bad that in 2018 Emma Kell even wrote a book to support her colleagues called *How to Survive in Teaching: Without Imploding, Exploding or Walking Away*. Two recent authors have pinpointed some ethical problems that affect teachers' practice—Andy Tharby (2018), a teacher, and Gert Biesta (2011, 2013, 2018), an educational philosopher. I suggest we might start with them in reconsidering and reconfiguring our understanding of education because they make some particularly pertinent points.

Teachers have to work with and assess all kinds of knowledge as a necessary part of their practice. There are fixed and shifting factual bodies of subject knowledge and there are *ways or modes of knowing*, for example, intrinsic knowledge or knowledge for knowledge's sake and instrumental knowledge or knowledge for practical use. Tharby explains that teachers should be aware of the value of intrinsic knowledge as well as the instrumental knowledge needed to pass goal-orientated prescriptive examinations (2018). In considering plasticity, we should also add tacit or implicit mode knowledge, the knowledge we learn mysteriously through the 'subconscious' activity of our bodies and minds and simply know (Claxton, 1998).

Intrinsic and/or tacit knowledge is often case specific, variable, subjective and open to interpretation. Although it can be clearly demonstrated by processes and products, it is harder to explain and measure by means of statistics. All these modes and ways of knowing interact within the plastic brain as it creates learning networks. In recent times, in the United

Kingdom there has been a general assumption of production line values and imposition of narrow results-based testing by politicians. This causes problems for practitioners, because not only does it reduce autonomy in personal professional development, it reduces the system's ability to adapt to changing factors and conditions. For example, updating the response to changes in subject knowledge, cultural and social issues within and outside school.

Biesta has investigated and argued to establish a fundamentally clearer view of the purpose, function and essential nature of education (2013). He believes education to be fundamentally 'uneducational' if it does not involve risk and points out that it can never be 100% safe. This is entirely consistent with new discoveries about human brain plasticity. We are finding that learning develops in a unique manner through constant reaction, creating unique variations and physical change. He talks about the beautiful risk of education and the fact that this involves 'weakness,' due to unpredictability, as a necessary ingredient. This must surely include another important, unpredictable element of the 'beautiful'—the possibility of unexpected discoveries and growth in knowledge.

Developing and increasing student knowledge is what most good and inspiring teachers consciously aim and plan for, even though they know their profession involves risk. Their work becomes more predictable as their experience of teaching in their area of expertise increases. However, *outcomes can never be a certainty.* Unforeseen learning occurs at all levels of ability and can improve learners' lives whatever their ability. Such gains are dependent on individual natural tendencies, readiness, motivation and opportunity. They sometimes lead to original discoveries and new knowledge.

What if we really wanted evidence-informed practice in the classroom? In 2018, at Institute of Education (IOE), University College London (UCL), Biesta delivered some enlightening assertions, insights, questions and answers in answer to this question. "What if.....we really wanted evidence informed practice in the classroom?" See Presentation Video Endnote.[17]

- Wasting money on educational research is not the same as spending money on educational research.
- The aim should not be to make education work but to make education meaningful for everyone involved so that teachers can be thoughtful in their actions.

- He is concerned about the interface between research and practice.
- Is medicine an evidence-based profession? Not really, surgeons admit they are often operating in the 'dark' when the unexpected happens. Teachers often also face uncertainty.
- Does knowledge about what works actually exist? Not really because knowledge about what works is based on what has worked in the past in concrete situations.
- Are advances in modern medicine and modern agriculture the result of the application of scientific knowledge about what works? No, scientific knowledge is not spread to farms; farms are included and developed within a nationally organised system (Latour, 1988).
- Is teaching an intervention and learning an outcome? No, because this type of language is based on an assumption of input and output and it ignores what happens in the middle—the interface between the two.
- Can we make education work? Yes, we can make education work, but not in the Newtonian, quasi-causal way that is only applicable to closed, deterministic and linear systems.
- What kind of system is education and how can it be meaningful? It is an open system, it has porous borders; semiotic, it runs on meanings, interpretations and communication—we talk and we try to make sense with and to our students. It is also a recursive system. The people in the system think and act on the basis of their thoughts creating feedback loops into the system. It is not closed, linear and deterministic.

How can we push such systems towards operating with predictability? We can do this through complexity reduction, but this is not advisable. Complexity reduction can be done and is in the process of being done in three ways: (1) we reduce the openness (buildings, classrooms, curricular and age groups); (2) we reduce interpretation through assessment; (3) we reduce divergent thinking/recursivity by training teachers to think as a teacher instructor and not to think in any way possible. Biesta says that if we continue down this route we will reach a tipping point when education turns into indoctrination and complete control. We have to ask the bigger question: what do we want education to work for?

What is the purpose of the whole endeavour of education? Qualifications (knowledge and skills), socialisation (orientation to the world as it is), philosophical subjectification (or individual responsibility) where students take responsibility for their freedom to live together well democratically.

Biesta says it is complicated, we can make education work but it comes at a price. What is the price we are willing to pay? It can cause stress and mental health problems. Do we still think of it as a life-enhancing process or are we giving up on the idea of education itself? He finds that the people who run education generally have good intentions, but he has an even bigger question for us. Why strive for equal educational opportunities in a society that has no interest in or place for real equality?

Historically there has been a cross-fertilisation of ideas across disciplines such as politics, economics, sociology, psychology, their 'philosophies' and schools of thought, which are generally informative but sometimes cause problems in relation to education. For example, traditionally psychologists believed in observation and measurement in order to create and test their theories on large groups of people in order to discover 'normative measures.' The feelings and emotions and own opinions of the chosen research 'subjects' were not recorded. The 'beautiful' aspects of education are life enhancing and improve ordinary lives in personal ways. They contribute to richer social communities and a feeling of well-being. For example, the factory worker who is an expert in classical music, the bus driver who plays in a band, the entrepreneur setting up a new business, people who collect antiques.

There are children who enjoy sport and go on to be football fans for life; senior citizens involved in physical activities and hobbies; those who investigate their ancestors, help charities, visit museums and art galleries, enjoy poetry, run credit unions, food banks, organise community activities, book clubs and co-operatives. *These are all knowledge-based learning activities that are educational.* Our love of and involvement in theatres, museums, art galleries, urban parks and national parks, sports facilities, architecture, large business corporations, transport infrastructures, governmental bodies and charities all require us to become educated and invest in social and cultural capital. Humans should not be reduced to a set of equations involving only the economics of input and output which is only valued by amounts of profit gained. A profitable life involves a great deal more than money—*it should also be measured in human happiness.*

This type of personal life fulfilment starts at home *and* at school within self-confident people and communities—feelings and emotions, talents discovered and developed are very much involved in creating this type of valuable commodity! Some past knowledge about learning is obviously lost or unavailable to us. However, in educating ourselves in the process of our everyday lives, we continue to rediscover, adapt, find and invent new ways to learn. Clinical and neurocognitive psychologists are gradually gaining more information and a better understanding of how this happens. Theorists in all knowledge areas can learn from this new information, contribute data and help each other. Education, at all levels, can provide a rich vein of practitioner evidence to complement psychological research. Better understandings arise particularly as the result of teacher research into a particular problem in a particular setting working with individual children towards a positive solution to achieve acceptable outcomes.

7.6 Knowledge, Human Brain Plasticity and the Use of Screen Technology

Screen technology presents us with both a huge advantage in our access to some kinds of knowledge and huge disadvantages by its very accessibility and possible susceptibility to the distortion of facts. The social platform world on the internet has created a fake reality—in some respects a fake 'real' world—or perhaps its own 'real' world. The boundaries are blurred and yet this new world is dependent on one power source, which may not always be reliable—electricity. The way in which modern journalists talk about 'developing the story' implies that news events are not real. Unfortunately they can be manipulated and apparently they often are. Of course this has always happened, but at a slower pace. Jaron Lanier (2014, 2017, 2018), creative inventor of virtual reality, Silicon Valley expert, musician and mathematician, says that technological discoveries have created new dangers.

It turns out we are highly stimulated by negative emotions and this means that when algorithms recycle popular messages, large numbers of people are alerted, become concerned and feed these messages back into the system. We are social animals and it is instinctive to want to share alerts with our social family. This creates a useful communication loop for those who sell data, goods or political ideas. The result has been that some journalists have become trauma tourists looking for stories that attract attention and readers. They present disasters far removed from the audience as story dramas which sometimes disrespect and distort our view of human suffering as it unfolds. Our knowledge of the world can be improved or reduced depending on the quality and balance of news reports. People who can't cope with or accept the truth now have the apparent option of making and spreading fake news, which denies truth. This is a problem for seekers of knowledge.

Lanier points out that reaping the benefits of the internet inevitably involves facing and dealing with these problems. It involves learning the skills to balance our immediate self-interest with the need to keep opportunities alive in our natural world environments. He believes, if we act promptly and wisely, we can learn to understand technologies managing them intelligently to benefit all humanity. His book *Who Owns the Future?* asks that we identify, pre-empt and solve the ecological and social catastrophes we face in this fourth (perhaps last) modern industrial revolution (2014). Lanier challenges us to understand human nature, values, temptations and realistic needs. This puts responsibility on individuals and governments to create appropriate administrative, political and social systems with embedded fail/safe mechanisms. Education is fundamental to choice and freedom if humanity is to have a worthwhile future. Everyone has a stake in this process.

We do have a huge body of knowledge already, but education is a changeable creative process. There is always more important information and knowledge to be gained, practised and developed. For example some possible research questions might be:

- How should we teach knowledge in different settings?
- How can we engage learners' emotional intelligence in different subjects?
- In the 'Post Truth' era, how do we sort fact from fiction? Is all truth potentially 'fake news'? On what do we base our judgements? How can we check information?
- How can we counteract the human tendency to misunderstand other people's experiences?
- How can we avoid believing incomplete or prejudiced information?
- What kinds of systems, checks and balances can help us solve problems? How can we test and trust our judgements?
- How can we empower ourselves, learners and each other to optimise intelligent inquiry?
- What learning strategies and systems can be devised to preserve our planet for future generations?
- Are any of these questions particularly useful or unhelpful and what hidden challenges do we face?

The implications of exploring and researching these questions are wide, but some of the answers could improve our lives as a species. We will have to choose wisely, adapt, invent and reinvent questions to suit different situations. There are some great challenges ahead as we face up to life in the third millennium (Davis, 2019). Human brain plasticity has already adjusted to the use of screen technology. New ideas and social trends have always occurred and often then disappear. Neural connections become dormant or wither away as we change our ideas and behaviours and lose interest. As a teacher who tends to be wary of controlling versions of 'brave new worlds' I find this quite reassuring, but it does not remove the problems of conditioning and coercion. We need to learn ourselves and help our students to learn about how to deal with these new forms of knowledge.

Computers are limited to sorting, comparing and matching in binary mode, but humans can operate as appropriate in binary, complex and multi-switching modes. They make decisions in new situations empowered by 'emotional' reasoning influenced by multisensory inputs. Algorithms are set and fixed by technicians and tend to be limited in their adaptability to react appropriately to unexpected circumstances that affect other humans. They are set to act upon predictions based on what was previously factored in, but become a problem when they persist and are unmanageable in the event of new evidence and unforeseen events. They do not have any *living emotional intelligence*. Currently checks and balances are not always sufficiently built in and a regulation-free internet is a very real danger to humanity. In fact such a system promises a dystopian future where humans are forced to think more like algorithms. We are now becoming more aware of the difficulties arising when 'logical' algorithms operating on the internet distort human behaviours.

One of those behaviours is the habit of responding to problems in a binary way as simply right or wrong. This habit dumbs down thinking and reduces intelligence. Another behaviour is the habit of instant gratification with very little pause for thought. We have replicated an easy way to gain information, but need to learn the art of questioning facts using wisdom and judgement. Algorithms used 'logically' to attract people on social media platforms have automatically reinforced thoughtless as well as 'valuable' interests and behaviour. Social platforms prompt people to seek the easy acceptance and acclaim provided by social media. They exploit our cognitive processes and habits as data about individual users is given freely and sold to advertisers. We have been 'conditioned' and 'groomed' to collect likes, followers and to post self-congratulatory idealistic 'selfies.'

Corporations, advertisers and political campaigners have discovered ways to use algorithms to influence us through subconscious misdirection, sensational announcements, phoney reward systems and manufactured (fake) news. This is knowledge which may be harmful. It can cause unhealthy preoccupations and distortions that affect behaviour, mental well-being and judgement even involving child abuse, suicide, self-harm and genocide. Democratic governments are currently responding to protests about social media practices by creating protective regulations. Whistle-blowers are revealing the truth about social media companies and the way they have acted against the interests and well-being of ordinary people to create wealth and power (Levy, 2020; Kaiser, 2019).

Teachers around the world are in the process of readjusting their thinking to deal with these effects on their learning environments. Other ideas, cultures and frameworks were also 'logical' for their times at various points in history. They helped human civilisations survive and make meaning. They came to be embedded in social and religious groups, regarded as set in stone, necessary to the status quo and externally logical, for example, religious doctrines and scientific approaches. Historically groups of people have accepted them and assumed that they would continue to be useful going forward. However, they may or may not always apply and need to be reimagined for different purposes by each generation for humanity to 'progress.' All idea frameworks are working prototypes from which we need to learn essential and 'appropriate' truths.

This is a skill particularly needed in our age of mass communication and the habit of researching, collecting alternative points of view needs to be taught. Sometimes we can understand and explain feelings and behaviour in context and sometimes we cannot. We are affected subconsciously and consciously by our own unique predictions and at times this causes difficulty and confusion for all of us on a daily basis—*experts included*. The 'intelligent human brain' is constantly attempting to analyse itself, while at the same time believing it is objective. Unfortunately it is not possible for any human to be completely impartial in any area of inquiry however much information we gain. Neuroplasticity can help describe thinking in new terms as neuroscience research brings a greater level of objectivity to the subject. It shows us that feelings and emotions are essential to human thought and physical survival and how each emotional event is unique (Feldman Barrett, 2017). See Chap. 2.

This information has implications for improving learning. The extraordinary achievements the human brain can achieve still have to be practised and learned. They could be dumbed down if children's and adults' neural nets come to rely on the internet too much for entertainment, distraction and information. This is especially the case where invisible gamers use reinforcing behaviourist mechanisms and propaganda promoting algorithms for their own ends. A rich humanist education and a creative, questioning attitude are key. Science is not the only form of knowledge and getting instant answers via the World Wide Web is not a substitute for a proper education. There is also the problem of preserving all of this knowledge. Who is going to decide what to delete, when we run out of enough electricity to support data storage for the transport, bitcoin and cloud systems we have created?

How will knowledge be valued and preserved for the future? On the upside, a few years ago when looking after my grandson, I asked him what he was watching online. "Funny men," he said. "What are they doing?" I asked peering over his shoulder curiously and a little apprehensively. "They are doing something clever with a thing called a Hadron Collider," he said. Fortunately he was not being 'groomed' by 'bad actors' but was fascinated by science documentaries. However, to my consternation his brother, aged 13, on being asked what he had learned from his technology class at school answered that he had learned and decided for himself that the human race had become so clever, it was going to destroy itself. This is a profound and worrying concern for us all, but particularly for his generation. It is an unfair responsibility and we all have to do something about it as the work of teenage Swedish activist Greta Thunberg demonstrates (2019, 2020). See Video Endnote.[18]

7.7 Conclusion

Human knowledge has its roots in ancient mathematical skills, astronomy, civil engineering, agriculture, crafts, alchemy, medicine and architecture. Historians of 'traditional' science have identified that some of its approaches of *apparent* 'dispassionate objectivity' and categorisation began in sixteenth and seventeenth century Europe. We are now in a position to look back, appreciate and reflect upon how scientific research and technological inventions have benefited and impacted our lives. *However, the 'scientific' conviction of the unimportance of emotions to learning has affected ordinary people in negative ways.* It is evident this attitude has perpetuated some forms of ignorance that cause us to abandon and refuse to engage positively with new forms of knowledge, discouraging learning. This approach is now proved completely wrong—not only by the wisdom of all kinds of humans in their everyday activities, but also by electrophysiological imaging technology.

Brain plasticity has made disbelief in feelings and emotions on a par itself with such beliefs as 'the Earth is flat'; 'global warming is untrue' and that 'literal views in old religious texts that justifying mayhem and the murder of fellow humans' are justified. The refusal to engage with any potentially unpleasant emotional discussion or disagreeable fact by university students has become a recent trend. For example, currently they have been refusing to engage with new 'emotional' information. This is popularly called 'no platforming' at universities with the justification that it may be potentially upsetting (Lukianoff & Haidt, 2018). However, we now have authentic and overwhelming proof that 'exploring challenging and innovative ideas'; 'reflecting upon the consequences and effects of our emotions'; and 'recognising feelings' which engage us in thoughtful reflection are an important part of our fundamental and very human intellectual skill set.

Notes

1. The Editors of Encyclopaedia Britannica (Aug 9, 2019) *Science* https://www.britannica.com/science/science Accessed 19/08/2019
2. Websites: https://www.historyandheadlines.com/top-ten-greatest-civilizations/ https://en.wikipedia.org/wiki/Civilization
   https://www.ancienthistorylists.com/ancient-civilizations/10-oldest-ancient-civilizations-ever-existed/ Accessed 04/04/2019.
3. Jarvis, P. (2019) 'Not Just 'Once' upon a Time' Institute of Childhood and Education, Leeds Trinity University, UK; Published in: Genealogy 2019, 3, 44; doi:https://doi.org/10.3390/genealogy30300441 – Link: www.mdpi.com/journal/genealogy – Licensee MDPI, Basel, Switzerland. This article is an open access article distributed under the terms and conditions of the Creative Commons Attribution (CC BY) license (http://creativecommons.org/licenses/by/4.0/). https://www.mdpi.com/2313-5778/3/3/44/htm Accessed 01/01/2020.
4. Mandela Foundation News Lessons from Nelson Mandela on reconciliation, reparation, and the path to prosperity Published 15 June 2020 https://www.mandelarhodes.org/ideas/mandela-rhodes-foundation-statement/ Accessed 06/01/2021.
5. Special Data Contribution

   My Education Journey by Muhammad Shehu Shuaibu M.A. Applied Economics, Sharda University, 20 January 2021.

   My name is Muhammad Shehu Shuaibu and I was born in Kano state, Nigeria. I have completed my primary school and Junior secondary school at Jarkasa Special primary and junior secondary school Kano. Then I proceeded with my senior secondary school at Aminu Kano community commercial college and graduated in 2012.

Upon the completion of my high school, I then started looking for admission into the university, and finally I was admitted into Northwest University, Kano, and graduated with first-class Honours in Bachelor of Economics in 2017. After completing my one-year national youth service corps as economics teacher then I got a Masters scholarship to study M.A. Applied Economics at Sharda University India through Kwankwasiyya Development Foundation.

Growing up in Africa and adopting the changes in the steps of life is not easy as well. I had started struggling to make a living when I was a kid. Immediately when I finished primary school then I established a small business and opened a small provision shop at the age of 12. I also lost my dad at that stage and continue living with my mother and 3 siblings.

After finishing my junior high school then I got another job at the age of 16 as a shopkeeper in one boarding secondary school within Kano-Nigeria where I spent almost 4 years doing that job. Around 2013 when I got admission into the University, I had to leave my shop-keeping job due to the lack of enough time for my study, but I continued with a part-time job selling garments and clothes in my previous place of work.

What I find difficult always is balancing between my studies and work. I had to work just to be self-reliant and also I had to study for my career development. It's not easy to grow up without much support from within or outside the community, government and other peer groups. When you were born in Africa the chances to succeed are less compared to other continents, and these as a result of a lack of good institutions, laws, and order, infrastructure, good governance, excess corruption, etc. These are among the factors that dragging us back and making life difficult in the African continent.

I have a keen interest in research, academics, traveling, demography, and finance. I have served in various community development volunteering projects for my society, I want to help my people to the best of my ability in the future. I want to be a full Ph.D. graduate and contribute with researches as the basis for solving my community's problems.

6. Robert J. Baker & Ronald K. Chesser (2000, 2009) 'The Chernobyl nuclear Disaster and Subsequent creation of a wildlife preserve' https://doi.org/10.1002/etc.5620190501 Journal-Environmental Toxicology and Chemistry https://setac.onlinelibrary.wiley.com/doi/full/10.1002/etc.5620190501 Accessed 25/04/2021.
7. Andrew Zimmerman Jones (updated 2019) 'How Quantum Mechanics Explains the Invisible Universe Quantum Physics Overview' 16 August 2019 Andrew Zimmerman Jones is a science writer, educator, and researcher. He is the co-author of *String Theory for Dummies*. https://www.thoughtco.com/quantum-physics-overview-2699370 Accessed 10/04/2019.

8. Kalid Azad A Visual, Intuitive Guide to Imaginary Numbers https://betterexplained.com/articles/a-visual-intuitive-guide-to-imaginary-numbers/ Accessed 10/04/2019.
9. Hern, A. (2019) *Myspace loses all content uploaded before 2016: Faulty server migration blamed for mass deletion of songs, photos and video* @alexhern https://www.theguardian.com/technology/2019/mar/18/myspace-loses-all-content-uploaded-before-2016 Accessed 04/04/2019.
10. The Basics of Philosophy Website *What is Philosophy* https://www.philosophybasics.com/general_whatis.html accessed 20/08/2019.
11. "Kircher & David (2003) listed five different areas, where human 'consciousness' is researched. Even though one might question the use of the word consciousness, this list shows complex ways human intelligence can be studied. I have edited these and added a sixth—philosophical biological studies about cognition within nature—as follows:

    - Phenomenological research is about mental states and what they mean.
    - The social sciences are about how we see ourselves and others, social roles and psychology and changes over time.
    - Cognitive science builds models with computer simulations and real time imaging and explains how the brain works researching into healthy, brain-damaged and disturbed subjects.
    - Neuroscientists are engaged in relating mental behaviour to the physical brain using brain imaging, and electrophysiological technology.
    - Traditional clinical neurosciences are concerned with psychopathology, mental health difficulties, treatments and therapies.
    - Philosophical studies by investigating cognition in other living creatures, considering how sentience, intelligence and consciousness have developed (Godfrey-Smith, 2016)"

    Page 5, Hawkins, J. A. Feelings and Emotion-based Learning; A New Theory. (2017). New York: Palgrave.

    Godfrey-Smith, Peter (2016) Other Minds: the octopus and the evolution of intelligent life. London: William Collins.

    Kircher, T. & David, A. S (2003) *The Self in Neuroscience and Psychiatry* Cambridge: Cambridge University Press.

    Hawkins, J. A. (2017) *Feelings and Emotion-based Learning; A New Theory* New York: Palgrave, P 5.
12. Accard, G. (2014) *Psychological Assessments – from Confucius to Modern Times* by Dr Giuseppe (Joe) Accard (17/12/ 2014) http://www.consultinghealth.com/psychological-assessments-confucius-modern/ Accessed 30/05/2019.

13. Kabeer, N. (2019) *The Human Element: Why Randomised Control Trials need Mixed-methods Approaches* https://blogs.lse.ac.uk/impactofsocialsciences/2019/06/10/the-human-element-why-randomised-control-trials-need-mixed-methods-approaches/ Accessed 19/06/2019.
14. Chivers, T. (2019) *Some star psychologists don't disclose in research papers the large sums they earn for talking about their work. Is that a concern?* https://www.nature.com/articles/d41586-019-02041-5 Accessed 02/07/2019.
15. Sarah Marsh and Amanda Boateng (2018) "Calls for action over UK's 'intolerable' child mental health crisis: Teachers and campaigners say young people under pressure but support is underfunded" Guardian Fri 31 August 2018 @sloumarsh https://www.theguardian.com/society/2018/aug/31/calls-for-action-over-uks-intolerable-child-mental-health-crisis Accessed 12/03/2019.
16. Titcombe, Roger (2019) 'Schools should be genuinely child-centred Part-2 Why this is always threatened by marketisation' https://rogertitcombelearningmatters.wordpress.com/2019/07/21/schools-should-be-genuinely-child-centred-part-2-why-this-is-always-threatened-by-marketisation/ Accessed 06/08/20.
17. Biesta, Gert J. J. (2018) What if… we really wanted evidence-informed practice in the classroom? 1 February 2018 Institute of Education (IOE) Public Debates, University College London (UCL) https://www.youtube.com/watch?v=Jg7onT_WUdw&t=2s Accessed 01/02/2018.
18. Intimate documentary about a young girl who has become the voice of a generation. The story of teenage climate activist Greta Thunberg is told through compelling, never-before-seen footage. British Broadcasting Corporation -BBC-UK https://www.bbc.co.uk/iplayer/episode/p090xz9z/i-am-greta

References

Accard, G. (2014). *Psychological Assessments – From Confucius to Modern Times* by Dr Giuseppe (Joe) Accard (17/12/ 2014). http://www.consultinghealth.com/psychological-assessments-confucius-modern/. Accessed 30 May 2019.

Andrew Zimmerman Jones. (updated 2019, August 16). *How Quantum Mechanics Explains the Invisible Universe Quantum Physics Overview*. Andrew Zimmerman Jones is a Science Writer, Educator, and Researcher. He is the Co-author of "String Theory for Dummies." https://www.thoughtco.com/quantum-physics-overview-2699370. Accessed 10 Apr 2019.

Areas of human 'consciousness' research in psychology.

Baker, R. J., & Chesser, R. K. (2000, 2009). The Chernobyl Nuclear Disaster and Subsequent Creation of a Wildlife Preserve. *Journal-Environmental*

Toxicology and Chemistry. https://doi.org/10.1002/etc.5620190501. https://setac.onlinelibrary.wiley.com/doi/full/10.1002/etc.5620190501. Accessed 25 Apr 2021.

Beresford Ellis, P. (1994). *The Druids*. Constable and Company Ltd.

Bernstein, D. A., Teachman, B. A., Olatunji, B. O., & Lilienfeld, S. O. (2020). *Introduction to Clinical Psychology: Bridging Science and Practice*. Cambridge University Press.

Biesta, G. (2011). *Good Education in an Age of Measurement: Ethics, Politics, Democracy* (Interventions: Education, Philosophy, and Culture). Paradigm Publishers.

Biesta, G. J. J. (2013). *The Beautiful Risk of Education*. Routledge/Taylor & Francis.

Biesta, Gert J. J. (2018). *What if... We Really Wanted Evidence-Informed Practice in the Classroom?* Institute of Education (IOE) Public Debates, University College London (UCL). https://www.youtube.com/watch?v=Jg7onT_WUdw&t=2s. Accessed 01 Feb 2018.

Bruner, J. (1996). *The Culture of Education*. Harvard. [Google Scholar].

Caesar, J. (1982). *The Conquest of Gaul* (Penguin Classics, Eds. Jane Gardner, et al.). Penguin.

Chase, P. (1999). Symbolism as a Reference and Symbolism as a *Culture*. In R. Dunbar, C. Knight, & C. Power (Eds.), *The Evolution of Culture* (pp. 34–49). Edinburgh University Press. [Google Scholar].

Chivers, T. (2019). *The AI Does Not Hate You: Superintelligence, Rationality and the Race to Save the World*. Orion.

Clark, R. W. (1987). *Freud, the Man and the Cause*. Paladin.

Claxton, G. (1998). *Hare Brain, Tortoise Mind*. Fourth Estate Limited.

Data Contribution Shehu Shuaibu, Muhammad. (2021, January 20). My Education Journey.

Davis, E. (2019). *Post Truth: Why We Have Reached Peak Bullshit and What We Can Do About It*. Little Brown.

Feldman Barrett, L. (2017). *How Emotions Are Made; the Secret Life of the Brain*. Pan Macmillan.

Foer, F. (2017). *World Without Mind: Why Google, Amazon, Facebook and Apple Threaten Our Future*. Penguin Random House.

Fry, H. (2018). *Hello World: How to be Human in the Age of the Machine*. Penguin Random House.

Gladwell, M. (2002). *The Tipping Point: How Little Things Can Make a Big Difference*. Abacus.

Gottschall, J. (2012). *The Story Telling Animal*. Mariner. [Google Scholar].

Gough, D. (1990). The Principle of Relevance and the Production of Discourse: Evidence from Xhosa Folk Narrative. In B. Britton & A. Pellegrini (Eds.), *Narrative Thought and Narrative Language* (pp. 199–218). Laurence Erlbaum.

Hatfield, E., Cappioco, J., et al. (1994). *Emotional Contagion (Studies in Emotion and Social Interaction Series)*. Cambridge University Press.
Hawking, S. (1988). *A Brief History of Time: From Big Bang to Black Holes*. Bantam Books.
Hawkins, J. A. (2017). *Feelings and Emotion-Based Learning; a New Theory*. Palgrave.
Heide, K. G. (2006). *Landscapes of Emotion: Mapping Three Cultures of Emotion in Indonesia (Studies in Emotion and Social Interaction)*. Cambridge University Press.
Hern, A. (2019). *Myspace Loses All Content Uploaded Before 2016: Faulty Server Migration Blamed for Mass Deletion of Songs, Photos and Video* @alexhern https://www.theguardian.com/technology/2019/mar/18/myspace-loses-all-content-uploaded-before-2016. Accessed 04 Apr 2019.
History Websites.: https://www.historyandheadlines.com/top-ten-greatest-civilizations/ https://en.wikipedia.org/wiki/Civilization, https://www.ancienthistorylists.com/ancient-civilizations/10-oldest-ancient-civilizations-ever-existed/. Accessed 04 Apr 2019.
Jarvis, P. (2006). Rough and Tumble Play, Lessons in Life. *Evolutionary Psychology, 4*, 268–286. [Google Scholar].
Jarvis, P. (2019). Not Just 'Once' upon a Time Institute of Childhood and Education, Leeds Trinity University, UK; Published in: *Genealogy, 3*, 44; https://doi.org/10.3390/genealogy30300441 https://www.mdpi.com/2313-5778/3/3/44. Accessed 16/09/2021.
Jarvis, P. (2021). Book Review: The Psychology of Family History. *Genealogy, 5*(2), 39. https://doi.org/10.3390/genealogy5020039. Accessed 25 Apr 2021.
Jones, J. (2018). *Sensations: The Story of British Art from Hogarth to Banksy*. Laurence King.
Jung, C. G. (1991). *The Archetypes and the Collective Unconscious (Collected Works of C. G. Jung1865–1961)*. Routledge.
Jung, C. G. (1995). *Memories, Dreams, Reflections: An Autobiography*. Fontana Press (1875–1961).
Kabeer, N. (2019) *The Human Element: Why Randomised Control Trials need Mixed-methods Approaches*. https://blogs.lse.ac.uk/impactofsocialsciences/2019/06/10/the-human-element-why-randomised-control-trials-need-mixed-methods-approaches/. Accessed 19 June 2019.
Kaiser, B. (2019). *Targeted: My Inside Story of Cambridge Analytica and How Trump, Brexit and Facebook Broke Democracy*. Harper Collins.
Kalid Azad A Visual, Intuitive Guide to Imaginary Numbers. https://betterexplained.com/articles/a-visual-intuitive-guide-to-imaginary-numbers/. Accessed 10 Apr 2019.
Kell, E. (2018). *How to Survive in Teaching: Without Imploding, Exploding or Walking Away*. Bloomsbury.

Kelly, L. (2017). *The Memory Code: The Secrets of Stonehenge, Easter Island and Other Ancient Monuments*. Pegasus Books Ltd..

Kircher, T., & David, A. S. (2003). *The Self in Neuroscience and Psychiatry*. Cambridge University Press.

Lanier, J. (2014). *Who Owns the Future?* Penguin.

Lanier, J. (2017). *Dawn of the New Everything: A Journey Through Virtual Reality*. The Bodley Head.

Lanier, J. (2018). *Ten Arguments for Deleting Your Social Media Accounts Right Now*. The Bodley Head.

Latour, B. (1988). *The Pasteurization of France*. Harvard University Press.

Lukianoff, G., & Haidt, J. (2018). *The Coddling of the American Mind: How Good Intentions and Bad Ideas Are Setting Up a Generation for Failure*. Allen Lane Penguin Random House UK.

Lutz, C. A. (2009). *Language and the Politics of Emotion (Studies in Emotion and Social Interaction)*. Cambridge University Press.

Lyle, S. (2000). Narrative Understanding: Developing a Theoretical Context for Understanding how Children Make Meaning in Classroom Settings. *Journal of Curriculum Studies, 32*, 45–63. [Google Scholar].

Mandela, N. (1995). *Long Walk to Freedom: The Autobiography of Nelson Mandela*. Abacus.

Marsh, S., & Boateng, A. (2018, 31 August). Calls for Action Over UK's 'Intolerable' Child Mental Health Crisis: Teachers and Campaigners Say Young People Under Pressure but Support is Underfunded. *Guardian*. @sloumarsh https://www.theguardian.com/society/2018/aug/31/calls-for-action-over-uks-intolerable-child-mental-health-crisis. Accessed 12 Mar 2019.

Menzies, G. (2009). *1434: The Year a Chinese Fleet Sailed to Italy and Ignited the Renaissance*. Harper Collins.

Mitchell, L. (2009). *Language, Emotion, and Politics in South India: The Making of a Mother Tongue (Contemporary Indian Studies)*. Indiana University Press.

Niemark, N. (Ed.). (2011). *The Environmental Debate: A Documentary History, with Timeline, Glossary, and Appendices*. Grey House Publishing.

Pascoe, B. (2018). *Dark Emu*. Scribe Publications UK Ltd..

Pinker, S. (2018). *Enlightenment Now: The Case for Reason, Science, Humanism and Progress*. Allen Lane.

Rogers, C. R. (1951). *Client-Centred Therapy*. Constable.

Rovelli, C. (2016). *Reality Is Not What it Seems: The Journey to Quantum Gravity*. Penguin Science.

Skinner, B. F., Ferster, C. B., Cheney, C. D., & Morse, W. H. (1957). *Schedules of Reinforcement*. The Skinner Foundation.

Syed, M. (2019). *Rebel Ideas: The Power of Diverse Thinking*. John Murray.

Tharby, A. (2018). *How to Explain Absolutely Anything to Absolutely Anyone: The Art and Science of Teacher Explanation*. Crown House.

The Basics of Philosophy Website. *What is Philosophy*. https://www.philosophybasics.com/general_whatis.html. Accessed 20 Aug 2019.

The Editors of Encyclopaedia Britannica. (2019, August 9). *Science*. https://www.britannica.com/science/science. Accessed 19 Aug 2019.

Thunberg, G. (2019). *No One Is Too Small to Make a Difference*. Penguin Random House.

Thunberg, G. (2020). Intimate Documentary About a Young Girl Who Has Become the Voice of a Generation. The Story of Teenage Climate Activist Greta Thunberg is Told Through Compelling, Never-Before-Seen Footage. *British Broadcasting Corporation* -BBC-UK. https://www.bbc.co.uk/iplayer/episode/p090xz9z/i-am-greta. Accessed 10 Mar 2021.

Titcombe, R. (2015). *Learning Matters*. FGI Publishing.

Titcombe, R. (2019). *Schools Should Be Genuinely Child-Centred Part-2 Why This is Always Threatened by Marketisation*. https://rogertitcombelearningmatters.wordpress.com/2019/07/21/schools-should-be-genuinely-child-centred-part-2-why-this-is-always-threatened-by-marketisation/. Accessed 06 Aug 2020.

Ungunmerr-Baumann, M. R. (1988). *Dadirri: Inner Deep Listening and Quiet Still Awareness*. Available online: https://www.miriamrosefoundation.org.au/about-dadirri/dadirri-text. Accessed on 1 Aug 2019.

Ungunmerr-Baumann, M. R. (2002). *Dadirri—A Reflection*. Available online: http://nextwave.org.au/wp-content/uploads/Dadirri-Inner-Deep-Listening-M-R-Ungunmerr-Bauman-Refl.pdf. Accessed on 1 Aug 2019.

Website Article. (2020, 15 June). *Mandela Foundation News Lessons from Nelson Mandela on reconciliation, reparation, and the path to prosperity* Published 15 June 2020. https://www.mandelarhodes.org/ideas/mandela-rhodes-foundation-statement/. Accessed 06/01/2021.

CHAPTER 8

Politics, Economics, World Outlooks and Influences

8.1 Introduction

It was never realistic to consider human intelligence to be fixed even when the world changed at a slower pace. We now have scientific and clinical proof that our brains have plasticity. They are active and interactive not blank slates to be written upon (Pinker, 2002). However, there are environmental problems outside school, which directly affect education and the lives of both child and adult learners. This chapter looks at some of the pressures, mindsets and social prejudices in the United Kingdom and world society that impact education. We need to be aware of these in order to recognise and deal intelligently with the way they affect us. Physical conditions, intellectual information, social attitudes and fashionable behaviours affect our lives and environment. These include incidental events, outside influences, emotional models, history and other changeable aspects impacting our thinking.

There are many skilful ways of living and they can be combined and adapted to different circumstances—nowadays many people have almost instant access to a huge range of ideas. Even so there are situations, inventions and ideas of which we may be generally unaware. For example, there

Special thanks to Sifu John Lobb for 'Confidence: Anything you can do once, you can do again better. Learning to invest through failure', Wing Chun martial arts master, March 2019.

J. A. Hawkins, *Brain Plasticity and Learning*,
https://doi.org/10.1007/978-3-030-83530-9_8

are disenfranchised first nation peoples and historical cultures being 'rediscovered,' who have a fund of wisdom to share about healthy approaches to life, medicines and ways of conserving and respecting the environment. In some cases 'educated people' have not only denied other peoples' feelings and different ways of knowing, but the fact that they have similar brain potential and capacities with similar variations to their own. Instead of learning from them, we have assumed we know best, sometimes repeating mistakes and behaving ignorantly. This made our own situation worse and sometimes prevented us from learning how to appreciate our own lives—for example, the Earth's environmental crisis.

At the same time as it teaches a body of factual information, education can help us with this. Our brains have a capacity to shift frames of reference, outlooks and perceptions. This is a great strength but only if we listen to others, learn to reflect and weigh up our options. The reality of lives includes neurodivergence, disability, frailty, opportunity, unforeseen setbacks and tragedy, imagination, deception and delusion, ignorance, unforeseen insights, discovery and invention. We have to learn to navigate them by learning to see where others are 'coming from' and collaborate appropriately with subject and community experts, colleagues, parents and students themselves. Our adult world nowadays involves sifting more facts. This world is extremely complex, even more so than in the past. Good education supports and gives rise to human improvement beyond school.

It requires skill and expertise to teach learners how to 'learn to learn' in ways likely to motivate and support them into the future. We all need to learn not to take information at face value. We need to know how to draw conclusions collaboratively in practice. A teacher needs a grounded, balanced view of educational aims. Teachers need to adapt curricular approaches and take account of the diverse experiences of real lives. Teachers and learners bring their outside lives into school and it is more important than ever that we relate to the world outside as we work with students. In this chapter I explore some of the political and societal pressures which form a backdrop and give a larger context to our work, all of which tend to involve feeling and emotion on some level (Freire, 2014; Bronfenbrenner, 1979).

8.2 How Economics Impacts All of Us: Our Long-Term Existence and Our Quality of Life

When we look at historical events, we see the sorts of economics problems which influence lives and education systems. The Great Depression in 1929 was the longest, deepest and most widespread depression of the twentieth century. Lasting until 1941 it shows how the world's economy can suddenly go into a catastrophic decline. It started in the United States after a major fall in stock prices that began in September 1929 and soon became worldwide. The stock market crash on 29 October of that year was known as Black Tuesday. People famously committed suicide and catastrophic privations, emotional and fiscal depressions of all kinds were widespread. Gross domestic product (GDP) fell by about 15% between 1929 and 1932 (whereas in our recent Great Recession [2007–2009] it fell by less than 1%).

For many Western and other economies the negative effects of the Great Depression lasted until the beginning of the Second World War. Everybody, at all levels of society, suffered devastating shocks because incomes, tax revenue, profits and prices dropped and international trade plunged by more than 50%. Unemployment rose in the United States to 25% and in some countries rose to as high as 33%. The Second World War (1939–1945) affected the global economy and ordinary people. The deaths of family and fellow workers, disease, destruction of property, disruption of employment, transport and infrastructure reduced everybody's standard of living. Feelings and emotions were strong indeed! This led to a general appreciation of the need for a recovery leading to a better life for all, through greater fairness, stability and social cohesion.

Paul Roberts (2014) explains how we have coped with past waves of industrialisation due to technological progress. Governments stepped in to regulate industry improving conditions for ordinary people. In the West a period of regulated and managed market economies followed (1945–1980). After such a traumatic time, privately owned industries were generally more inclined to accept responsibility. Union rights were accepted. Many took an active role in engaging with government interventions during recessions and depressions. Some Western industrialists thought social values and a sense of duty towards workers helped their businesses to succeed. People tended to thrive and their standard of living and aspirations grew. Historical examples include the development of the car by General Motors when fashion-based advertising and quick turnover incentives contributed to successful marketing and large profits.

Manufacturing started to drive economies and yet Keynesian economists advised these should still be managed and regulated by government. John Maynard Keynes (1883–1946) wrote a book called *The General Theory of Employment, Interest and Money*, published in 1936. Keynes advocated the use of fiscal and monetary policies to mitigate the adverse effects of economic recessions and depressions. This was the standard economic model in developed nations during the latter part of the Great Depression, the Second World War and the post-war economic expansion (1945–1973). People's lifestyles, general health and well-being benefited but depending on personal circumstances some had less fulfilling lives in other ways, for example, the 'treadmill' of factory assembly lines, time and motion study methods.

Manufacturers favoured neoclassical economics arguing that free markets would automatically provide full employment, as long as workers were flexible in their wage demands. Neoclassical economics is a free market approach based on production, consumption and pricing of goods and services according to supply and demand. Just as back in Aristotle's day there were important issues around human feelings, emotional responses and who are entitled to succeed and lead fulfilled lives (Hall, 2018). In the United States even people who are disadvantaged believe in the 'winners and losers American dream' rhetoric. Some people become embittered and some people discover that material wealth without social engagement and responsibility is not *entirely sufficient* for happiness and well-being. The interplay and effect of different emotional and social models on humanity is evident. More realistic and universally beneficial value models are needed.

The Great Recession financial crisis of 2007–2009 had some similar effects as the Depression—reviving the popularity of Keynesian economics. It was caused by unsecured loans in the money markets in the United States. When the optimistic property values 'bubble' burst in North America, finances in the banking sector collapsed. This affected the world economy when large banks were bailed out by governments. Some countries, for example, China and India, suffered less impact than those in the West. This was due in part to the fact that large Western companies use these economies to manufacture their goods where labour and materials are cheap (globalisation). Manufacturers have pursued short-term 'efficiency' leading to unemployment in their home countries, the decline of traditional industries and a reduction in people's spending power.

As a result, the workforce in the United States, used to a comfortable lifestyle, have found themselves vulnerable to debt. The election and popularity of President Donald Trump by an American workforce with declining work opportunities and standards of living is due to some of these factors. Other Western economies have been affected by similar economic events. In the United Kingdom, teachers have become one group of convenient scapegoats used by fearful politicians as economic circumstances have impacted on society and on education. Education is an emotive issue which engages the electorate. Schools have been made responsible for remedying difficulties and foreseeing situations which are beyond their control. The implication is that not only is it the duty of education to train any future workforce, but schools are to blame because they cannot read the future accurately (Mazzucato, 2018).

Benjamin Doxtdator (2017) (teacher and blogger) points out that since the 1950s there have been frequent claims circulated by economists and academics that schools have failed in their duty to prepare the future workforce for future jobs, the '*jobs that haven't been invented yet.*' At the same time it is acknowledged that many current technological jobs did not exist when the current teaching workforce were at school. This provided an argument of convenience with which to criticise education. The solutions put forward are paradoxical and often both inflexible and simplistic creating a 'catch 22'situation for teachers. They hardly ever allow for professional development and research by experienced teachers. They often overburden trainee students with learning more technical knowledge, cutting down the induction time into professional teaching they need. See Blog Article Endnote.[1]

8.3 Different Interpretations of Liberalism: Social, Classical and Neoliberalism

Liberalism is a term which appeals to different political groups with very different moral philosophies. Liberals generally espouse civil rights, democracy, secularism, gender and race equality, internationalism and freedom of speech, the press, some religions and free markets. The word is based on liberty and equality but there is a problem of semantics caused by the differences in how this term is interpreted. This can be summed up by the popular saying that 'all people are equal but some are more equal than others.' Different groups have different frames of reference about other people and *unless they have been educated to respect others may discount, disagree intolerantly with, dislike and fear their difference.* This is an unfortunate consequence of brain plasticity we are all prone to, especially where upbringing encourages a 'fixed mindset.'

Social liberalism is different to classical liberalism and neoliberalism, the latter sometimes involve an extremely strong belief in a free market economy, which does not hamper one's own interests and profits. Social liberalism includes a political system which depends upon a regulated economy with a proscribed civil, political rights and human rights legal system and programmes which address social issues such as poverty, health care and education. Classical liberalism is similar to classical economics. Problems of understanding and frames of reference arise when people perceive and experience conflict between the social good and the 'freedom' and 'self-interest' of the individual. For example, many of the socialist dictators and their governments initially brought real improvements to the lives of citizens, but occasionally became violent and oppressive in order to maintain their own power and wealth, for example, the Soviet Union, Chile and China.

Different groups of people have a very clear view of meanings for themselves but there are groups within groups and splinter groups who adopt labels and define them differently. The terms 'conservative,' 'socialist,' 'right' and 'left' have become hard to define just as 'progressive' and 'traditional' used both in politics and in education cause a great deal of confusion. For example, some classical liberalists believed in civil liberties under the law and economic freedom, but argued for individual rights over interest group rights and limited government interference. They believed in property rights and the rule of law. These arguments provide and continue to provide a useful argument of justification for the wealthy. They have drawbacks since proponents may have a tendency to dislike democratic systems and, although ultimately in their interest, may not want to contribute to integrated social well-being. They may be resistant to independent ideas and may not cope well with unforeseen events like climate change or pandemics.

The term neoliberalism has currently been a popular one. The idea is based in a resurgence during the last 20–30 years of the neoclassical ideas above, which originated during the industrial revolutions of the eighteenth and nineteenth centuries. These ideas were derived from the historically popular philosophies of 'laissez-faire' and 'economic liberalism.' Definitions and uses of both terms have changed over the decades. Currently the following words, ideas and terms are associated with neoliberalism and are often currently bandied around by political journalists discussing government policy:

- *Economic liberalisation or deregulation* which can lead to oppressive and aggressive practices towards the workforce and consumers, for example, cartels and monopolies.

- Privatisation aimed at increasing the role of the private sector in the economy and society. Perhaps with the notion of sharing the burden of paying for social infrastructure necessary for all, but with the inherent difficulty of the need to regulate commercial self-interest and raise taxes. For example law and order, government, education, defence, roads, railways and health care.
- Austerity (public service cuts) measures are used by governments that find it difficult to pay their debts. It is generally believed that austerity policies increase unemployment as government spending falls.
- Globalisation (which has generally been used by the West) is a term referring to the manufacture of goods in distant countries where labour is cheap.
- Hidden economy (cash in hand) and short-term contracts.
- Free trade, which may or may not involve protecting or not protecting home industries.
- *Gross domestic product (GDP)* is a measure of the market value of a country's goods and services produced during a quarterly or yearly period. The measure is used to determine the economic performance of a country in order to make international comparisons. It does not explain underlying differences in the cost of living and the causes of all of the different living standards in different countries, for example, climate, genocide, war, natural catastrophes, social cohesion, quality of life and cultural history.

Neoliberalism is an economic philosophy which has its roots in Europe in the 1930s as scholars sought to encourage a market economy that would benefit society. Roberts (2014) explains the rise of neoliberalism in the United States and some of the effects of industrialisation on the world. It seems to have had its roots in ideas taken, adapted and developed from classical liberalism which emerged out of the Industrial Revolution in the nineteenth century advocating civil liberties under the law and included greater economic freedom. Adam Smith's book *The Wealth of Nations* (1776) was a source of many of these ideas (1723–1790). Currently the concept of 'neoliberal,' however, has been interpreted by some of those in power as an argument of justification amongst their peers to mean that a person should be allowed to do as they please.

This is done in the name of efficiency, aka for profit, even directly or indirectly at the expense of others (Silicon Valley technology companies) and ultimately even against their own and their families' long term interests. These ideas have been linked to the theories of economists Milton Friedman and James Buchanan and politicians Ronald Reagan and Margaret Thatcher. Philosopher and writer Ayn Rand with her libertarian philosophy has been linked to the education and commercial rationale of the big tech giant corporate executive officers. Since the global financial crisis of 2008–2009 the social phenomenon of neoliberalism has been increasingly discussed by scholars. A school of thought has arisen that criticises neoliberalism and seeks to develop policy alternatives.

Our current unashamedly self-serving consumer society has resulted in some of the following problematic situations for society:

- A waste of resources, for example, the destruction of wildlife habitats.
- Problems of excessive pollution, for example, built-in obsolescence creating unrecyclable waste.
- A greed-driven economy leading to the constant dissatisfaction of those who cannot afford to keep up, for example, cars which are out of date as soon as they are sold.
- Problems of social pressure to achieve status—even to the point of getting into debt and repeatedly being bankrupted.
- Consumer credit which tends against appreciating possessions through saving for them, for example, more people can instantly gain prestige for buying a sought after and admired object or service (Mazzucato, 2018).

The effect on education of a neoliberal culture was reported on by Pam Jarvis, British chartered psychologist and historian. She discusses her concerns about the 'datafication' of English education in a newspaper article called 'A Journey Through the Looking Glass' (2017). She points out that those in charge have started to see education as existing in a 'looking glass world'—forgetting that as parents and grandparents themselves they would not want their own children to be viewed unsympathetically and unrealistically in this way. She says that this distorted inhumane approach has been caused by the currently dominant neoliberal economic philosophy which defines human beings including children as ultimately of value only as generators of profit, valued or discounted from a distant 'othering' perspective.

In some cases in the apparent strive for 'efficiency' the vulnerable have been undervalued and education settings reconstructed as private businesses, even though they often ultimately rely on taxpayer subsidies.

Children with special educational needs are seen as inferior as prospective clients being thought likely to reduce national league tables' results leading towards a tendency to favour exclusion. She believes that this approach to education has been used to justify operational deceptions of purpose in order to achieve personal wealth and power at the expense of others.

While some of these schools are run as innovative ethical institutions—the political culture has created opportunities for others to engage in dubious 'business practices' so that school investors and managers can preserve and enhance their own incomes. This philosophy encourages the idea that 'deregulation' produces a society which is more efficient, but of course this depends what is valued and why. A business-based mantra about the 'survival of the fittest' appropriated from Darwin's evolutionary research is also often connected to these arguments. See Article Endnote.[2]

8.4 Organising Collaborative Responsibilities for Democracy and Equality

Democracies are based on the ideal of human equality with the government in the power of citizens who vote for representatives in elections. In any democratic country the government oversees the education system paid for by citizens' taxes. Education is necessary for democracy to be maintained. We need to learn to consider facts and events that affect our world in order to vote wisely and we have to be taught to collaborate and give way to the majority vote. Two problems of democracy are the human habit of self-interest and the problem of what 'equality' actually means. Some people seek status in office and power in general to become rich and evade taxes. Others work for the common good and see that this includes their own, their families' and society's long-term best interests.

It requires everybody's effort to maintain a democracy. Much depends upon the intentions and motivations of those we elect and the needs and wishes of the wider electorate, our personal and group emotion models. Aristotle apparently believed that the 'status quo' should be revised systematically so that written laws can be regularly updated and improved because it is unlikely state structures can be framed and fixed correctly for all time (Hall, 2018:18). An idea of some world problems involving democratic issues can be gained by looking at those reported upon by the Open Democracy Foundation for the Advancement of Global Education. See Website Endnote.[3] This organisation, based in the United Kingdom, describes itself as an independent global media platform, publishing articles, reports and analyses on social and political issues across the world on issues of rights, equality and justice.

Here in plain terms is a list of some of the current issues the website reports on:

- Dictatorships in Europe, West Asia, the Middle East and North Africa.
- Digital liberties and data infringements—surveillance and human rights.
- 'Democratic' dictatorships that are reducing human rights and freedoms, for example, Russia, Belarus, India, Brazil, Lebanon, China and Myanmar.
- Campaigns for inclusive democracy, for example, social justice, gender equality and women's and LGBT human rights.
- Telling the stories of peaceful protestors who are attempting to democratise their countries.
- Investigative journalism and storytelling to provide intelligence and ammunition to people working for policy change.
- Evidence-based advocacy to unveil exploitation, structural political, economic and social root causes of global trafficking and slavery.
- Publishing in a range of languages to provide a global platform for different national voices.
- Hosting discussion groups and forums, for example, on health, education and the media.

In 2011 Open Democracy UK published an e-book by Andrea Westall, a researcher and economist, called *Time to revisit* Associative Democracy? (2011). It documents and draws together the ideas and thoughts of a group of the organisation's leading associates from around the world. They met in 2010 to discuss Paul Hirst's book Associative Democracy: *New Forms of Economic and Social Governance* (1994). Discussing this not long after the Great Recession (2006–2008), they hoped a new framework could be worked out by bringing together social groups engaged in competition and collaboration. The group wondered how we might get more ***democratic co-operation***, co-ordination and collaboration into our economies, democracies, public services, and lives and made some of the following points. The democratic 'visions' of both of the main political parties in the United Kingdom were examined.

They found that neither of the main political parties, Labour and Conservative, on left or right fully understood, what had gone wrong in society and the economy. The left did not recognise the 'ill-effects of the social liberalism of the 1960s' and had not developed 'a coherent response to cultural pluralism' or recognised the limits of economic liberalism. The right had not admitted the social disasters brought about by the neoliberal experiment of the 1980s and 1990s or recognised the 'brokenness' beneath from the top of the society. The language and policies of three decades of neoliberalism had gone so deep with politicians, local government and citizen behaviour that change would be difficult. They concluded that we had become constrained by an economic set of beliefs and practices that were no longer sufficiently collaborative and humane enough to continue to improve our well-being.

The environmental crisis has proved them to have been correct in their predictions. The trend of short-term gain, economic profligacy (loan culture) and the breakdown of social cohesion was and still is a problem. In the West there were problems of escalating immigration, debt and 'austerity' with less government support for social infrastructure and increasing job insecurity. Looking back over the subsequent years this still appeared to be the case when the pandemic started in 2020 (Runciman, 2018; Murray, 2017; Roberts, 2014). Hirst had an 'ideal' vision that can now be supported by a better understanding of our own brain plasticity—our weaknesses and strengths and our ability to reflect, surmise and decide wisely. He thought associational forms of democracy or collaborative decision-making between ordinary people could rebalance the central role of the state versus big business interests.

Hirst wanted to distribute decision-making to groups of 'interested' people who would come together to collaborate and make decisions—to in effect action research the future. Formal networks of multi-stakeholders and cross-sectoral partnerships, coordinated regulatory institutions and economic and social governance groups would have to be set up. This could enable more effective, informed and appropriate actions and increase the flow of accurate information, accountability, trust, inclusion and choice. He wanted to enable individual empowerment and freedom. It would reduce the need for complex top-down regulation and more effectively share wealth and security, potentially solving mistrust and social disintegration and supporting communities.

This would require ongoing cross-party political interest in the establishment of distributed systems to create more resilient and effective local and regional economies. Associative spaces would be considered an integral part of government policy. Politicians would need to recognise the benefits of more democratic spaces in local government with partnerships of different players. It would tend to improve regulation through more involved mutual responsibility in a better system of public scrutiny and accountability. There would have to be better co-ordination, collaboration and *cooperation* across sectors in local and national government overlapping, interacting and supporting small democratic self-governing groups of people doing similar jobs.

Different types of co-operatives with co-ordination and collaboration between economic actors using joint services would be needed. For example training and research and development at local level between small firms; peer groups or unions of the self-employed; third sector businesses, unions, mini-trusts, friendly societies, credit unions, direct citizen participation, cross-sectoral and social-public partnerships. Some simplistic solutions suggested by politicians have found support amongst the general population—they are propagated through crowd hysteria, linked to popular ideas and promulgated by self-interest. People and/or voters who have not had the privilege of a good education may not consider the long-term effect of their actions. Innocent people are sometimes subjected to propaganda by others with vested interests and/or an irresponsible outlook on life. Outrage and blame sometimes fuel and drive a chaotic an unhelpful complaint culture in politics instead of contributing practical ideas and solutions.

Human beings are vulnerable to social and monetary influences. There have even been those in the education community who 'provide research support' for inefficient, inappropriate and unworkable government policies for their own financial gain. A co-operative associative democratic process would tend towards a natural system of checks and balances and could also play an important role in finance. For example, through membership of professional bodies, codes of practice and mutual financial bodies. Multi-stakeholder partnerships or governance which bring together customers, stakeholders and suppliers can help us to innovate more effectively to meet need and change culture and practice. Such democratic stakeholder microanalysis would enable citizens to register, complain and help to regulate capital markets. A way of responding to a complex, self-reinforcing marketing system that constantly pushes boundaries beyond reasonable current taxation and regulation limits.

Associative Democracy's expert discussion group in 2010 agreed that associative, collaborative or *co-operative* democracy systems could help citizens forge relationships with the aim of creating a better life for everybody. Westall says these ideas deserve serious scrutiny, in order to complement and renew, but not replace existing direct forms of democracy (2011). See Website Note.[4] The co-operative movement operating worldwide, which often consists of businesses owned and managed by the people who use their services offers us a model for this type of approach. It is a model that millions of people in 96 countries around the world understand and participate within on some level (World Watch Institute, 2011). These organisations are autonomous and run on the principle of benefiting everyone who contributes their labour or money on an equal basis. A cooperative is "an autonomous association of persons united voluntarily to meet their common economic, social, and cultural needs and aspirations through a jointly owned enterprise."

Co-operative Democracy through Citizen's Assemblies at all levels of society based on an agreed interpretation of the principle of equality might be a useful concept going forward. All of this brings us back to education of citizens at all levels, ages and conditions in society. This is necessary because neoliberal emotion models are blocking our ability to develop collaborative ways of renewing our democracy, humanising and introducing more fairness into the economy. The current world dominance of profit focussed, short-term 'business as usual' thinking appears to be problematic for education as well as our economy and government organisations, for example, the UK Office for Standards in Education (Ofsted). This form of system reversal and democratising process through negotiation with stakeholders is currently being suggested for the education system and is starting to be researched and considered by education experts. It is popularly referred to as 'flipping the system' (Rycroft-Smith & Dutaut, 2018; Evers, 2015).

8.5 Problems and Benefits of Taking Personal Responsibility for Democratic Equality

Once a government has been elected, citizens may be prevented by them from complaining about their ongoing performance and management. Good intentions can be subverted if system checks and balances on local and national government decisions are circumvented. Insufficient regulation of institutions, corporations and powerful private individuals tend to threaten people's human rights, freedom and happiness. For example, the tactic of Silicon Valley monopolies to 'move fast and break things' using 'disruption' without responsibility. On the plus side, communication via the internet helps us research causes, crimes, talk to strangers, gain knowledge, solve problems, glitches and breakdowns and communicate our needs and wishes quickly to better effect. It is the human element that is problematic, for example, through subversion and suggestible compliance.

Jonathan Taplin wrote a book called *Move Fast and Break Things: How Facebook, Google and Amazon Have Cornered Culture and What It Means for All of Us*. He reveals the *faceless* gaming tactics of corporate internet platforms such as Facebook, Uber and Airbnb. It is vital for the human species' independent survival and development that the World Wide Web remains a public arena that is properly regulated. The idea that if we 'treat others as we want to be treated' we would get a better world for ourselves, seems to be a difficult lesson for us all to learn. A country's citizens can become too complacent and selfish for their own good. After extended periods of peace they can become naïve, blinkered, corrupted and unaware of potential catastrophe, for example, the Roman Empire. This is the downside of brain plasticity.

People too easily forget and reject the lessons of history that the basic physical challenges of survival are never far beneath the surface and ultimately apply to everyone. For example, national social services, if appropriate and fair, benefit the whole population because when poor people have a safety net everybody's general social and physical environment improves. Ultimately we always have to 'navigate' the universal human condition or as Pinker (2018: 17) puts it "*to fight back the tide of entropy and carve out refuges of beneficial order.*" In modern times and in some cases surprisingly recently, new democracies spread and became established after the Second World War, particularly in the West.

They continue to become established, evolve, weaken, disappear, adapt, re-establish and redevelop. David Runciman (2018) believes they have the potential for improvement, but in his book *How Democracy Ends*, warns of the possibilities for gradual decline. Education can teach that we must all work to achieve a good democratic system at different levels and this starts by listening to, questioning our own and others' feelings about facts and opinions and checking them out. Our plastic brains do give us the potential to consider and solve problems in order to (if possible) avoid their further creation and remedy errors made. Are our systems wise enough to enable this process in practice?

Democracy at all levels of our lives is always going to be a work in progress because we and our world are always changing. Nancy Bermeo (2016a, b) points out that democracy is now under attack from six kinds of democratic backsliding (2016a). She finds that this can lead to either complete democratic breakdown or to the serious weakening of existing democratic institutions. When these reasons are hidden from public scrutiny, democratic principles are obscured, the public can be deceived more easily and democracy becomes 'particularly difficult' to defend.

Democratic backsliding involves a deliberate turning away from the democratic ideal—some types of which can be identified as:

- *Coup d'etat*—a surprise take-over of the state machinery of government including all the administrative organisations before the general population has had time to resist (fait accompli).
- *Executive coups*—freely elected leaders suspend the constitution in order to amass power in one swift take-over often using their military, for example, Ferdinand Marcos (1965–1986), the Philippines; Alberto Fujimori, Peru in 1992; Levon Ter-Petrosian, Armenia in 1995; Alyaksandr Lukashenka, Belarus in 1995; Frederick Chiluba, Zambia in 1996 and René Préval, Haiti in 1999 (removal of leader accountability).
- *Election-day vote fraud*—where votes are miscounted to create a false result (cheating).
- *Promissory coups*—where a take-over promises to be temporary before a future election to redress wrongs (pie in the sky and/or deferment).
- *Executive aggrandisement*—when governments and their departments subvert democracy from within—(bureaucratic paperwork, denial, delay and obfuscation).
- *Strategic electoral manipulation or gerrymandering*—a range of actions which influence voters and tilt the 'electoral playing field' in an underhand manner, that is, are unlikely to be observed, criticised or perceived as fraudulent. These include (1) hampering media access; (2) using government and/or private funds inappropriately; (3) preventing opponents from standing; (4) hampering voter registration; (5) harassing and falsely discrediting opponents; and (6) changing electoral rules and voting areas.

These final two points contain some of the greatest current hidden dangers, for example, the ongoing problems in the courts following the 2020 presidential election in the United States. The first three had declined in popularity but some have been re-invented from the extreme violence of genocide to the establishment of gerrymandering restrictions on voting by politicians promoted by bogus news reports. As Runciman (2018) points out complacency is an ever-present danger now that many democracies have become so established their citizens will not easily give up comfortable lives to resist an assault on freedom. Particularly now the manipulation of communications via algorithm has made it difficult for ordinary citizens to fully understand the implications of their conditioned responses. For example, in the United Kingdom, people are apparently having to drop their regional accents when talking to online voice recognition assistants—an example of technology initiating, imposing and re-enforcing an unrequested behaviour (Field, 2018).

Bermeo (2016b) says that often democratic backsliding takes the form of gradual change. This is not so likely to cause obvious regime change but gives us political systems that are 'ambiguously democratic or hybrid.' There is a danger of citizens losing control, personal freedoms and the power to challenge corrupt government practices. Arbitrary restrictions may be imposed covertly with the loss of individual freedom and agency. Education has an important role in teaching how to use old and new knowledge and power responsibly and to inform at the 'popular' level. Education about past human mistakes, disasters, wars, genocide, historical events, great literature, heroic and scientific progress can give a more balanced view of the possibilities ahead.

We can use historical and current narratives to demonstrate how democracy works. In establishing future democracies that self-correct, it may be useful to consider a range of experiences, emotional models and conceptual points of view. Dambisa Moyo thinks our current democracies need 'urgent reform' before it is too late (2018). Angry citizens are demonstrating from Belgium via Belarus, to Lebanon via Mexico to Brazil demanding a greater say in their political and economic future, better health care, education and living standards. She says 'equalities' are eroding around the world and governments must take more responsibility for their citizens' welfare and protection from private-sector monopolies. The pandemic has highlighted the need for co-operation and we need to reset the world economy.

Moyo says education is critical because humanity will only survive if people themselves take responsibility for democracy. Madhukar Pai, a writer on health, says that "people power, not powerful people, can help end inequality." He interviewed Ben Phillips, who has written a book called *How to Fight Inequality (and Why That Fight Needs You)* (2020). He tells us Philips is an international anti-inequality activist and advisor to the UN, governments and civil society organisations, who has written a 'thoughtful and inspiring' new book. After the serious inequalities that the Covid-19 pandemic revealed, he explains that we need to build a better and fairer world, which will require a commitment from ordinary individuals. It is an approach which has worked in the past and he believes it can work again. Pai says justice will have to be fought for and won because those who benefit from inequality will not easily concede their positions of power. See Article Endnote.[5]

With insightful honesty, Julie Fellmayer, a college tutor in Brussels, says administrators and teachers generally do not want to embrace the concept of freedom in education if it means 'rocking the boat.' It means they would have to stop talking, listen and promote marginalised people's voices in education. In the West, white women teachers are in the majority with mostly white men in positions of power. Fellmayer has thought about herself in relation to other voices. She cites Gloria Jean Watkins (bell hooks) (1994, 2003, 2010) who says that teachers should 'self-actualise' for their own benefit in order to empower students. She explains her difficulty as an entitled white woman. Even as a woman, she presently has little to struggle against. She says it is hard to even be aware of the full extent of her privilege and as an 'entitled' white woman myself I relate to her position. I see people of colour who are valuable, they are in my family and I respect and revere them, but it is hard for me to fully understand their difficulties. However, all of us need to fight the 'status quo' together because all of us are needed.

It is hard to visualise yourself accurately from a position of advantage, but ironically an acceptance of one's flaws and vulnerabilities is necessary for personal empowerment. Fellmayer honestly details her current advantages. She says she already has everything on Maslow's Hierarchy of Human Needs. White society normalises and supports her accomplishments, choices and physical appearance giving her the gift of self-esteem. She benefits from white supremacy, and from cis-het supremacy, and from the class system and she is not disabled. Her mother tongue is spoken all over the world. She feels that a person must acknowledge their privilege in order to correctly understand Watkins' (bell hooks') concept of self-actualisation. If she wants to be truly self-actualised she feels it is an obligation in reaping the rewards of an unequal society to use her privilege to act to disrupt the status quo. She is right. See Blog Article Endnote.[6]

8.6 The Dangers of Screen Technology Versus the Freedom to Think Independently

The Information Technology Revolution is taking us through some similar processes to that of the Industrial Revolution of the nineteenth century. In the sense that our addiction to instant gratification through our devices is creating a new treadmill that limits freedom. At times it has been overly driven by neoliberal commercialism and consumerism, for example, online shopping, computer games, social media, gaming and gambling addictions. This means we now have a greater need to understand and manage our emotions, feeling sensations and impulses. Some understanding of how brain plasticity works and is affecting us can help us understand these problems. Not by suppression, but by learning to work with and on ourselves. If we can learn to do this constructively, we can develop the emotional intelligence that we pay lip service to but do not always understand and practice sufficiently (Goleman, 1996).

There have always been thoughtful individuals who have understood the power of the mind. Individuals developing their own ideas and creations by discovering methods and systems of thought that helped them achieve. We have meditative schools of thought and physical practice. For example, yoga and meditation; particular forms of dancing and music with practice regimes and performance repertoires; gymnastics and ice skating and all kinds of performance and creative arts; creative and philosophical literature; mathematical theories with sequences of theorems and algorithmic designs to name but a few. Teacher Sifu John Lobb is a martial arts expert who is constantly developing his physical and mental strength and teaching his students how to manage their expertise responsibly. He says, "Wing Chun draws from physics, mathematics, and geometry to augment physical form and guide human choices for efficient violent exchange." He has written a piece on confidence, which explains his perspective on mental effort. See Special Data Contribution.[7]

Computer applications are structuring our lives and limiting our decisions. Many choices are created and presented to us and for us, based upon the prescriptive systems and needs of Silicon Valley's financiers and shareholders. These people have been clear about their dislike of democracy and its institutions. They are proud of their ability to cause social disruption in pursuit of material gain. There has been a recent spate of books by former internet inventors, financiers and computer engineers disillusioned with the sector. Billionaire chief executive officers have typically made huge profits and taken no responsibility for any detrimental effects on our world. They have no interest in learning the lessons of history and are proud of their part in building a value free society. They take a 'buyer beware' attitude, but are now facing possible regulation as people become aware of the implications (Lanier, 2017, 2018).

In his book *Who Owns the future?* Jaron Lanier (2014) points out that humans have always faced annihilation in various forms. A mathematician, artist and musician, 'virtual reality' is a world Lanier was instrumental in creating. He now warns against its improper use. He says in the past scientific discoveries and technological developments brought change and risks we learned to manage, but our current impulse to want quick-fix solutions is dangerous. Listening to other people's points of view even when we don't immediately agree can teach and help us cope better with our new 'virtual' world. Lanier (2018) says if we decide to use social media we should behave unpredictably in order to avoid being tracked and trained in our responses by algorithms.

Social media uses behavioural training as described in B.F. Skinner's (1904–1990) research with animals (rats) and humans (Skinner et al., 1957). Behavioural training methods are researched and screen technology applications carefully designed by technicians to achieve the maximum number of hits, for example, Twitter and Facebook sell advertising space based on 'customer' numbers and their preference data. Skinner's theory that human free will is an illusion has been discounted, but it is certainly informative about the way in which humans have unwittingly interacted with social media. His research into human behaviour found if consequences are bad, there is a probability an action will not be repeated. If the consequences are good, there is a strong probability of repetition. Skinner called this the principle of reinforcement. Social media platforms have used computer algorithms to encourage human responses and predict likely behaviour with the aim of making sales.

They use continuous reinforcement (CRF)—each time a specific action is performed, the subject receives a reinforcement, which links the target behaviour to a reward. They are also rewarded with free use. We have been behaving like rats. Our iPads and smartphones have become a kind of operant conditioning chamber we carry around with us. Lanier warns us these can become a kind of portable cage which prevents us from interacting with the real world (2018). They use what Skinner called 'operant conditioning' in order to strengthen behaviour. The process is set up to earn money by 'helping' sellers to advertise their services to customers likely to buy goods. They find out people's interests by 'harvesting' their data, which they sell to advertisers and political influencers (Kaiser, 2019).

It is the latter that have caused most problems by publishing and encouraging fake news and conspiracy theories. The rewards we get from social media include feelings of satisfaction as we express anger and gain support. Our comments are recycled in a feedback loop by the system to promote attention and encourage us. Comments are easily misinterpreted and the long-term effect on individuals encourages negativity—in some cases violence. Lanier explains that negative emotions have a deeper impact upon us than positive ones. This is probably because they activate an adrenalin rush as our bodies react with an involuntary 'fight or flight' electrochemical response. He says that negative feedback has turned out to be the 'bargain' feedback for social media systems where automatic systems encourage responses. They constantly feedback news and comments, stimulating more.

As Lanier says currently we are subjected to an 'over amplification' of negative emotions. They are caused remotely by siren servers whose creation is affecting our well-being in negative ways, for example, fake news, stolen identities, criminal activity via the dark web, terrorist infiltration and political manipulation. The internet has opened up a new environment, which we have to learn to negotiate safely. Just as in our physical life encounters in the 'real' world we have to learn to interact using the communication benefits while avoiding the dangers. We need systems of law, tools, skills and education in order to understand the difficulties. This brings us back to positive aspects of Skinner's behavioural research, because that research shows us how we can train our own brains to be smarter (1957).

We can reward ourselves constructively improving our lives by thinking about more enjoyable ways to achieve our aims in life. The answer to many of our problems is around being more aware. Brain plasticity tells us that this is possible. We need an awareness of those factors that affect our ability to think and make wise choices. Many of these phenomena are contradictory, insistent and confusing. Our neuroplastic brains have evolved to respond to the world at a variable, but generally more measured pace than we are currently experiencing in everyday life. There is a massive escalation in distracting 'white noise' because there are many more lines of communication than in the past. This is the information revolution. Technicians have even tried speeding up human speech in television dramas to give more time for advertising on television! Virilio has written a book about 'dromoscopy' (the logic of speed) in social, commercial and political communications (2008).

Steve Redhead (2011) has critiqued Virilio's work, taking up and developing associated ideas. Interruptions and distractions to our thinking time are triggered by, for example, television, iPads, advertisements, emails, mobile phones and social media. It is harder to comprehend, negotiate and deal with the world in which we live than in the past. Social beliefs and norms have been slowly established and absorbed by us over time. Sometimes we accepted them as mental straitjackets, but now we are inundated with distractions. Successful social media actors and entrepreneurs have learned how to attract our attention by manipulating emotions and feelings. Political manipulators and bad actors have spread misinformation, conspiracy theories, dangerous and destructive ideas. They have blurred the difference between fact and opinion. Humans have to learn to manage this new environment.

The owners of social media platforms have created systems that enabled disruption. At the same time they have refused to take responsibility for the consequences of the services they provide. This is a form of exploitation governments have been far too slow to regulate. The consequences such as the 6 January 2021 insurrection on Congress in the United States and the genocide in Myanmar have shocked the world (Taplin, 2017). Such events are triggered on social media platforms by self-serving ideas, inherited fixed mindsets and prejudices based in ignorance and a lack of an open-minded education. For example, anti-vaccine propaganda; when old jobs become obsolete and we are reluctant to be trained in new jobs or when we hark back to a 'golden age' (Pinker, 2018; Roberts, 2014). Politicians and employers may wrongly assume us to be unable to adapt to change and old ideas are perpetuated.

In 2017, Walter Veit was inspired by the adult animated series *Rick and Morty* and *BoJack Horseman* to challenge the absurd. He writes that elements of Friedrich Nietzsche's life-denying philosophy of Existential Nihilism are based on the idea we are living in a meaningless universe devoid of purpose. Existentialism was a philosophy promoted by Jean-Paul Sartre emphasising the isolation of human experience in a hostile and indifferent universe. Although these ruminations seem far removed from everyday concerns of ordinary people, such ideas are worth considering since they can impinge upon our well-being when they are adopted and become a justification for extremists. They resonate because they do sometimes express our feelings of depression about human situations. See Cartoon Endnote.[8]

People nowadays get upset without a smartphone and constant internet access. These have become an extension of ourselves. The apparent omniscience of the 'self image' has seduced us, but it is a spurious world—an alternative reality. People have come to consider the phone to be an extension of their brain, a data bank and a magic mirror, for example, you can change your image in a world of hits and likes and instantly buy whatever goods you want. It is easy to be conditioned by virtual media. It makes you godlike and omnipotent in your own 'virtual' world. This outlook on the world encouraged by social media builds on and serves our natural self-interest and fear of adversity. If we don't want it to be true, we can change it and such an attitude is proving dangerous to our safety in the real world and damaging to our mental health. We search for safety and comfort and status in the face of a challenging world and we often feel and are inadequate. It is time to re-educate ourselves and our children (Lukianoff & Haidt, 2018).

8.7 The Environment, Emotion Models, Brain Plasticity and Educated Transformation

Before it ceased *to exist* in 2011 the Worldwatch Institute asked how education could prepare 9.5 billion people for life in the Anthropocene, the current geological age so badly affected by human activity. How could people thrive in this uncharted and much more chaotic future? They pointed out that on Earth education is traditionally defined and confined by specific topics such as ecoliteracy, outdoor education and environmental science, but these are really interconnected. The present century is on track to be the warmest in human history. The changing planet will affect every aspect of human life. Drought, food shortages, emerging diseases and waterlogged cities are just some of the unprecedented challenges students will face. They found researchers, students and teachers looking for answers all around the world. They were sharing 'innovative approaches to teaching and learning in a new era.' They hoped to inspire us to prepare people not just for the problems ahead but to become the next generation of sustainability leaders.

According to Jeff Sachs, co-creator of the World Happiness Report and a professor at Columbia University, the Nordic countries have rather a different view of modern life than some other countries. He explains that people in Finland and Denmark are generally happier than in the United States. The United Nations' World Happiness Report Index measures life satisfaction in six areas—income, generosity, freedom, social support, trust and healthy life expectancy. Since the report began in 2012, Nordic countries, including Denmark, Finland, Norway, Sweden and Iceland, plus the Faroe Islands, Greenland and Aland consistently turn up at the top of the list. It is not a measure of celebrity, power or material possessions, but about 'satisfaction' with the overall way a person's individual life is working out. *The United States is typically placed in 18th or 19th place.* See Video Endnote.[9]

This is an example of useful scientific informative qualitative research to show a different emotional outlook. At base it is about lifestyles, cultural aspirations, what people want and what they are able to achieve. Aspiring to lifestyles advertised on screen has affected people's values and satisfaction with their lives. Many of us have lost touch with nature and stopped appreciating the simple pleasures of life. Governments and community leaders must understand they need to legislate so that everyone can contribute and take responsibility for making our lives more humane in this area. In fact, at this moment in history all of our real lives and existences depend on it from billionaires to immigrants. *Now that we understand how our brain plasticity makes us vulnerable, we have to create social checks and balances.*

School councils can demonstrate democracy in action, for example, questioning local business owners and politicians. Currently, especially ideas about democracy have been clouded by social media where giving voice without evidence and gaining an audience in the 'post truth' era can undermine individual happiness. There are huge emotional issues here about learning how to respond with reflection, wisdom and integrity, to infer and reserve judgement. For example, what is said online is not necessarily authenticated as true, but when it is believed to be and action is taken, death, imprisonment and catastrophe may follow, for example, the attack on the World Trade Centre in 2001 orchestrated by terrorists (9/11), Iraq War (2003–2011); the Arab Spring demonstrations (2010), the Afghan War (2001–2021), the genocides in Myanmar up to the present day and the MMR and Covid-19 vaccination disputes. Particularly as children, but also in adulthood our brain plasticity makes us vulnerable.

Teachers can draw upon examples from the past of how educational methods and approaches have served, disrupted and taught us important lessons. We can be proactive in reframing, expressing and adapting these values and suggestions to framework the curricula of the future. For example, Martin Robinson suggests we teachers adapt our curriculum according to the tried and tested tenets and wisdom of the ancient Greeks. The Trivium was their way of checking and balancing their learning system. They had a check list—'Grammar' or Knowledge, which involves learning about the way things were or are; Dialectic, around questioning the way things were or are and Rhetoric, showing how things are or could be. They give us three different positions or frames of reference in learning how to learn to think about the world (Robinson, 2019). See Chap. 4 Sect. 4.4.

Education can help us to take a 'longer' and 'wiser' view interpreting our choices and living our lives differently, but much more collaboratively to save ourselves and planet Earth. We can use real issues to teach and reach curricular objectives using facts and opinions on and off line that can be researched, discussed and questioned. Dropping the arts from the curriculum is not an option because they are a necessity for human emotional well-being. Even when we reject their products, they can help to keep us sane and balanced by helping us explore and assess reality. They are a way of expressing, challenging and balancing our perceptions ideas and dreams. Imaginative dialogic discussion, creativity and listening to diverse voices are key not only for equality, but also for human survival.

Education is instrumental in helping us build and update concept frameworks. These, together with emotional learning experiences, can help us to adjust our personal emotion model of our world. We are complicated primates and brain plasticity enables our survival process, but it has disadvantages too. There are many ways to be human and much of our character is formed in childhood, but it is not *necessarily* fixed. Characters are capable of growth and adjustment. They can deteriorate and improve.

A variety of books by previous insiders have been published about the rise of the oligarchs of Silicon Valley. The drive to make money and wield power has been very much a part of their competitive group emotion model. It explains much of their unethical ambivalence towards the rest of the world and the serious power imbalances their products have enabled.

It seems likely these entrepreneurs were influenced by the 'American dream,' inculcated in early childhood. American children are sometimes taught America is better than other cultures. There childhoods are typically affected by the constant use of computer games in which fast responses to gain money or tokens gives one the ability to beat other players. In these there is no penalty for being irresponsible, anything goes because you are *in it to win it*—you are a winner or a loser. For example, apparently Mark Zuckerberg was particularly motivated and fascinated by these. His maths was exceptional, but his own social education may have been wanting. Like many Americans Zuckerberg has been attracted by the idea of the Roman Empire, which we know used psychological propaganda to dominate other countries (McNamee, 2019).

It is not hard to see some of the influences on Zuckerberg's apparent emotion model of the world. The evidence is Facebook itself. It turned into a big game in which compliance with regulation applied only when rules could not be circumnavigated. It appealed to many people and they happily joined in. We are only now realising some of the downsides of this behaviour. Computer games, although not necessarily bad in themselves, may be responsible for helping to create emotion models that can be destructive for humanity. They have become more and more realistic and can be played without any personal consequences other than failure. They are now used to train drone operators, making war a 'remote game' with the use of the chilling expression 'collateral damage' when referring to civilian casualties in modern warfare (Levy, 2020).

Nelson Mandela was a well-respected world leader who had a difficult life. However, he learned from mistakes and become an example of the advantages of lifelong learning, insightful reflection and adjustment. He took an open-minded approach to those who did him harm and he developed his knowledge through reflection and self-regulation in prison. He came out of prison and set his people on a path to freedom and democracy, which is still being worked out and developed. Brain plasticity shows us that change and development are possible. If we are motivated, we can change ourselves, make progress, reparations and become reconciled to some of those with whom we disagree. Whether or not you believe in a religion, there is a possibility that your brain plasticity can help you change, make progress and 'be forgiven' enough to make improvements to your thinking and behaviour. You may also 'deteriorate,' but many of us will hopefully have some choice. See Article Endnote.[10]

When we are considering the research evidence given by teachers and learners about how learning develops and how intelligence works; it may be helpful to bear in mind that our encounters with other humans are affected by the following:

- The need to survive as external and internal conditions affect our physical bodies.
- Trial and error—reflecting and working things out over time is important. A poor idea may lead to a better one.
- 'Logical' emotions (schema memories) are stimulated by new events which cause true and false interpretations (connections) (Bennett-Goleman, 2001).

- Not understanding or agreeing with each other—having variable communication skills. We react quickly and automatically—jumping to conclusions.
- Semantics are important—the understanding of words. For example, in giving verbal instructions, I have found as a teacher that repeating the same words may not result in understanding or compliance, but rephrasing or using alternative words quite often does.
- Subconscious memories affect how we feel about new people, environments and events.
- Feelings, both physical and mental, are meaningful to individuals and should be considered—whether or not we agree with them and whether or not they are 'correct.'
- We all need to encounter new ideas, surface old ones and revise patterns of reasoning to avoid being trapped by habitual thoughts.
- Acknowledging feelings during and after learning helps the thinking process, may solve learning difficulties and explain confusing thoughts and behaviour—this process helps us to revise, create, update and embed ideas.
- The human brain is capable of adjustment throughout life. Its inherent plasticity operates in superficial and fundamental ways—these may be viewed as 'beneficial' or as 'detrimental,' depending on your point of view.

In explaining the work of John Abbott (2009), Luba Vangelova (2014) says that in order to advance education, we must reimagine our current aims for society. He says we have not been able to reform education to help the disadvantaged because we have suffered from a malaise in society. He asks us to consider whether we want to be populations of responsible people who respect each other and thrive in interdependent communities, or in a world where we are 'customers' who rely on goods, services and authority figures, who we expect will look after us. See Article Endnote.[11]

We're all part of the human family and our brain plasticity makes us all unique. It therefore makes sense, especially in our current problematic world environment situation, to learn about world cultures—to be confident in our own cultures and interested in other people's. The human family together have the power to move forward and create an interesting, wise and kind new world for us all. Learning involves acquiring, exploring and interpreting knowledge. Our emotional experiences are key as we respond to information, share useful ideas, interact, inform and leverage new knowledge. Now that this process has been speeded up by technology, we can improve our skills, systems and strategies of intelligent emotional awareness and foresight. Education is a good arena in which to study how to sharpen our intelligence. This requires research—by everyone personally and socially; intimately and widely—at every level—in a constant cycle of activity, collaboration and revision.

As evidenced throughout history, humans may not apprehend the full meaning and implications of their thoughts and behaviour until further reflection at a later date. Brain plasticity developed to give us the powers we need for survival, but again these can veer towards unsocial, self-destructive behaviours. We are super social animals, ultimately group dependent and our physical biology is adapted accordingly. Our brains have evolved to respond at high speed to the many threats and challenges we face through life. For example, embedded tendencies become extreme when a person seeks more and more material things. Power is a strong motivator that 'develops logically' from the basic survival instinct as such phenomena as power positioning, war, crime, conspicuous consumerism and racial and minority group scapegoating have shown.

Our feelings about our emotional intelligence are ambivalent, but they perform a fundamental function of the brain. They are in effect a communication system between all parts of this multi-thought biological generator. It is foolish to ignore or suppress them because our body-brain through our central nervous system is our own personal information highway. Although we have created a technology system that can 'serve' us with facts, we should never allow it to prevent *our own more sophisticated thinking from being operational.* In all our thoughts we need to learn to be able to connect to our own personal and external physical reality *and manage ourselves better.* Good schooling can deliver this, but in recent years politicians have taken a rather negative view of education requiring instant answers, wanting a fixed structure, control, power and to deny *the need for complexity.*

Such approaches have actually broken parts of the system, for example, the teacher retention crisis. They have caused some people to abandon hope (mental health difficulties, unsupported young people turning to crime). An 'overwhelming' number of teachers in the United Kingdom have suffered either physically or mentally because of their jobs, a study has suggested. The research, commissioned by the charity Education Support Partnership, indicated 75% of teaching staff in schools and colleges experienced symptoms stemming from their work. Physics teacher

Pranav Patel told the BBC that although he was passionate about teaching the "sheer amount of workload" had caused him to have bouts of depression, anxiety and sleeplessness. See Video Endnotes [12] and [13].

The evidence is that we can too easily become trapped in destructive cycles of thought. 'Tell me the answer I don't need to think about it' has become our habitual cry as we turn to the internet. Our brain capability and plasticity may actually be in jeopardy if we abandon elaborated thought and extended thinking. Perhaps we have become overly used to separating out our information into discrete subject areas of knowledge as we search for facts, for example, Psychology, Sociology, Anthropology, Education and Neuroscience. We have become very good at deconstructing, categorising, sorting children into types, listing, metric collecting, mechanical measurement, producing spreadsheets, graphs, technology and bureaucracy.

We have been seduced by the expedient notion that simple logical mechanisms can supply easy answers and successful outcomes without doing the work of reserving judgement, checking complex facts, finding humane solutions and developing wisdom. This has become our cultural environment and it is undermining our progress and happiness. We have rich and practical areas of expertise around the ways we think, philosophise and feel about the world. However, they are disparate and unconnected. In education we have often failed to co-ordinate our thoughts and create a comprehensive overview of the 'spiritual' aims of learning that involves considering fundamental human values.

8.8 Conclusion

My research is one teacher's phenomenological research experience. I am not trained in the sciences, but they appear to me to back me up to some extent. I have found the debate fascinating and learned many new things by listening to a range of viewpoints. I reached these conclusions after considering the evidence I collected from 2018 to 2021. I gained a lot of information by looking at experts', teachers', parents' and learners' concerns and listening to their narratives, opinions and evaluations. It has been a wide canvas and I hope the areas of interest and information I have found will help others to start their own disciplinary action research, to discuss, agree with, or to argue against going forward. As a monolinguist, I use English written or translated sources as most accessible and understandable to me. My exploration is inevitably informed by my own personality, interests, learning and teaching experience in the United Kingdom.

As I write I am relating current neuroscience and psychological research findings to see what they can contribute to my personal teacher's professional understanding. My conclusions are that teachers, since they work with varied learners, settings, knowledge sets and skills, should be expected to do the following:

- Since with training, good teachers have proved themselves capable of it, justify their work both to themselves and others in a variety of ways.
- Those who develop specific skill sets and experience in middle management, such as master crafter teachers (who mentor and advise) should be more listened to by politicians and more in charge of education!
- The profession would then be in a better position to self-regulate, reporting upwards and instructing for the development of an adaptable workforce.
- Education can contribute to a happier, healthier, more contented society where people are supported to learn and develop interests throughout their lives.

I invite you to consider how you might similarly look around you and make your own cultural, personal and subject contributions. I find the researching journey with its experiences, downs and ups along the way can turn out to be more rewarding than any final conclusion! I invite you to enter with me into a fairly new hall of mirrors and look at your own knowledge, theories and problems from some different angles in the light of neuroplasticity. National diversity enriches world culture. Beneficial aspects of all our current earth's cultures give us opportunities to combine and build a new richer, better, *whole world civilisation and body of knowledge*.

The evidence is telling us we have to bring the humane back, appreciating our human skills and abilities on the ground and making sure there are checks and balances that recognise the way we actually think. We need flexible systems that involve a dialogic approach to determining meaning and human purpose. We need to develop our skills of inference, compassion and the divergent pursuit of wisdom for the betterment of humanity. We need to train children to challenge their assumptions, reserve judgement and understand who they and others are, if they are to solve their own and world problems into the future. They need schooling that promotes freedom to learn and is thought provoking, intellectually challenging, enlightening, knowledge-rich, adaptable and creative. The late Sir

Ken Robinson (1950–2020) in his famous TED talk on creativity in 2006 said that updating our understanding about ourselves is urgent. See Video Endnote.[14]

Robinson continued to stress *the need to reconceptualise intelligence in education* throughout his career. His last long presentation in 2019, just eight months before his death is based on 'real life' anecdotes and observations after a lifetime working in education. It touches on many of the key issues in this book. Today an ability to reconceptualise intelligence for learning through education is even more urgent, because our environment is presenting us with serious challenges in new ways. We have to learn how to innovate to cope with the effects of science and information technology in our changing world and work towards forms of restoration, reconciliation, reparation and development. Currently we face environmental pollution, materialistically motivated politics, globalisation, runaway consumption, genocide, dysfunctional social systems, democratic backsliding, tyrannical regimes and a world pandemic. We must change our behaviour and remediate our world. See Video Endnote.[15]

On the positive side, the 'fourth industrial revolution' of the computer age gives us opportunities to share data. Many more of us can access old knowledge and share new knowledge about our various cultures and situations over the internet. Perhaps we can solve problems, acknowledge our human failings, address our mistakes and misdemeanours and learn wisdom. We will need to change, adapt and develop better thinking strategies in order to deal with these difficulties. Our children need to be taught to research, infer, surmise, interpret, create and deduce. A rich and humane education can offer solutions. 'Feelings phenomena' and narrative data can be recorded both in science and education. They are increasingly researched and acknowledged. There is hope—the discovery and fact of neuroplasticity is changing views—its very existence could force a change in attitudes throughout our world.

Solving the kinds of world problems listed above, and recently in the Covid-19 pandemic, were an impossible dream in the past but we now have new technical capabilities combined with urgent imperatives and have no excuse not to make the effort. There are more 'educated' people and many opportunities. We can co-operate and create concepts of value to improve our environment and our well-being. We can share our successes and expertise. We can develop learning strategies that ultimately tend towards improving our politics, environment and everyday life. All of this involves the incorporation of feeling and emotion within the intelligence model including adjustable and agreed value systems of reference and reflection, checks and balances. Again Ken Robinson tells us we need to make a better contract with nature and our own humanity. See Video Endnote.[16]

Groups of people in every area in societies around the Earth can make value judgements and become collaborative arbiters. Understanding human nature and possible motivations for our behaviour is fundamental to this endeavour. As educationalists and learners, modern life requires us to make personal investigations and listen with more attention to others. We have an opportunity to be more connected to other people and to do it in a better way. We can look through different lenses aided by our digital inventions and listen to different points of view. We can now research collaboratively at a distance and together to find out how to create, conserve and promote a better world and a better reality for ourselves.

Notes

1. Benjamin Doxtdator (2017) A Field Guide to 'jobs that don't exist yet' 8 July 2017. http://www.longviewoneducation.org/field-guide-jobs-dont-exist-yet/. Accessed 06/04/2018.
2. Jarvis, Pam (2017) 'A Journey Through The Looking Glass' 2017, October 05 Huffington Post; Available at: https://www.huffingtonpost.co.uk/pam-jarvis/a-journey-through-the-loo_b_17903012.html. Accessed 19/07/2018.
3. Website Open Democracy an independent global media organisation https://www.opendemocracy.net/en/about/. Accessed 13/02/2019.
4. International Cooperative Alliance (2021) 'The Statement on the Cooperative Identity, Values & Principles' https://www.ica.coop/en/cooperatives/cooperative-identity. Accessed 04/04/2021.
5. Madhukar Pai (2021) 'People Power, Not Powerful People, Can Help End Inequality' an interview with Ben Phillips https://www.forbes.com/sites/madhukarpai/2021/01/01/people-power-not-powerful-people-can-help-end-inequality/?sh=4a9eac281717. Accessed 01/01/21.
6. Julie Fellmayer 'Disruptive Pedagogy and the Practice of Freedom' 11 October 2018 Critical Pedagogy https://hybridpedagogy.org/disruptive-pedagogy-and-the-practice-of-freedom/. Accessed 23/06/2020.
7. Special Data Contribution

   'Confidence: Anything you can do once, you can do again better. Learning to invest through failure' by Sifu John March 2019

   Sifu John has been practicing Kung Fu and Thai Boxing since 1991 with active teaching duties since 1994 teaching the Gary Lam Wing Chun system since 2008. He has been a regular contributor to Wing Chun Illustrated Magazine since 2012, and creator of a unique pressure tested method of application specifically for combative exchange. Website: https://warringtonkungfustudio.co.uk/

Confidence is experience, and experience is born through repetitive practice.
We live in a fast paced world, information and tasks are completed as quickly as possible.
The speed in which we are expected to complete tasks (in the modern world), does not always allow for reflection and absorption.
As a result, sometimes we may find that when revisiting the same problem a second time, we are not attacking it with useful information gleaned from the previous occasion in mind—we may not be approaching the problem with the level of confidence we should, as we have not taken the time to process and bank the previous experience intelligently.
Without filtering and prioritising positive knowledge gained from unsolved problems, we run the risk of stagnation or magnifying inadequacy.
Negative mindsets relating to learning new skills can prevent us from considering that some problems need to be tackled from a different angle or with better information gleaned from a new source.
Shying away from the potential to fail and change is not useful.
Failure is only the result, if you fail and stop. When you fail and try again, failure becomes part of a growth process.
Self-doubt or an unwillingness to try new things, blocks action and interaction necessary to attain and achieve all things new. We should strive to remember this at all times.
Failure and deliberation, should be understood as intrinsic to success. A call to action, a challenge still to be met.
Rather than build a culture of avoidance based on experiences that did not meet our preconceived expectations, we should be open to chance, both failure and success—and willing to work methodically and consecutively, to create opportunities for modification and growth.
The fear of judgement or failure is presently magnified by a society that shares its self so readily via social media.
With so much self-imposed attention we can create a great deal of apprehension.
It is easy for people get caught up in thinking about a process, or making assumptions about a result before they experience it for themselves.
It is also very easy (and sometimes lazy) and 'safe', to be guided by ideas held as popular opinion. It involves no risk.
Just because an idea or opinion is popular, does not mean its foundations are imbedded in truth or practicalities. Most importantly, if these ideas do not stem from our personal experience, they are but borrowed assumptions.
It is useful to attempt to qualify external opinions before they are accepted as truths, because sometimes they have nothing to do with the actuality of an activity or efficiency within it.

Limited perspective can sometimes make very simple things feel like complex puzzles.

Sometimes external opinions about something, do not relate to the capacity of an individual to be successful at a 'thing'. People are complex creatures. We are all very different, and usually capable of much more that we give ourselves credit for.

Self-knowledge rather than borrowed knowledge is firmness of mind.

To be truly confident, you must carry your own set of experiences.

Doing things is testing, exploring, refining.

Today when most people say 'I think', what they actually mean is 'I read' or 'somebody told me'. This is not thinking.

It is always useful to distinguish between retention and reflection.

It takes great honesty to divide the opinions we built from an external source and opinions we built through practice and reflection.

We have to recognise borrowed from built, assumed from learnt, belief from ascertained.

Confidence is self-knowledge through practice, and practice is confidence in self-knowledge.

Information may expand our knowledge base, but a lack of focused action does not invoke a confident character.

Not employing our 'doing' aspect is out and out loss.

Thinking is a function used to produce direction and ultimately choice. We should try to appreciate it as a device separate from our character.

The German word for character is Persönlichkeitsmuster. The literal translation is 'people pattern.' A 'people pattern' is an excellent way to perceive human character.

8. Wisecrack (2018) 'The philosophy of Bo Burnham asks the question "On a scale of 1–0 are you happy?"' Writer Mike Burns Sponsor BetterHelp. https://www.youtube.com/watch?v=W307nbJNK_E&list=PLghL9V9QTN0jve4SE0fs33K1VEoXyL-Mn. Accessed 26/07/2018.
9. Jeff Sachs (2020) 'Why Finland And Denmark Are Happier Than The U.S.' United Nations World Happiness Report CNBC 9 January 2020. https://www.youtube.com/watch?v=6Pm0Mn0-jYU. Accessed 06/08/2020.
10. Karen MacGregor (2015) 'Nelson Mandela as an exemplar of lifelong learning' 26 September 2015 University World News – The global window on higher education Africa Edition. https://www.universityworldnews.com/post.php?story=2015092514572690 06/08/2020.
11. Vangelova, Luba (2014) 'To Advance Education, We Must First Reimagine Society' NewsScienceArts & CulturePodcastsTVSupport KQED. https://www.kqed.org/mindshift/34658/to-advance-education-we-must-first-reimagine-society. Accessed 19/05/2019.

12. Pranav Patel (2017) 'Physics teacher, Pranav Patel, told the BBC that although he was passionate about teaching the "sheer amount of work-load" had caused him to have bouts of depression, anxiety and sleeplessness'
    Published 18 September 2017 Section Education BBC. https://www.bbc.co.uk/news/av/uk-england-41286104. Accessed 05/06/2018.
13. William Walton 'Frankie McCamley speaks to teacher William Walton who says some colleagues drop out due to pressures and the workload' Published 21 February 2017 Section BBC News Subsection Education & Family. https://www.bbc.co.uk/news/av/education-39045553. Accessed 03/04/2019.
14. Robinson, K. (2006) 'Do schools kill creativity?' [Monterey: TED Online] Available from: https://www.ted.com/talks/sir_ken_robinson_do_schools_kill_creativity. Accessed 04/10/2018.
15. Robinson, K. (2019) 'Finding Your Element' 18 December 2019 King Rose Archives, California. https://www.youtube.com/watch?v=17fbxRQgMlU. Accessed 25/08/2020.
16. Robinson, K. (2020) 'My thoughts for the Call to Unite' 7 May 2020, London. https://www.youtube.com/watch?v=QU4Q17t4muY. Accessed 25/08/2020.

References

Abbott, J. (2009). *Over Schooled But Under Educated: How the Crisis in Education Is Jeopardizing Our Adolescents.* Continuum.

Bartlett, J. (2017). *The Secrets of Silicon Valley UK: British Broadcasting Corporation.* Available at: https://www.bbc.co.uk/iplayer/episode/b091zhtk/secrets-of-silicon-valley-series-1-2-the-persuasion-machine. Accessed 28 Aug 2018.

Bartlett, J. (2018). *Secrets of Silicon Valley: Uncovers the Dark Reality Behind Silicon Valley's Glittering Promise to Build a Better World Open University.* https://www.open.edu/openlearn/tv-radio-events/tv/secrets-silicon-valley. Accessed 10 Feb 2018.

bell hooks. (2003). *Teaching Community: A Pedagogy of Hope* (No. 2 bell hooks Teaching Trilogy). Routledge.

bell hooks. (2010). *Teaching Critical Thinking Practical Wisdom* (No. 3 bell hooks Teaching Trilogy). Routledge.

bell hooks alias Gloria Jean Watkins. (1994). *Teaching to Transgress: Education as the Practice of Freedom* (No. 1 bell hooks Teaching Trilogy). Routledge.

Bennett-Goleman, T. (2001). *Emotional Alchemy.* Rider.

Bermeo, N. (2016a, January). On Democratic Backsliding. *Journal of Democracy, 27*(1). Johns Hopkins University Press

Bermeo, N. (Ed.). (2016b). *Parties, Movements, and Democracy in the Developing World* (Cambridge Studies in Contentious Politics). Cambridge University Press.

Bronfenbrenner, U. (1979). *The Ecology of Human Development*. Harvard University Press.

Cherry, Kendra. (2021). *The 5 Levels of Maslow's Hierarchy of Needs*. Reviewed by David Susman, PhD Updated on March 19, 2021. https://www.verywellmind.com/what-is-maslows-hierarchy-of-needs-4136760#. Accessed 28 Apr 2021.

Chivers, T. (2019). *The AI Does Not Hate You: Superintelligence, Rationality and the Race to Save the World*. Orion.

Christian, B., & Griffiths, T. (2016). *Algorithms to Live By: The Computer Science of Human Decisions*. William Collins.

CNBC. (2019, July 27). *Why Democratic Socialism Is Gaining Popularity in The United States*. In the Latest Explainer from CNBC We Try to Answer What Socialism Means to Different Parts of the Political Spectrum and Whether It's Fad or the Future of Politics in America. https://www.youtube.com/watch?v=BlrDpTGRREA. Accessed 10 Aug 2020.

Data Contribution Sifu John Lobb. (2019). *'Confidence' a Wing Chun Martial Art Instruction Piece About Focused Thinking*.

Davis, E. (2019a). *Post Truth: Why we Have Reached Peak Bullshit and What We Can Do About It*. Little Brown.

Decolonise the Curriculum Pranav Patel. https://theteacherist.com/category/decolonise-the-curriculum/

Doxtdator, Benjamin. (2017). A Field Guide to 'jobs that don't exist yet'. Jul 8, 2017. http://www.longviewoneducation.org/field-guide-jobs-dont-exist-yet/. Accessed 06/04/2018.

EarthEd. (2017a). *State of the World: Rethinking Education on a Changing Planet*. Island Press.

Enurture Network. *Promoting Young People's Mental Health in a Digital World a Network of Cross-disciplinary Professionals*. https://www.enurture.org.uk/

Evers, J. (Editor). (2015). *Flip the System: Changing Education from the Ground Up*. Routledge.

Facebook, Algorithms and News, Fake News. https://popular.info/p/the-dirty-secret-behind-ben-shapiros. Accessed 22 July 2020

Field, M. (2018, August 16). How to Make Alexa Listen: Drop Your Regional Accent. *The Daily Telegraph*, p. 11. Telegraph Media Group Ltd.

Foer, F. (2017). *World Without Mind: Why Google, Amazon, Facebook and Apple Threaten Our Future*. Penguin Random House.

Freire, P. (2014). *Pedagogy of Hope: Reliving Pedagogy of the Oppressed*. Bloomsbury Academic.

Fry, H. (2018). *Hello World: How to Be Human in the Age of the Machine*. Penguin Random House.

Goleman, D. (1996). *Emotional Intelligence: Why It Can Matter More Than IQ*. Bloomsbury.
Greenfield, S. (2004). *Tomorrow's People: How 21st-Century Technology Is Changing the Way We Think and Feel*. Penguin.
Greenfield, S. (2014). *Mind Change*. Penguin Rider.
Haas, R. (2020). *The World: A Brief Introduction*. Penguin Press.
Hall, E. (2018). *Aristotle's Way: How Ancient Wisdom Can Change Your Life*. Bodley Head.
Harley, D., Morgan, J., & Frith, H. (2018). *Cyberpsychology as Everyday Digital Experience Across the Lifespan*. Palgrave Macmillan.
Hedges, C. (2018). *America: The Farewell Tour*. Simon & Schuster.
Hirst, P. (1994). *Associative Democracy: New Forms of Economic and Social Governance*. Blackwell, Polity Press.
International Cooperative Alliance. (2021). 'The Statement on the Cooperative Identity, Values & Principles' https://www.ica.coop/en/cooperatives/cooperative-identity. Accessed 04/04/2021.
Irwin, W. (2015). *The Free Market Existentialist*. Wiley Blackwell.
Jarvis, Pam. (2017). 'A Journey Through The Looking Glass' 2017, October 05 Huffington Post; Available at: https://www.huffingtonpost.co.uk/pam-jarvis/a-journey-through-the-loo_b_17903012.html. Accessed 19/07/2018.
Joyce, R. (2013). Nihilism. In H. LaFollette (Ed.), *International Encyclopedia of Ethics*. Wiley-Blackwell.
Joyce, R. (2016). Moral Skepticism. In D. Machuca & B. Reed (Eds.), *Skepticism: Antiquity to the Present*. Bloomsbury.
Kaiser, B. (2019). *Targeted: My Inside Story of Cambridge Analytica and How Trump, Brexit and Facebook Broke Democracy*. Harper Collins.
Keynes, J. M. (2009). *The General Theory of Employment, Interest and Money and Essays in Persuasion*. Classic Books America.
Lanier, J. (2014). *Who Owns the Future?* Penguin.
Lanier, J. (2017). *Dawn of the New Everything: A Journey through Virtual Reality*. The Bodley Head.
Lanier, J. (2018). *Ten Arguments for Deleting Your Social Media Accounts Right Now*. The Bodley Head.
Levy, S. (2020). *Facebook: The Inside Story*. Penguin Random House UK.
Lukianoff, G., & Haidt, J. (2018). *The Coddling of the American Mind: How Good Intentions and Bad Ideas Are Setting up a Generation for Failure*. Allen Lane Penguin Random House UK.
Mandela, N. (1995). *Long Walk to Freedom: The Autobiography of Nelson Mandela*. Abacus.
Mazzucato, M. (2018). *The Value of Everything: Making and Taking in the Global Economy*. Penguin Random house.

McNamee, R. (2019). *Zucked: Waking Up to the Facebook Catastrophe.* Harper Collins.

Moghraby, O. S. (2017, February 7). Screen Time Balance – Is There a Healthy Diet? Kidcrafters UK. https://www.youtube.com/watch?v=kmRA7qZcV7w&feature=emb_logo. Accessed 10 Feb 2019.

Moyo, D. (2012). *How the West Was Lost: Fifty Years of Economic Folly—And the Stark Choices Ahead.* Penguin Books Ltd.

Moyo, D. (2018). *Edge of Chaos.* Little Brown Book Group.

Moyo, D. (2021). *How Boards Work: And How They Can Work Better in a Chaotic World.* The Bridge Street Press.

Murphy, K. (2020). *You're Not Listening: What Your're Missing and Why It Matters.* Penguin Random House.

Murray, D. (2017). *The Strange Death of Europe: Immigration, Identity, Islam.* Bloomsbury Continuum.

Open Democracy an independent global media organisation https://www.opendemocracy.net/en/about/. Accessed 13/02/2019.

PBS NewsHour. (2020, October 28). The Effects of Screen Technology-Mark Zuckerberg, Others Testify at Senate Tech Hearing. https://www.youtube.com/watch?v=VhjHtZoCv3w

Phillips, B. (2020). *How to Fight Inequality: (and Why That Fight Needs You).* Polity Press.

Pinker, S. (2002). *The Blank Slate: The Modern Denial of Human Nature.* Allen.

Pinker, S. (2018). *Enlightenment Now: The Case for Reason, Science, Humanism and Progress.* Allen Lane.

Rand, A. (1961). *The Virtue of Selfishness.* Penguin Group.

Redhead, S. (2011). *We Have Never Been Postmodern: Theory at the Speed of Light.* Edinburgh University Press.

Roberts, P. (2014). *The Impulse Society: What's Wrong with Getting What We Want?* Bloomsbury.

Robinson, K., & Aronica, L. (2015). *Creative Schools.* Penguin Random House.

Robinson, M. (2019). *Curriculum: Athena Versus the Machine.* Independent Thinking Press.

Runciman, D. (2018). *How Democracy Ends.* Profile Books.

Rycroft-Smith, L., & Dutaut, J.-L. (Eds.). (2018). *Flip the System.* Routledge.

Sharma, V. (2018, May 10). Machine Learning and Its Algorithms to Know. *Data Science Central.* https://www.datasciencecentral.com/profiles/blogs/machine-learning-and-its-algorithms-to-know-mlalgos. Accessed 10 Feb 2019.

Skinner, B. F., et al. (1957). *Schedules of Reinforcement.* The Skinner Foundation.

Smith, A. (1982). *The Wealth of Nations.* Penguin Classics.

Soboroff, J. (2020). *Separated: Inside an American Tragedy.* Harper Collins Publishers.

Sumpton, J. (2019a). *Trials of the State: Law and the Decline of Politics.* Profile Books.

Taplin, J. (2017). *Move Fast and Break Things: How Facebook, Google and Amazon Have Cornered Culture and What It Means for All of Us.* Pan Macmillan.

The Worldwatch Institute. (2011). *Vital Signs* From Economic Growth to Sea-level Rise, Vital Signs 2011 Documents the Trends That Are Shaping Our Future in Concise Analysis and Clear Tables and graphs.report. https://web.archive.org/web/20111210063613/http://www.worldwatch.org/taxonomy/term/38 https://web.archive.org/web/20111217061529/http://www.worldwatch.org/bookstore/vital-signs. Accessed 01 May 2021.

Thunberg, G. (2021). A Year to Change the World: Climate Activist Greta Thunberg Takes a Year off School to Explore the Science of Global Warming and Challenge World Leaders, Calling for Action on Climate Change. British Broadcasting Association – BBC-3 Episodes Available https://www.bbc.co.uk/iplayer/episodes/p099f58d/greta-thunberg-a-year-to-change-the-world. Accessed 25 Mar 2021.

Vangelova, L. (2014). To Advance Education, We Must First Reimagine Society. NewsScienceArts & CulturePodcastsTVSupport KQED https://www.kqed.org/mindshift/34658/to-advance-education-we-must-first-reimagine-society. Accessed 19 May 2019.

Veit, W. (2017). *Why Humeans Are committed to Moral Nihilism' & 'How Does Evolutionary Game Theory Explain?'* Conference Abstracts Depts of Philosophy & Economics, Bayreuth University: [Online] Available at: https://uni-bayreuth.academia.edu/WalterVeit. Accessed 09 July 18.

Virilio, P. (2008). *Negative Horizon: An Essay in Dromoscopy.* Continuum.

Westall, Andrea. (2011, January 10). *Time to Revisit Associative Democracy?* Open Democracy UK Revisiting Associative Democracy, an e-book, Draws Together the Ideas and Thoughts of a Group of People Who Met Last…https://www.opendemocracy.net/en/opendemocracyuk/time-to-revisit-associative-democracy/. Accessed 13 Feb 2020.

Ximena. *@Edk8R Primary Teacher.* Aotearoa New Zealand. http://apassion4learning.blogspot.com/

References

Abbott, J. (2009). *Over Schooled But Under Educated: How the Crisis in Education Is Jeopardizing Our Adolescents.* Continuum.

Allen, R., & Sims, S. (2018). *The Teacher Gap.* Routledge.

Ampaw-Farr, J. (2021). *Because of You, This Is Me.* Crown House Publishing.

Armstrong, S. (2019). *Borrowed Time: The Science of How and Why We Age.* Bloomsbury Sigma.

Arrowsmith-Young, B. (2013). *The Woman Who Changed Her Brain: How We Can Shape Our Minds and Other Tales of Cognitive Transformation.* Random House.

Bach-y-Rita, P. (1979). *Recovery of Function Following Brain Injury: Theoretical Considerations.* Academic Press Inc..

Ballantyne, et al. (2005). Measuring Environmental Education Program Impacts. *Australian Journal of Environmental Education, 21.* University of Queensland.

bell hooks. (1994). *Teaching to Transgress: Education as the Practice of Freedom* (No. 1 bell hooks Teaching Trilogy). Routledge.

bell hooks. (2003). *Teaching Community: A Pedagogy of Hope* (No. 2 bell hooks Teaching Trilogy). Routledge.

bell hooks. (2010). *Teaching Critical Thinking Practical Wisdom* (No. 3 bell hooks Teaching Trilogy). Routledge.

Beresford Ellis, P. (1994). *The Druids.* Constable and Company Ltd.

Bermeo, N. (2016, January). On Democratic Backsliding. *Journal of Democracy, 27*(1). Johns Hopkins University Press.

Bhopal, K. (2018). *White Privilege: The Myth of a Post-Racial Society.* Policy Press.

J. A. Hawkins, *Brain Plasticity and Learning*,
https://doi.org/10.1007/978-3-030-83530-9

Bickle, J., Mandik, P., & Landreth, A. (1999). The Philosophy of Neuroscience. *The Stanford Encyclopedia of Philosophy* (First Published Mon Jun 7, 1999; Substantive Revision Fall 2019 Edition), Edward N. Zalta (ed.). https://plato.stanford.edu/entries/neuroscience/. Accessed 07 May 2020.

Biesta, G. (2013). *The Beautiful Risk of Education*. Routledge, Taylor & Francis.

Bignell, S. (2018). Autism, Asperger's and ADHD. In *What You Need to Know. A Guide for Parents, Students and other Professionals*. My Child Services Publishing.

Blakemore, S.-J. (2018). *Inventing Ourselves: The Secret Life of the Teenage Brain*. Penguin Random House.

Bliss, T. V., & Lomo, T. (1973). Long-lasting Potentiation of Synaptic Transmission in the Dentate Area of the Anaesthetized Rabbit Following Stimulation of the Perforant Path. *The Journal of Physiology, 232*(2), 331–356. https://doi.org/10.1113/jphysiol.1973.sp010273. PMC 1350458. PMID 4727084. Accessed 03 Apr 2021.

Boldrini, M., Fulmore, C. A., Tartt, A. N., Simeon, L. R., Pavlova, I., Poposka, V., Rosoklija, G. B., Stankov, A., Arango, V., Dwork, A. J., Hen, R., & Mann, J. J. (2018, April 5). Human Hippocampal Neurogenesis Persists Throughout Aging. *Cell Stem Cell, 22*(4), 589–599.e5. https://doi.org/10.1016/j.stem.2018.03.015. PMC 5957089. PMID 29625071.

Bowlby, E. J. M. (1956, 2018). *Attachment: The Attachment and Loss Trilogy*. Pimlico.

Bronfenbrenner, U. (1979). *The Ecology of Human Development*. Harvard University Press.

Caesar, J. (1982). *The Conquest of Gaul* (Penguin Classics, eds Jane Gardner, et al.). Penguin.

Chatterley, G. (2020). *Building Positive Behaviour: Returning to Learning Using a Sequential Approach*. Email gchatterleyroasa@gmail.com Self-published: Amazon Kindle Direct Publishing

Chivers, T. (2019). *The AI Does Not Hate You: Superintelligence, Rationality and the Race to Save the World*. Orion.

Christian, B., & Griffiths, T. (2016). *Algorithms to Live By: The Computer Science of Human Decisions*. William Collins.

Clark, A. (2001). *Mindware: An Introduction to the Philosophy of Cognitive Science*. Oxford University Press.

Clark, A. (2008). *Supersizing the Mind: Embodiment, Action, and Cognitive Extension*. Oxford University Press.

Clark, R. W. (1987). *Freud, the Man and the Cause*. Paladin.

Claxton, G. (1998). *Hare Brain, Tortoise Mind*. Fourth Estate Limited.

Claxton, G. (2002). *Building Learning Power: Helping Young People Become Better Learners*. TLO.

Cobb, M. (2020). *The Idea of the Brain: A History*. Profile Books Ltd.

Collard, P. (2010). *The Little Book of Mindfulness: 10 Minutes a Day to Less Stress More Peace*. Octopus Publishing.

Cox, T. (2018). *Now You're Talking: The Story of Human Conversation from the Neanderthals to Articicial Intelligence*. Vintage Random House.

Cozolino, L. (2013). *The Social Neuroscience of Education: Optimizing Attachment and Learning in the Classroom*. Norton.

Crehan, L. (2016). *Clever Lands: The Secrets Behind the Success of the World's Education Superpowers*. Unbound.

Crick, F., & Koch, C. (1990). Towards a Neurobiological Theory of Consciousness. *Seminars in the Neurosciences, 2*, 263–275.

Cromby, J. (2007). Feelings and Embodied Subjectivity. In V. van Deventer, M. Terre-Blanche, E. Fourie, & P. Segalo (Eds.), *Citizen City: Between Constructing Agent and Constructed Agency* (pp. 232–240). Captus University Press.

Cromby, J. (2015). *Feeling Bodies: Embodying Psychology*. Palgrave.

Davis, E. (2019). *Post Truth: Why We Have Reached Peak Bullshit and What We Can Do About It*. Little Brown.

de Thierry, B. (2019). *The Simple Guide to Attachment Difficulties in Children: What They Are and How to Help*. Jessica Kingsley.

Dean Hobbs, R. (2016). *The Global Child: How Experts Would Change Education. Research-Based Acquisition of Languages*. Lap Lambert Academic Publishing.

Dennett, D. C. (1991). *Consciousness Explained*. Penguin Random House.

Dennett, D. C. (2017). *From Bacteria to Bach and Back: The Evolution of Minds*. Penguin Random House.

Desautels, L. (2020). *Connections Over Compliance: Rewiring Our Perceptions of Discipline*. Wyatt-MacKenzie Publishing.

Desautels, L., & McKnight, M. (2019). *Eyes Are Never Quiet: Listening Beneath the Behaviors of Our Most Troubled Students*. Wyatt-MacKenzie Publishing.

Dewey, J. (1976). *The Collected Works of John Dewey (1859–1952)*. Southern Illinois University Press.

Dix, P. (2017). *When the Adults Change, Everything Changes: Seismic Shifts in School Behaviour*. Independent Thinking Press.

Doidge, N. (2007). *The Brain That Changes Itself: Stories of Personal Triumph from the Frontiers of Brain Science*. Penguin.

Doidge, N. (2015). *The Brain's Way of Healing: Stories of Remarkable Recoveries and Discoveries*. Penguin.

Doxtdator, Benjamin. (2017). A Field Guide to 'jobs that don't exist yet'. Jul 8, 2017. http://www.longviewoneducation.org/field-guide-jobs-dont-exist-yet/. Accessed 06/04/2018.

Dweck, C. (2006). *Mindset: Changing The Way You think To Fulfil Your Potential*. Random House.

Eaton, H. (2010). *Brain School: Stories of Children with Learning Disabilities and Attention Disorders Who Changed Their Lives by Improving Their Cognitive Functioning*. Glia Press.

Eaton, H. (2018). *The Brain Pioneer: The True Story of How Barbara Arrowsmith-Young Used Brain Science to Help Children with Learning Disabilities*. Barnes and Noble.

Elder, Z. (2012). I. Gilbert (Ed.), *Full on Learning: Involve Me and I'll Understand*. Crown House Publishing.

Feldman Barrett, L. (2017). *How Emotions Are Made; The Secret Life of the Brain*. Pan Macmillan.

Finch, C. (2007). *The Biology of Human Longevity: Inflammation, Nutrition, and Aging in the Evolution of Lifespans San Diego*. Academic Press-Elsevier.

Florance, C. (2004). *A Boy Beyond Reach*. Simon & Schuster UK Ltd.

Fodor, A. J. (1983). *The Modularity of Mind*. MIT Press-A Bradford Book.

Foer, F. (2017). *World Without Mind: Why Google, Amazon, Facebook and Apple Threaten Our Future*. Penguin Random House.

Freire, P. (2014). *Pedagogy of Hope: Reliving Pedagogy of the Oppressed*. Bloomsbury Academic.

Freud, S. (2010). *Sigmund Freud Collected Works: The Psychopathology of Everyday Life, The Theory of Sexuality, Beyond the Pleasure Principle, The Ego and the Id, and The Future of an Illusion*. Freud (1856–1939). Pacific Publishing Studio.

Fry, H. (2018). *Hello World: How to Be Human in the Age of the Machine*. Penguin Random House.

Gadsby, C., & Evans, J. (2019). *Dynamically Different: Create Spaces that Spark Learning*. Independent Thinking Press.

Gardner, H. (1993). *Frames of Mind: The Theory of Multiple Intelligences*. Fontana.

Garvey, P. (2019). *Talk for Teaching: Completely Rethinking Professional Development in Schools*. John Catt Educational Ltd.

Geary, D. C. (2005). *The Origin of Mind: Evolution of Brain, Cognition, and General Intelligence*. American Psychological Association.

Gilbert, I. (2015, 2018). *There Is Another Way: The Second Big Book of Independent Thinking*. Independent Thinking Press.

Gilbert, I. (Ed.). (2018). *The Working Class: Poverty, Education and Alternative Voices*. Independent Thinking Press, Crown House Publishing.

Gladwell, M. (2002). *The Tipping Point: How Little Things Can Make a Big Difference*. Abacus.

Gladwell, M. (2005). *Blink: The Power of Thinking Without Thinking*. Penguin.

Glasser, M., Coalson, T., Robinson, E., et al. (2016, July 20). Academic Paper 'A multi-modal parcellation of human cerebral cortex'. *Nature, 536*, 171–178. https://doi.org/10.1038/nature18933. Accessed 12 Feb 2019.

Godfrey-Smith, P. (2016). *Other Minds: The Octopus and the Evolution of Intelligent Life*. William Collins.

Goleman, D. (1996). *Emotional Intelligence: Why It Can Matter More Than IQ*. Bloomsbury.
Goodall, J., & Weston, K. (2018). *Engaging Parents: 100 Ideas for Primary Teachers*. Bloomsbury.
Greenfield, S. (2004). *Tomorrow's People: How 21st-century Technology Is Changing the Way We Think and Feel*. Penguin.
Greenfield, S. (2014). *Mind Change*. Penguin Rider.
Gropius, W. (1955). *The New Architecture and the Bauhaus*. Faber and Faber Limited.
Haas, R. (2020). *The World: A Brief introduction*. Penguin Press.
Hale, I. (2018). *Asperger's, Autism and You: Is the Epidemic Real? Are Vaccines the Cause?* Institute for Education, Research and Scholarships.
Hall, E. (2018). *Aristotle's Way: How Ancient Wisdom Can Change Your Life*. Bodley Head.
Harley, D., Morgan, J., & Frith, H. (2018). *Cyberpsychology as Everyday Digital Experience Across the Lifespan*. Palgrave Macmillan.
Hatfield, E., Cappioco, J., et al. (1994). *Emotional Contagion* (Studies in Emotion and Social Interaction Serie). Cambridge University Press.
Hattie, J., & Yates, G. (2014). *Visible Learning and the Science of How We Learn*. Routledge.
Hawkes, N. (2014). *From My Heart: Transforming Lives through Values*. Independent Thinking Press.
Hawkes, N., & Hawkes, J. (2018). *The Inner Curriculum: How to Nourish Wellbeing, Resilience and Self-leadership*. John Catt Educational.
Hawking, S. (1988). *A Brief History of Time: From Big Bang To Black Holes*. Bantam Books.
Hawkins, J. A. (2017). *Feelings and Emotion-based Learning; A New Theory*. Palgrave.
Heathcote, D., & Bolton, G. (1995). *Drama for Learning: The Mantle of the Expert Approach to Education*. Heinemann Educational.
Hedges, C. (2018). *America: The Farewell Tour*. Simon & Schuster.
Hejlskov Elven, B., & Wiman, T. (2017). *Sulky, Rowdy, Rude? Why Kids Really Act Out and What To Do About It*. Jessica Kingsley Publishers.
Henden, J. (2017). *What It Takes To Thrive: Techniques For Severe Trauma And Stress*. World Scientific Publishing Company.
Hirst, P. (1994). *Associative Democracy: New Forms of Economic and Social Governance*. Cambridge.
Hohwy, J. (2013). *The Predictive Mind*. Oxford University Press.
Holt, J. C. (1982). *How Children Fail*. Perseus Books.
Jackson, P. Z., & McKergrow, M. (2006). *The Solutions Focus: Making Coaching and Change Simple*. Nicholas Brealey International.

James, G. (2016). *Transforming Behaviour in the Classroom: A Solution-focused Guide for New Teachers*. Sage.

James, G. (2019). *Solutions Focused Coaching: A Workbook for Educators*. Singular Publishing.

James, W. (1918). *The Principles of Psychology – Vols. 1–2 (Illustrated Works of W. James 1842–1910)*. Pantianos Classics.

Jarvis, P., Swiniarski, L., & Holland, W. (2016). *Early Years Pioneers in Context: Their Lives, Lasting Influence and Impact on Practice Today*. Routledge.

Jeewon, O., Chopik, W. J., & Kim, E. S. (2019). The Association Between Actor/Partner Optimism and Cognitive Functioning Among Older Couples. *Journal of Personality*. https://doi.org/10.1111/jopy.12529; Michigan State University, USA: "A happy partner leads to a healthier future." *ScienceDaily* 11 February 2020; Available at: www.sciencedaily.com/releases/2020/02/200211092555.htm. https://www.sciencedaily.com/releases/2020/02/200211092555.htm. Accessed 12 Feb 2020

Johnson, M. H., & de Haan, M. (2015). *Developmental Cognitive Neuroscience: An Introduction*. Wiley & Sons Ltd..

Jones, J. (2018). *Sensations: The Story of British Art from Hogarth to Banksy*. Laurence King.

Joyce, R. (2016). Moral Skepticism. In D. Machuca & B. Reed (Eds.), *Skepticism: Antiquity to the Present*. Bloomsbury.

Jung, C. G. (1991). *The Archetypes and the Collective Unconscious (Collected Works of C. G. Jung 1865–1961)*. Routledge.

Jung, C. G. (1995). *Memories, Dreams, Reflections: An Autobiography*. Fontana Press (1875–1961).

Kaiser, B. (2019). *Targeted: My Inside Story of Cambridge Analytica and How Trump, Brexit and Facebook Broke Democracy*. Harper Collins.

Karmiloff-Smith, A. (1996). *Beyond Modularity: A Developmental Perspective on Cognitive Science*. MIT Press.

Kell, E. (2018). *How to Survive in Teaching: Without Imploding, Exploding or Walking Away*. Bloomsbury.

Kelly, L. (2017). *The Memory Code: The Secrets of Stonehenge, Easter Island and Other Ancient Monuments*. Pegasus Books Ltd..

Kidd, D. (2014). *Teaching: Notes from the Front Line*. Independent Thinking Press.

Kidd, D. (2020). *A Curriculum of Hope: As rich in Humanity as in Knowledge*. Independent Thinking Press.

Kincheloe, J. L., Steinberg, S. R., & Villaverde, L. E. (1999). *Rethinking Intelligence: Confronting Psychological Assumptions About Teaching and Learning*. Routledge.

Klein R. G. with Edgar, B. (2002). *The Dawn of Human Culture*. Wiley & Sons Ltd..

Lampert, M. (2001). *Teaching Problems and the Problems of Teaching.* Yale University Press.

Langer, S. K. (1953). *Feeling and Form: A Theory of Art.* Routledge Kegan Paul.

Langer, S. K. (1967). *Mind: An Essay in Human Feeling – Volume 1.* The John Hopkins University Pres.

Lanier, J. (2014). *Who Owns the Future?* Penguin.

Lanier, J. (2017). *Dawn of the New Everything: A Journey Through Virtual Reality.* The Bodley Head.

Lanier, J. (2018). *Ten Arguments for Deleting Your Social Media Accounts Right Now.* The Bodley Head.

Leader, D. (2016). *Hands: What We Do with Them – And Why.* Penguin Random House UK.

Leland, J. (2018). *Happiness Is a Choice You Make: Lessons from a Year Amongst the Oldest Old.* Sarah Chrichton books.

Levy, S. (2020). *Facebook: The Inside Story.* Penguin Random House UK.

Lukianoff, G., & Haidt, J. (2018). *The Coddling of the American Mind: How Good Intentions and Bad Ideas Are Setting Up a Generation for Failure.* Allen Lane Penguin Random House UK.

Luria, A. R. (1973). *The Working Brain: An Introduction to Neuropsychology.* Penguin Books.

Luria, A. R. (1976). *Cognitive Development: Its Cultural and Social Foundations Cambridge.* Harvard University Press.

Lutz, C. A. (2009). *Language and the Politics of Emotion* (Studies in Emotion and Social Interaction). Cambridge University Press.

Lytra, V., & Martin, P. (2010). *Sites of Multilingualism: Complementary Schools in Britain Today.* Trentham Books.

MacCarthy, F. (2020). *Walter Gropius: Visionary Founder of the Bauhaus.* Faber and Faber Limited.

Mandela, N. (1995). *Long Walk to Freedom: The Autobiography of Nelson Mandela.* Back Bay Books.

Mareschal, D., Johnson, M., Sirois, S., Spratling, M., Thomas, M., & Westermann, G. (2007). *Neuroconstructivism: Volumes I & II* (Developmental Cognitive Neuroscience). Oxford University Press.

Mazzucato, M. (2018). *The Value of Everything: Making and Taking in the Global Economy.* Penguin Random house.

McGill, R. M. (2017). *Mark. Plan. Teach.* London.

McNamee, R. (2019). *Zucked: Waking Up to the Facebook Catastrophe.* Harper Collins.

Menary, R. (2012). In A. Clark et al. (Eds.), *The Extended Mind: Life and Mind: Philosophical Issues in Biology and Psychology.* Massachusetts Institute of Technology (MIT).

Menzies, G. (2009). *1434: The Year a Chinese Fleet Sailed to Italy and Ignited the Renaissance*. Harper Collins.
Merzenich, M. (2013). *Soft-wired: How the New Science of Brain Plasticity Can Change Your Life* (2nd ed.). Parnassus Pub. ISBN 978-0989432825.
Mitchell, L. (2009). *Language, Emotion, and Politics in South India: The Making of a Mother Tongue* (Contemporary Indian Studies). Indiana University Press.
Montessori, M. (1991). *The Montessori Method*. Schocken Books Inc.
Montessori, M. (1965). *Dr. Montessori's Own Handbook: A Short Guide to Her Ideas and Materials*. Schocken Books Inc..
Moskowitz, M. A. (2010). *Reading Minds: A Guide to the Cognitive Neuroscience Revolution*. Routledge.
Moskowitz, M. H., & DePolo Golden, M. (2013). *Neuroplastic Transformation* Workbook Spiral-bound. The Bay Area Medical Associates.
Moskowitz, M. H., & DePoloGolden, M. (2013). *Neuroplastic Transformation Workbook: How to Use the Power of Your Mind to Heal Your Brain, Your Pain, and Change Your Life LLC: Neuroplastic Partners*. https://elitemanmagazine.com/heal-your-brain-your-pain-dr-michael-moskowitz-dr-marla-golden/. Accessed 24 Oct 2019.
Moyo, D. (2012). *How The West Was Lost: Fifty Years of Economic Folly – And the Stark Choices Ahead*. Penguin Books Ltd..
Moyo, D. (2018). *Edge of Chaos*. Little Brown Book Group.
Moyo, D. (2021). *How Boards Work: And How They Can Work Better in a Chaotic World*. The Bridge Street Press.
Munro, E. (2011). *Independent Report-Munro Review of Child Protection: A Child-Centred System-Report Setting Out Reform Proposals to Enable Professionals to Make the Best Judgements to Help Children, Young People and Families*. Published 10 May 2011. Department for Education. https://www.gov.uk/government/publications/munro-review-of-child-protection-final-report-a-child-centred-system. Accessed 15 Sept 2021.
Murphy, K. (2020). *You're Not Listening: What You're Missing and Why It Matters*. Penguin Random House.
Murray, D. (2017). *The Strange Death Of Europe: Immigration, Identity, Islam*. Bloomsbury Continuum.
Myatt, M. (2018). *The Curriculum: Gallimaufry to Coherence*. John Catt Educational.
O'Mara, S. (2019). *In Praise of Walking: The New Science of Why We Walk and Why It Is Good For Us*. Penguin Random House.
Ostby, H., & Ostby, Y. (2018). *Adventures in Memory: The Science and Secrets of Remembering and Forgetting*. Greystone Books.
Pascoe, B. (2018). *Dark Emu*. Scribe Publications UK Ltd..
Pearce, S. (2005). *You Wouldn't Understand: White Teachers in Multi-ethnic Classrooms*. Trentham Books.

Pepper, J. D. (2012). *Reverse Parkinson's Disease*. John Douglas Pepper Website: www.reverseparkinsons.net

Perry, B. D., & Szalavitz M. (2006). *The Boy Who Was Raised as a Dog: And Other Stories from a Psychiatrist's Notebook*. Hachette Book Group.

Phillips, B. (2020). *How to Fight Inequality: (And Why That Fight Needs You)*. Polity Press.

Pinker, S. (1994). *The Language Instinct*. Penguin.

Pinker, S. (2002). *The Blank Slate: The Modern Denial of Human Nature*. Allen.

Pinker, S. (2018). *Enlightenment Now: The Case For Reason, Science, Humanism And Progress*. Allen Lane.

Pinker, S. (2015). *The Village Effect: Why Face-to-Face Contact Matters*. Atlantic Books.

Plomin, R. (2018). *Blueprint: How DNA Makes Us Who We Are*. Penguin Random House.

Porges, S. W. (2011). *The Polyvagal Theory: Neurophysiological Foundations of Emotions, Attachment, Communication, and Self-regulation*. Norton Publishing.

Postman, N., & Weingartner, C. (2009). *Teaching as a Subversive Activity: A No-holds Barred Assault on Outdated Teaching Methods-with Dramatic and Practical Proposals on How Education Can Be Made Relevant to Today's World*. Delta.

Ramachandran, V. S. (2002). *Encyclopedia of the Human Brain: Volume 1–4*. Elsevier Science Ltd.

Ramachandran, V. S. (2003). *The Emerging Mind: The BBC Reith Lectures*. BBC/Profile Books Ltd.

Ratner, H. (2012) (with Evan George & Chris Iveson-contributors). *Solution Focused Brief Therapy: 100 Key Points and Techniques* Routledge.

Reay, D. (2017). *Miseducation: Inequality, Education and the Working Classes*. Policy Press.

Renner, M. J., & Rosenzweig, M. R. (1987). *Enriched and Impoverished Environments: Effects on Brain and Behavior*. Springer. ISBN 978-3-540-96523-7.

Roberts, D. (2017). *Empowering Generation Z: How and Why Leadership Opportunities Can Inspire Your Students*. John Catt Educational.

Roberts, H. (2012). In I. Gilbert (Ed.), *Oops! Helping Children Learn Accidentally*. Independent Thinking Press.

Roberts, P. (2014). *The Impulse Society: What's Wrong With Getting What We Want?* Bloomsbury.

Robertson, J. (2014). *Dirty Teaching: A Beginner's Guide to Learning Outdoors*. Independent Thinking Press Crown House.

Robertson, J. (2017). *Messy Maths: A Playful, Outdoor Approach for Early Years*. Independent Thinking Press Crown House.

Robinson, K., & Aronica, L. (2015). *Creative Schools*. Penguin Random House.
Robinson, M. (2013). *Trivium 21c: Preparing Young People for the Future with Lessons from the Past*. Independent Thinking Press.
Robinson, M. (2016). In T. Sherrington, S. Gorse, N. Wells, D. Hall, M. Grenier, N. Rose, C. Hendrick, & J. L. Taylor (Eds.), *Trivium in Practice*. Independent Thinking Press.
Robinson, M. (2019). *Curriculum: Athena Versus the Machine*. Independent Thinking Press.
Rogers, C., & Freiberg, H. J. (1983). *Freedom to Learn* (Rev. ed.). Prentice Hall International.
Rogers, C. R. (1951). *Client-Centred Therapy*. Constable.
Rogers, C. R. (1961). *On Becoming a Person: A Therapist's View of Psychotherapy*. Constable.
Rose, H., & Rose, S. (2012). *Genes, Cells and Brains*. Verso.
Rovelli, C. (2017). *Reality Is Not What It Seems: The Journey to Quantum Gravity*. Penguin Random House.
Runciman, D. (2018). *How Democracy Ends*. Profile Books.
Rycroft-Smith, L., & Dutaut, J.-L. (Eds.). (2018). *Flip the System*. Routledge.
Shalaby, C. (2017). *Troublemakers: Lessons in Freedom from Young Children at School*. New Press.
Shea, T. (2008). *Dismantling Language Barriers. HR Magazine, 53*(11), 1–4.
Sherrington, T. (2017). *The Learning Rainforest: Great Teaching in Real Classrooms*. John Catt Educational.
Sherrington, T. (2019). *Rosenshine's Principles in Action*. John Catt Educational.
Shore, S. (2001). *Beyond the Wall: Personal Experiences with Autism and Asperger Syndrome*. Autism Asperger Publishing Co.
Shotter, J. (1985). *Social Accountability and Selfhood*. Blackwell.
Shotter, J. (1993). *Conversational Realities: Constructing Life Through Language* (Inquiries in Social Construction Series). Sage Publications Ltd.
Skinner, B. F., Ferster, C. B., Cheney, C. D., & Morse, W. H. (1957). *Schedules of Reinforcement*. The Skinner Foundation.
Steiner, R. (1920). *Lectures in Stuttgart*. Teaching from Humane Knowledge translated by Pusch, R., Querido, R., & Amrine, F. Amazon Fulfillment.
Stenhouse, L. (1975). *An Introduction to Curriculum Research and Development*. Heinemann Educational.
Stevenson, B. (2015, 2016, 2017, 2018, 2019, 2020). *Just Mercy: A Story of Justice and Redemption*. Scribe Publications.
Sumpton, J. (2019). *Trials of the State: Law and the Decline of Politics*. Profile Books.
Syed, M. (2019). *Rebel Ideas: The Power of Diverse Thinking*. John Murray.
Szalavitz, M., & Perry, B. D. (2010). *Why Empathy Is Essential and Endangered: Born for Love*. Harper Collins.
Taplin, J. (2017). *Move Fast and Break Things: How Facebook, Google and Amazon Have Cornered Culture and What It Means for All of Us*. Pan Macmillan.

Taub, E., Burgio, L., Miller, N. E., Cook, E. W., III, Groomes, T., DeLuca, S., & Crago, J. (1994). An Operant Approach to Overcoming Learned Nonuse After CNS Damage in Monkeys and Man: The Role of Shaping. *Journal of the Experimental Analysis of Behavior, 61*, 281–293.

Taylor, T. (2016). *A Beginner's Guide to Mantle of the Expert: A Transformative Approach to Education*. Singular Publishing.

Tharby, A. (2018). *How to Explain Absolutely Anything to Absolutely Anyone: The Art and Science of Teacher Explanation*. Crown House.

The Academic Press Profiles in Science U. S. National Library of Medicine. https://profiles.nlm.nih.gov/spotlight/sc/catalog?utf8=%E2%9C%93&exhibit_id=sc&search_field=all_fields&q=Koch. Accessed 05 Apr 2021.

Thunberg, G. (2019). *No One Is Too Small to Make a Difference*. Penguin Random House.

Titcombe, R. (2015). *Learning Matters*. FGI Publishing.

Tomsett, J. (2017). *This Much I Know About Mind Over Matter: Improving Mental Health in Our Schools*. Crown House Publishing.

Townend, J., & Walker, J. (2006). *Structure of Language: Spoken and Written English*. Wiley.

Velasco, C., & Obrist, M. (2020). *Multisensory Experiences: Where the Senses Meet Technology*. Open University Press.

Vygotsky, L. S. (1896–1934). In R. W. Rieber & A. S. Carton (Eds.), *Collected Works – Volume 1*. Plenum, (1987).

Walker, M. (2017). *Why We Sleep: The New Science of Sleep and Dreams*. Allen Lane.

Walker, P. (2013). *Complex PTSD: From Surviving to Thriving: A Guide and Map for Recovering from Childhood Trauma*. Azure Coyote Publishing.

Ward, J. (2019). *On the Fringes: Preventing Exclusion in Schools Through Inclusive, Child-centred, Needs-based Practice*. Crown House.

Waters, M. (2013). *Thinking Allowed On Schooling*. Independent Thinking Press.

Watson, L. (1999). *Jacobson's Organ and the Remarkable Nature of Smell*. Allen Lane Penguin.

Wenger, E. (1998). *Communities of Practice*. Cambridge University Press.

Wenger, E., McDermott, R., & Snyder, W. (2002). *Cultivating Communities of Practice: A Guide to Managing Knowledge*. Harvard Business School; McGraw-Hill.

Westall, A. (2011, January 10). *Time to Revisit Associative Democracy? Open Democracy UK Revisiting Associative Democracy, an e-book, Draws Together the Ideas and Thoughts of a Group of People Who Met Last…* https://www.opendemocracy.net/en/opendemocracyuk/time-to-revisit-associative-democracy/. Accessed 13 Feb 2020.

Whitebread, D. (2017). *Quality in Early Childhood Education: The Contribution of Developmental Psychology* (pp. 319–334). International Handbook of Early Childhood Education (Springer).

Whitehead, A. N. (1929). *Process and Reality: An Essay in Cosmology.* Macmillan Company.
Williams, R. (1958, 2017) *Culture and Society.* Penguin Vintage.
Willingham, D. T. (2009). *Why Don't Students Like School? A Cognitive Scientist Answers Questions About How the Mind Works and What It Means for the Classroom.* Wiley.
Winfrey, O., & Perry, B. (2021). *What Happened to You? Conversations on Trauma, Resilience, and Healing.* Bluebird Publications.
Zacks, J. M. (2015). *Flicker.* Oxford University Press.
Zadina, J. N. (2015). The Emerging Role of Educational Neuroscience in Education Reform *Tulane University School of Medicine, New Orleans, U.S.A.*: Published by Elsevier España, S.L.U. on Behalf of Colegio Oficial de Psicólogos de Madrid. This Is an Open Access Article Under the CC BY-NC-ND License.
Zlatic, M. (2004). Establishment of Connectivity in the Embryonic Central Nervous System of Drosophila. cam.ac.uk (PhD thesis). University of Cambridge. OCLC 890159518. British Library EThOS uk.bl.ethos.615935. https://ethos.bl.uk/Home.do;jsessionid=687EDE6B3BFA4FEC9CD134C7FF1AD231. Accessed 05 Apr 2021.

Index[1]

[1] Note: Page numbers followed by 'n' refer to notes.

J. A. Hawkins, *Brain Plasticity and Learning*,
https://doi.org/10.1007/978-3-030-83530-9

The manufacturer's authorised representative in the EU is Springer Nature Customer Service Centre GmbH, Europaplatz 3, 69115 Heidelberg, Germany. If you have any concerns regarding our products, please contact ProductSafety@springernature.com

Printed and bound by CPI Group (UK) Ltd, Croydon, CR0 4YY
30/09/2026
02225191-0001